Telecommunication Network Economics

Presenting a balance of theory and practice, this up-to-date guide provides a comprehensive overview of the key issues in telecommunication network economics, as well as the mathematical models behind the solutions. These mathematical foundations enable the reader to understand the economic issues arising at this pivotal time in network economics, from business, research, and political perspectives. This is followed by a unique practical guide to current topics, including app stores, volume-based pricing, auctions for advertisements, search engine business models, the network neutrality debate, the relationship between mobile network operators and mobile virtual network operators, and the economics of security. The guide discusses all types of players in telecommunications, including users, access and transit network providers, service providers (including search engines, cloud providers and content delivery networks), content providers, and regulatory bodies. Ideal for graduate students, researchers, and industry practitioners working in telecommunications.

Patrick Maillé has been with the Networks, Security and Multimedia Department of Telecom Bretagne (Institut Mines-Telecom) since 2002. He has written or co-written more than 60 papers on game theory and economic concepts applied to telecommunication ecosystems.

Bruno Tuffin has been with Inria in Rennes, France, since 1997. He has written or co-written more than 100 papers and two books on Monte Carlo and quasi-Monte Carlo simulation techniques for the performance evaluation of telecommunication systems, and on developing new Internet-pricing schemes and telecommunication-related economic models.

Telecommunication Network Economics

From Theory to Applications

PATRICK MAILLÉ
Telecom Bretagne

BRUNO TUFFIN
Inria

CAMBRIDGE
UNIVERSITY PRESS

CAMBRIDGE
UNIVERSITY PRESS

University Printing House, Cambridge CB2 8BS, United Kingdom

Published in the United States of America by Cambridge University Press, New York

Cambridge University Press is part of the University of Cambridge.

It furthers the University's mission by disseminating knowledge in the pursuit of education, learning and research at the highest international levels of excellence.

www.cambridge.org
Information on this title: www.cambridge.org/9781107032750

First published 2014

Printed in the United Kingdom by TJ International Ltd. Padstow Cornwall

A catalogue record for this publication is available from the British Library

ISBN 978-1-107-03275-0 Hardback

Contents

Preface

Network economics is a very hot topic, at the same time from a research point of view (with several conferences devoted to the theme, plus a devoted section in most of the other main telecommunication conferences), from a political point of view (as highlighted by the network neutrality debate, the increasing discussion on volume-based pricing, etc.), and of course from a business point of view (encompassing advertisement pricing definition, spectrum selling and sharing, bundling of offers, etc.). We believe that it is very good timing to release a book on the issue, describing both the theory and key specific applications, because of all the economic issues that constantly pop up in telecommunications. It is probably a cornerstone time for a redefinition of the Internet business model. While writing (scientific) papers on network economics, we also had the feeling that there was some room for a (new) book in the area. Indeed, there was to the best of our knowledge no recent work mixing mathematical theory and deep analysis of the economic stakes that had appeared in telecommunications. As illustrations of the issues we have in mind, we can mention the network neutrality debate, the relations between mobile network operators and mobile virtual network operators, the management of application stores, the economics of security, auctions for advertisements on different media (content pages, applications, or search engines), etc. In all those cases the interactions among their components need to be described, together with a solid scientific foundation, leading towards a careful analysis. Our book is designed to have a balance between theory and practice. Finally, after around ten years of research experience in the area, we think that this book is an opportunity to put together all the pieces of our activity and to push the analysis one step further.

This book is a monograph, but we believe that it can be also considered as a textbook at the Master level, for instance, because it will include all the necessary theoretical material.

We would like to thank Mia Balashova, Samantha Richter and Phil Meyer from Cambridge University Press for their assistance, encouragement, and patience during the preparation of this book. We also express our gratitude to Dr Steven Holt for his careful reading and the numerous improvements he suggested, and to Arindam Bose for helping implement them. Finally, we thank Maria Maillé for providing the cover picture. Any mistake, error of judgement, or treatment imbalance in the book is our sole responsibility.

About the authors

Patrick Maillé graduated from the Ecole Polytechnique and Telecom ParisTech, France, in 2000 and 2002, respectively. He has been with the Networks, Security and Multimedia department of Telecom Bretagne since 2002, where he obtained his PhD in applied mathematics in 2005, followed by a six-month visit to Columbia University in 2006. His research interests are in game theory and economic concepts applied to telecommunication ecosystems (resource pricing, routing, consequences of user selfishness for network performance). He has authored or co-authored more than sixty papers on those topics.

Bruno Tuffin received his PhD in applied mathematics from the Université de Rennes 1 (France) in 1997. Since then, he has been with Inria in Rennes. He spent eight months as a postdoc at Duke University in 1999. His research interests include developing Monte Carlo and quasi-Monte Carlo simulation techniques for the performance evaluation of telecommunication systems, and developing new Internet-pricing schemes and telecommunication-related economic models. He has published more than one hundred papers on those issues. He has also led or participated in several French and European projects, and co-organized several conferences. He is currently Associate Editor of *INFORMS Journal on Computing*, *ACM Transactions on Modeling and Computer Simulation*, and *Mathematical Methods of Operations Research*. He has written or co-written two books devoted to simulation: *Rare Event Simulation Using Monte Carlo Methods*, published by John Wiley & Sons in 2009, and *La simulation de Monte Carlo* (in French), published by Hermes Editions in 2010.

1 Introduction: telecommunications evolution and the set of actors

1.1 The evolution of telecommunications and the associated economic models

Telecommunication networks are occupying an increasing role in our daily life: almost everything is now available from the Internet (possibly via a mobile phone), and getting this kind of access has even become compulsory for some administrative operations, without mentioning the social pressure to be part of the trend. While the telephone network started commercially in 1877 following the birth of the Bell Telephone Company, and its development slowly democratized, mobile networks have since the 1970s quickly come to occupy a major place, with an estimated 5.3 billion mobile subscriptions worldwide at the end of 2010. It is also estimated by the International Telecommunication Union (ITU), in its report "The World in 2010," that 90% of the global population can reach a wireless network. In the same report, the Internet is also shown to be an increasing market worldwide, with the number of subscribers having doubled between 2005 and 2010, reaching more than two billion at the end of 2010 (1.6 billion having access at home). To highlight this growth, the average household telecommunications budget has considerably increased, counting wired and wireless telephony subscriptions (with often a wireless subscription per member of the family), Internet access, television, etc., with a role that is becoming compulsory for operations such as access to information and to documentation, and making declarations (taxes, etc.). E-commerce, namely the buying and selling of products or services over electronic systems such as the Internet and other computer networks, is also exploding: according to Goldman Sachs, global Internet sales are growing at about 19% a year, and are expected to reach almost one trillion dollars in 2013; see Figure 1.1. Also, the average time spent on the Internet keeps increasing: comScore Media Metrix estimates that the average American spent 32 hours per month online in 2010, and this is now surpassing the time spent watching TV.

While the above illustrates the increasing importance of telecommunications from the users' and content points of view, the networks themselves have evolved considerably [238]. The Internet is the key network of our analysis, being the one over which all networks converge. Although we do not intend to provide a complete description of the Internet's history, we wish to give a broad idea of it, since its economic model is highly related to the evolution of the network structure. The general idea of the Internet started in the early 1960s, with the principle of connecting computers in order to share

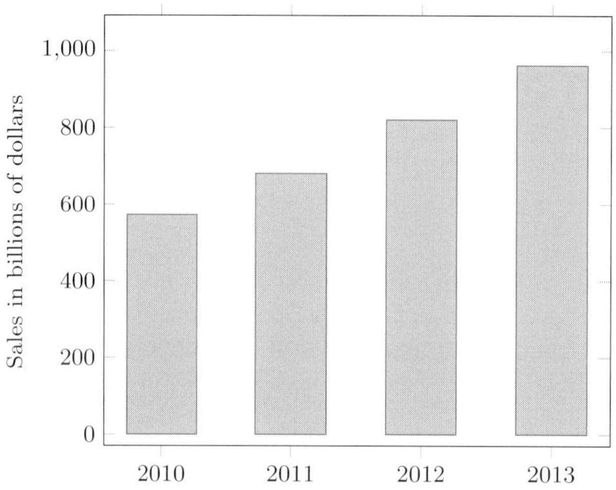

Figure 1.1 Expected evolution of e-commerce sales (Goldman Sachs data).

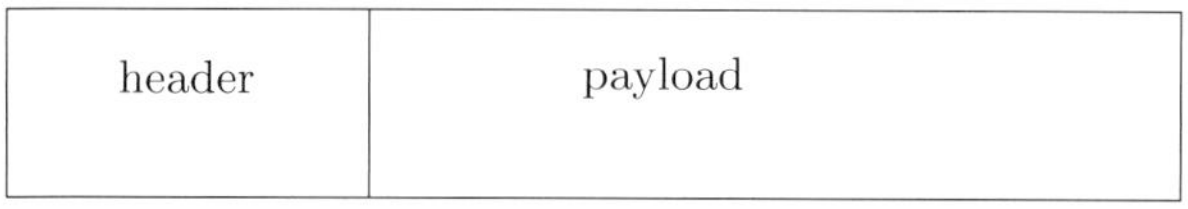

Figure 1.2 Packet description.

information and resources for research, but also in order to provide a robust network to the US military. The project was led by J. C. R. Licklider, and developed by the Defense Advanced Research Projects Agency (DARPA). The key **packet-switching** method for transmitting data consists in cutting the information into **packets** sent one after the other. Basically, a packet is made of two parts: the information (also called *payload*) and the **header**; the header contains fields such as source and destination address, data length and type, etc.; see Figure 1.2. See [154, 295] for more details. This is in contrast to the **circuit-switching** method used in telephony, for which a **circuit** is a communication channel along one line, which is fully used by the conversation. Packets have the advantage of being easily built and transmitted over the telephone network, and of allowing more reliable transmission, thanks to retransmission if needed. Two computers were effectively connected between the Massachusetts Institute of Technology and California in 1965 using dial-up telephone lines and packet switching, and the network sponsored by DARPA was called ARPAnet. ARPAnet allowed universities and research labs to share the cost of long-distance telephone lines. Still-in-use applications (such as FTP for example) and routing and addressing protocols were then imagined. Owing to the success of the network, other universities wanted to get connected, and created their own networks using the same protocols, so that the networks could interoperate. Thanks to the creation of Ethernet, which permitted one to cheaply interconnect computers on campuses, the network could expand, but was still just for non-commercial use. In the early 1990s, the Internet was opened to commercial organizations, which stepped in by

Table 1.1 IP traffic worldwide, measured in petabytes per month (data from Cisco Systems)

Year	Total global IP traffic	Fixed Internet traffic	Mobile IP traffic
1990	0.001	0.001	–
1995	0.18	0.17	–
2000	84	75	–
2005	2,426	2,055	0.9
2010	20,197	14,929	256
2011	27,483	20,634	597

building their own networks (especially telecommunication operators) and/or using the existing network. As illustrations, we have network operators that propose access at home to end-users, and connect them to the core network; content providers, which propose some content and get revenue by direct sales or through advertising; and more recently content delivery networks that provide resources to content providers for delivering their content on the network, social network applications such as Facebook, etc.

We can remark that, in most countries, telephony was initially a government monopoly, a situation that has changed almost everywhere. This analogy with the Internet and its interconnection of academics not interested in a business model, and now transformed into a commercial network and supply chain (from content creation to delivery to users), can be investigated to understand the pitfalls to avoid. In terms of business models for communications, it is interesting to note that Bell System had already compared the advantages and drawbacks of flat rate and metered rates for telephony in the 1880s, and preferred metered rates in large cities over flat rates, to cover the high marginal costs. Even if this situation is not necessarily true for the Internet now, studying the most "appropriate" charging scheme can lead to changes in the economic vision of the network.

We have mentioned the telephone network and the Internet, but wireless telephony (and communications) has grown and continues to grow significantly, and represents a prominent part of the economic business and technological works currently developed. Wireless communications are made through cellular networks (first 2G, then 3G encompassing data and Internet), WiFi, and now 4G/LTE transmissions. Those technologies will also be discussed in the book.

The success of the Internet and wireless networks, with an increasing number of subscribers but also because of more and more demanding applications in terms of bandwidth and resources, has seen a tremendous increase of traffic worldwide, as illustrated by Table 1.1. The traffic growth is often considered to be exponential, though it is estimated by Cisco that the yearly Internet traffic growth in the USA will fall from 42% in 2010 to 18% in 2014, and from 42% to 30% worldwide (a smaller reduction due to the later uptake in developing countries). To better check the evolution of this traffic increase, we plot the volume of traffic year by year from 1990 to 2011 in Figure 1.3. One can see that, even though the traffic is still increasing, the slope is slowing down (except for mobile communications, which are still in their infancy in terms of data

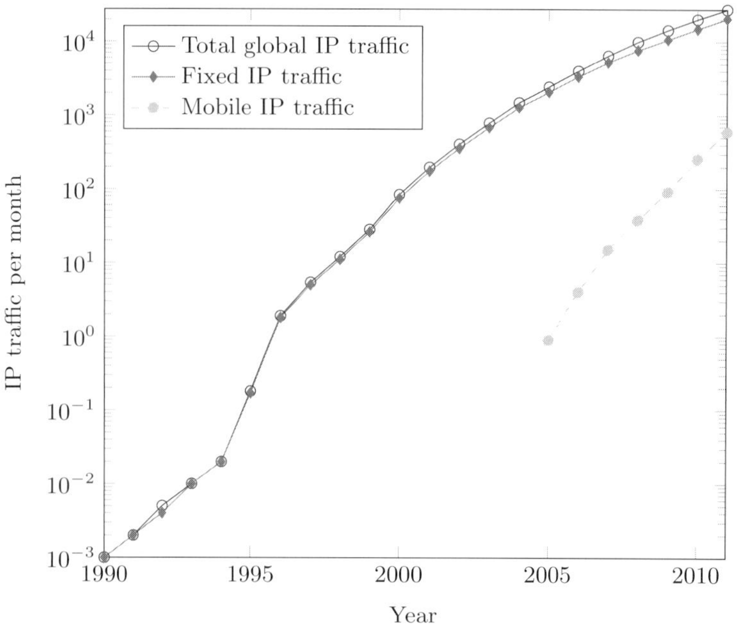

Figure 1.3 IP traffic evolution in petabytes per month, on a logarithmic scale (Cisco data).

transfer). The generally accepted statement that there is a doubling of traffic every year is no longer valid.

As a consequence of all that we have described, the initial (actually, non-existent) business model of the Internet, with free interconnection between academics, which still partly applies, is more and more being questioned: some ways to make revenues have to be defined, in order to at least cover the capacity expansion costs. Another complication is that the economics of the Internet are very related to the economics of (wired) telephony. Indeed, the Internet has experienced an easy development, and a rapid success, thanks to the use of the telephony infrastructure, even with its bandwidth restrictions. Therefore, a change in the business model of the Internet has to take into account the history of pricing mechanisms of the Internet, but also of telephony, in order to seek a better acceptance from users.

Among the many questions that the telecommunication network actors need to answer, we can non-exhaustively mention the following.

- *Determining the most relevant and profitable network access pricing scheme for end-users*, in a competitive context between network providers, also called Internet service providers (ISPs).

  One of the main aspects that could explain the success of the Internet is the fact that users just pay a subscription fee to the network, and can freely use it as much as they want, the so-called *flat-rate* pricing scheme. This scheme is/was very attractive to users, since they know exactly what they are going to pay and do not have to bother about usage. But, due to the tremendous development of traffic volume, users started to complain around the 1990s about data transfer times becoming too large without there being any possibility to improve this situation by themselves, by paying more, for instance. The problem arose from the congestion on communication links, and many people thought that link capacity expansion was not a viable solution because the network must respond to an increasing demand, and they claimed that experience had shown that demand for bandwidth has always been ahead of supply. Furthermore, in this context of congestion, flat-rate pricing is considered unfair since users with a huge consumption level pay the same amount as "light" users, while congesting the network and lowering the quality of service for all users. It was therefore suggested that one should replace the flat-rate fees by usage-based fees. Both flat-rate and usage-based schemes have their own respective advantages: first, flat-rate pricing is easy to implement, not requiring any costly measurement, and well-accepted by users. On the other hand, usage-based pricing allows better control of network usage, and makes it possible to differentiate services for potentially higher revenues and user satisfaction. There was a really strong research activity in the 1990s and early 2000s directed towards the design of various usage-based pricing mechanisms (some of which will be described in Chapter 3), but this activity slowed down afterwards, especially with the over-provisioning of the core network with optic fiber. However, usage-based pricing is coming back because of the increase of data traffic in wireless networks, as highlighted by Figure 1.3, for which bandwidth is scarce and more difficult to increase. Similarly, there is a push (mainly from Internet service providers) towards usage-based pricing: for instance in Canada in early 2011, usage-based pricing was enforced by the regulator, the Canadian Radio-television and Telecommunications Commission (CRTC), following the request of providers Bell, Rogers, Videotron, Cogeco, and Shaw, which were expecting to earn more. A lot of protests from user associations and content providers followed. Indeed, there is a strong public preference for flat-rate pricing [89], since users are more comfortable with this principle even though it is often said that usage-based pricing would mean a lower overall bill. Indeed, people always think about the possibility of consuming more, even if they will eventually not do so.

Another issue arising from the pricing scheme definition is related to network convergence: now wired and wireless telephony, Internet, and television are provided by the same operators. Because of competition, those operators merge the services into single offers, called *bundles*. Triple play offers that combine telephony, television over Internet, and Internet access, have become the norm, wireless telephony being added, too, for a quadruple play offer. Those offers and their associated prices have to be studied, taking care that a bundle offer would not cannibalize other potential offers, at the expense of the provider's profits.

- *Determining the best investments for network providers.*

  Network service providers have several strategic decisions to make, among which are those concerning investment in new technologies, and capacity and infrastructure expansions, but also participation in spectrum auctions. Those decisions are very important ones, because of the enormous costs involved, and the consequences of bad strategies can be economically very damaging for a company.

Indeed, providers should not only decide carefully the access price they will impose (the topic of the previous item) but also on which technologies to operate. They need, for example, to decide whether to invest in terms of infrastructure in new technologies such as the third generation (3G) of wireless networks (antennas, routers), and/or the fourth generation (4G) such as LTE (Long Term Evolution) or WiMAX. These investments should allow them to attract customers in order to get new sources of revenue. The LTE technology which is starting to be deployed is expected to lead to new broadband services, encouraged by mobile devices with larger screens, better batteries, and better performance. For instance, standard e-mails and SMS are moving towards photo messages, instant messaging, and video messaging, as well as social networking thanks to Facebook and Twitter, among others. As a consequence, investing should foster customers' subscriptions, but it has to be investigated whether the expected gains exceed the costs. Looking at the full picture, taking into account potential investments of competitors, would an investment by other providers reduce the market share too much? Will it help to segment the market and propose specific offers? Is it worth it for an operator to pay a license and devote resources to infrastructure in order to be present in a new technology? Will the return on investment be sufficient, and will it not be at the expense of other technologies already implemented and with limited maintenance costs?

Again, and to emphasize this issue, infrastructures are not the only required investment when dealing with wireless networks. Radio spectrum is indeed sold in most countries through auctions. Bidding to get a part of this spectrum is a costly and very strategic decision that has to be taken with caution. The auction run in 1999 for the radio spectrum in the USA was considered a disaster and had to be re-run after being declared null and void, because all the major bidders defaulted and declared bankruptcy (they could not bear the costs): the industry was far too speculative. Designing a fair auction mechanism is a key issue for regulators, in order for the governments to get as much revenue as possible but also to make sure that the providers' operations will not be endangered. Another complicated situation occurred also in France in 2009, when the country opened a fourth 3G license in order to foster competition (this license was won by the operator called Free). The price and conditions had to be decided in such a way as to let the entrant survive in confrontation with incumbents with existing infrastructure and licenses. This new license was offered at a lower cost than the initial ones, but the decision had to be as fair as possible for all actors.

Investing in new technologies is not the only strategic question. When a technology has already been implemented, providers have to decide whether or not, and, if so, when, to invest in capacity expansion, and when to cease a service. This decision has to forecast the evolution of demand and of capacities, and ponder the available options.

- *Defining the economic relations between network operators.*

  There are different types of network operators, organized in tiers and defining a hierarchical Internet. The characterization of an operator is not very easy, but it is considered that a Tier-1 operator is an operator network that is able to reach any other portion of the Internet without paying settlements. In other words, it peers with all other Tier-1 networks. A Tier-2 operator, on the other hand, is a network that peers with some networks, but pays transit fees to reach at least some portion of the Internet. Finally, a

Tier-3 operator is a network that is only buying transit fees to reach the Internet. The transit costs between those providers have to be carefully computed, and here too the regulator can intervene to ensure that a dominating network does not impose its rules on others at the expense of competition and users. These charging agreements structure the Internet in a tiered architecture, at the top of which Tier-1 operators dominate the market.

Among new questions, there is now a huge asymmetry of traffic exchanges due to some content providers, such as Google via YouTube, producing a non-negligible proportion of total Internet traffic. As a consequence the standard peering agreements are now less relevant. The economic model between providers is now being rethought. The network providers also think about which other provider they should be connected to. For instance, Tier-2 providers are starting to connect to each other to avoid the costs of going through Tier-1 providers. Similarly, big content providers are starting to become network providers to avoid paying the connection fees. Therefore we are leaning towards a flatter (or more meshed) Internet with more direct interconnections because there is more traffic originating from very few content providers or content delivery networks such as YouTube and Akamai (with the increasing importance of video streaming). This may reshape the Internet economy.

- *Understanding the relations between content providers and ISPs.*

  The previous item naturally leads to the question of relations between competitive and profit-seeking content and network service providers. Up to now, because of the historic non-profit organization of the Internet, the main principle driving the network has been the *universal access* principle, meaning that all consumers are entitled to reach meaningful content, whatever the technical limitations of their service. Here also, the fact that there are large and bandwidth-consuming content providers (e.g., YouTube) connected to a limited number of ISPs means that the competitive ISPs to which they are not connected, which therefore are not directly getting money from them, are starting to wonder why distant content providers should not be charged by them, with the threat that their traffic will not be delivered if they refuse to pay [168]. YouTube, for instance, is accessed by all users while being hosted by a single Tier-1 ISP. This issue was the starting point of the *network neutrality* debate which was launched at the end of 2005 by Ed Whitacre (CEO of AT&T) saying that content providers should also be charged by ISPs to which they are not directly connected:

  > How do you think they're going to get to customers? Through a broadband pipe. Cable companies have them. We have them. Now what they would like to do is use my pipes free, but I ain't going to let them do that because we have spent this capital and we have to have a return on it. So there's going to have to be some mechanism for these people who use these pipes to pay for the portion they're using. Why should they be allowed to use my pipes? The Internet can't be free in that sense, because we and the cable companies have made an investment and for a Google or Yahoo! or Vonage or anybody to expect to use these pipes free is nuts.

  The underlying concern is that investments are made by ISPs, but content providers appropriate a large part of the dividends. The revenue arising from online advertising (i.e. showing graphical ads on regular web pages) is estimated at approximately

$24 billion in 2009, while textual ads on search pages have led to a combined revenue of $8.5 billion in 2007, with those figures increasing every year. Meanwhile, transit prices – which constitute the main source of revenues for transit ISPs – are decreasing and predicted to be under $1 per Mbps (per month) by 2014. ISPs argue that there is insufficient incentive for them to continue to invest in the network infrastructures if most benefits go to content providers. Another behavior of ISPs has been to lower the quality of traffic coming from distant content providers. For instance, Comcast, one of the main ISPs in the USA, started in 2007 to block P2P applications such as BitTorrent, using the argument that P2P is mostly used to share illegal content. Advocates of neutrality argued here that P2P has legitimate uses and that other types of initiatives should be imagined. In all cases, the goal of ISPs is to change the current behavior of the Internet, where all users (of whatever type) have full access to the network with the same quality at a flat-rate fee.

This threat of modifying the current Internet model has led to a lot of protests from content providers and user associations arguing that charging for content or blocking some types of flow is an impingement of freedom of speech and/or human rights that will impact the network development. The relevance of arguments from both sides needs to be investigated, to see whether service differentiation should be allowed, and, if the answer is in the affirmative, at which level and how far it should be implemented. This issue has already been a subject of debate at the legal and political level. In the USA the trend was first to go with imposing network regulation on ISPs to ensure neutrality, but it is not clear who, if anyone, has the authority to regulate the Internet (for instance, the Federal Communications Commission (FCC) lost a lawsuit where the Supreme Court found that it (the FCC) lacked the authority). The Federal Trade Commission (FTC) released in 2007 a report not supporting neutrality constraints, increasing the debate at the political level. This debate is also active in the European Union, as illustrated by the open consultation on network neutrality launched in 2010 at the EU level but also in each country.

- *Defining the economic model of content and (application) service providers.*

  Content and (application) service providers have two main options for getting revenue: either they charge users for access to the content, or the access is free, but advertisements are inserted thanks to a banner on the displayed web pages, and it is the advertisers who remunerate the content provider. A typical example of a service provider with an access charge is Netflix, an American provider of on-demand Internet streaming media that allows users to watch an unlimited amount of movies and TV episodes over the Internet for a monthly fee. On the other hand, we have, for instance, many newspapers, which are used to provide articles freely on the web, but display ads on their web pages. However, that trend is changing, as some newspapers start to charge for the content found on their web sites. A trade-off (as applied by the French newspaper *Le Monde*) is to publish a limited amount of information for free, while the full newspaper is available for a price (15 euros per month in 2012). According to Peter Barron of Google, "the future is going to be a mixture of paid-for content behind pay walls and free content." As we said, ads were inserted in horizontal banners, but now you can find them on other areas of the websites, even within the content of the page, in vertical banners (which are popular among advertisers because they are permanently seen even when the reader is scrolling down) and on small buttons. There are tools to optimally choose the ads that are displayed, the most well-known being Google AdSense, but the competition is organizing itself, like Microsoft with AdCenter. Those tools are easy to use. For instance, webmasters can place Google AdSense JavaScript code on their web pages in order to allow Google's servers to show context-sensitive advertisements (Google Adwords). The history of users' browsing can be used to target ads even more; this is especially the case for social networks.

A similar issue occurs for search engines, such as Google, Yahoo! and Bing. Search engines are used to reach desired content when its address or even its existence is not known by end-users. Those search engines propose a ranked list of web sites corresponding to the keywords that have been typed. But, being commercial entities, they make money by additionally presenting advertisement links, usually at the top and/or on the right of the page. The advertisement links are selected from an auction to which an advertiser can submit a bid if she wants her ad to be displayed, because she believes the keyword search may result in a financial transaction when the user sees the ad. Search engine advertising has become an important business, with the combined revenue of the two main actors in the area, Yahoo! and Google, amounting to more than $11 billion in 2009, this business being expected also to count for about 40% of total advertising revenue.

Generalizing to all types of content/applications, when ads are thought to correspond to specific content and users (for web sites), or keywords (for search engines), the selection of the displayed ads is made through auctions. There are different ways to perform an auction. The advertisers submit bids, but the highest bids are not necessarily selected, since the selection depends also on the paying mechanism. Several principles can be applied: pay-per-view, such that the advertiser has to pay each time the ad is displayed; pay-per-click, where it pays each time the ad is clicked through; and pay-per-transaction, where it pays each time the click is transformed into a sale. There are also different possibilities regarding the amount to be paid: the first-price principle, according to which the advertiser pays its bid; the second-price principle, where it pays the bid of the second-highest bidder, etc. All this will be detailed in the book, with the respective interests and properties of the different alternatives.

Those items clearly illustrate some of the stakes for all actors in the telecommunications business. Since telecommunication networks keep evolving, the question about the most favorable economic models always comes up. Some other recent economic questions that we could have developed here, but will describe more later, are, non-exhaustively, app stores business models, grids/clouds economy, content delivery networks (CDNs) and caching, etc.

1.2 The need for modeling and analysis

There are therefore a lot of questions to be answered in terms of optimal decisions for the actors, but we wish now to highlight that mathematical modeling and analysis is an important way to avoid pitfalls that can have dramatic consequences. We aim at giving examples of such situations and of some typical paradoxes that may occur if a model is not properly defined and analyzed.

1.2.1 The tragedy of the commons

An argument that is often advocated for changing Internet pricing is the so-called *tragedy of the commons*. The basic idea is that several individuals acting independently, rationally, and selfishly can actually deplete a shared limited resource. This may seem surprising, since at first sight it is not in the interest of any of the individuals to let this happen. This type of counter-intuitive outcome can be understood, and maybe avoided, thanks to modeling and analysis through *game theory*.

A *common* is a resource owned by no one, but to which all have access, i.e., it is in some sense a public good. The word *tragedy* follows here the definition of the philosopher Whitehead: "The essence of dramatic tragedy is not unhappiness; it resides in the solemnity of the remorseless working of things."

The tragedy of the commons has been highlighted by Hardin [129], taking as an example the case of herders sharing a parcel of land on which they can let their cows graze. In this situation, it is in each herder's interest to put as many cows as possible on the land, with the outcome that the quality of the grass decreases, because of overgrazing. Why is that the best strategy for a herder? This comes from the fact that the herder gets all the benefits from each additional cow put on the land, whereas the degradation is shared by all of the herders.

There are several other examples of the tragedy of the commons to consider, before discussing telecommunications. A second example we can mention is fishing, for which each fisher has an interest in fishing as much as possible, but this results in overfishing and the resource depletion which can be observed. Global warming is another striking example, with countries having a selfish interest in developing their industry and over-exploiting the soil, at the expense of the Earth's climate.

There are two commonly accepted solutions for solving this problem in general: (i) management of the common goods by a regulatory authority, such as a government; and (ii) privatization of the common goods. Regulation allows one to control the common resources by limiting their use through rules defined to drive the system to a socially optimal outcome. Converting the common resources into private ones (when possible; it is hardly possible when one is talking about oxygen, for example) is another way to produce incentives to use those resources efficiently. Regulation is used for fishing, for example, by limiting the amount which can be fished, with the risk of fines if the rule is not obeyed. The same happens for the amount of pollutant that can be released. Privatizing the goods allows one to avoid the outcome that the negative effect of using a resource is shared by all individuals, hence privatization provides an incentive not to overuse it. Interestingly, the Internet network, which was initially public, has been privatized, and the question is how to define the associated business model.

Though it is interesting, we need to mention that Hardin's work has been argued to be inaccurate by other researchers, claiming that the common land was effectively managed to prevent overgrazing and that self-interested individuals often find ways to cooperate.

It is also often argued that regulation by distant bureaucrats who lack expertise may lead to a commons in worse shape than when it remained uncontrolled.

Looking at telecommunications, unlicensed (wireless) radio spectrum is shared by all users. It has been advocated by the US regulator, the Federal Communications Commission (FCC), that humans might be vulnerable to the overuse of high-power transmitters. This is especially the case for WiFi (802.11). As a consequence, the FCC has defined a maximum transmission power for devices. Another related problem is spam in e-mails, which degrades the interest of the service, and benefits only a few individuals, or free Internet access in coffee shops, where, depending on the number of demanding customers, the quality can be low and even getting a connection can be difficult. However, a "free Internet" sign attracts customers and the gain for the coffee shop is from sales of beverages and food, with the Internet being there only for its power of attraction.

But the most important concern is about Internet use as a whole and its future: it is often seen as an almost free resource shared by users. Indeed, thanks to the flat-rate pricing, there is no additional cost if you increase your consumption as soon as you have paid your subscription. Therefore, there is a risk of ending up with an over-congested network, similar to what can be observed with overfishing, or overgrazing for Hardin's initial example. It is the opinion of the 2000 Nobel Prize winner in Economics, Daniel McFadden, that "the commons that is likely to have the greatest impact on our lives in the new century is the digital commons" and that the "management of the digital commons is perhaps the most critical issue of market design that our society faces," even if the free Internet were a motor for permanent innovation. The next step would be for providers to find ways to earn money without harming customers or innovation. This is related to the *network neutrality debate* and openness of the Internet that we will discuss in Subsection 1.2.4. For many, such as Professor Lessig from Stanford University, any attempt to legislate or regulate would be a real tragedy for the Internet's development and innovation, but, for others, we can no longer rely on self-regulation.

We therefore emphasize the need for modeling and analysis to help to understand the benefits of the different business models for each actor and anticipate the resulting situations. Appropriate mechanisms can then be designed to avoid bad outcomes, such as networks increasing their exploitation of the resource if no immediate high cost is experienced.

1.2.2 The Braess paradox

Another paradox illustrating the need to be careful when taking decisions, especially at the governmental/regulation level, is the so-called *Braess paradox*. This paradox has been highlighted for road traffic, but a similar concern can be raised for data networks, packets being seen as cars and links as roads. The basic idea is that adding extra capacity to a network when the entities selfishly choose their route can reduce the overall performance of the network. This seems highly counter-intuitive, and can be explained thanks to game theory. Similarly, closing roads can sometimes increase efficiency and allow drivers to travel faster thanks to a better repartition of routing over

Figure 1.4 An illustration of the Braess paradox: Cheonggyecheon River in Seoul, before and after the highway was closed (image taken from www.lafoundation.org).

the network. This will be explained mathematically in the next chapter. In words, the reason is that adding a new link in a user-optimized network changes the equilibrium flows, but this might result in a higher overall cost, implying that users were better off without that link: time is lost due to individuals opting for strategies that maximize their own personal interest rather than the social optimum, which are not the same in most cases. Closing those roads may force them to take the (socially better) initial path.

There have been striking examples of the Braess paradox in real life. The most well-known one is probably the Cheonggyecheon restoration project in Seoul, South Korea. In 2002, Mayor Lee Myung-bak invested $380M to tear down a six-lane road built directly over the Cheonggyecheon River and carrying on average 160,000 cars per day. This road was usually jammed. In addition to a nicer and quieter landscape, it turned out that the overall traffic flow improved after the road removal. The Cheonggyecheon river before and after restoration can be seen in Figure 1.4. Similar experiences have been observed, for example, in 1969 in Stuttgart, Germany: there was public investment in a new route into the city center to ease traffic, but the situation actually got worse until a section of the newly built road was closed for traffic again. In New York City in 1990, the closing of 42nd Street reduced the amount of congestion in the area. This type of situation in transportation networks is likely to occur more and more frequently, with drivers following the indications of their GPS devices computing the (selfish) shortest paths from source to destination.

Other funny illustrations of the Braess paradox can be found in the literature: it has been used to explain the so-called *Ewing theory*,[1] under which the basketball teams of Patrick Ewing (a player at Georgetown University and then in the New York Knicks team in the NBA) were surprisingly better when Ewing was not playing, though Patrick Ewing was one of the most renowned and efficient players. As a brief description, this is because all of the other players were going too much through him (as an added capacity), "forgetting" about the other options to get to the basket. Also, similarly to the examples involving roads, in an electric circuit, removing wires can sometimes increase the conductance.

[1] See http://gravityandlevity.wordpress.com/2009/05/28/braesss-paradox-and-the-ewing-theory/.

As we have described, the Internet can be abstracted as a transportation network, with packets corresponding to cars and links to roads. Selfish routing can lead to a non-optimal situation, and adding links or expanding capacities could lead to worsened situations if not done properly. The possibility of experiencing the Braess paradox in the context of wireless networks has been highlighted in [13]. The paradox stems from the possibility for each mobile to transmit information to more than one base station at a time thanks to multi-homing.

1.2.3 Spectrum auctions

The radio spectrum is a scarce resource that has to be shared by the network providers to serve their (wireless) customers. The governments are looking for ways to earn money from this spectrum use, and it has commonly been admitted that *auctions* are the best way to assign scarce spectrum resources. There exist other possibilities to assign the spectrum, such as "beauty contests," in which a hearing is organized and the most attractive providers are selected, and lotteries, selecting the winners randomly. But they have quickly been sidelined because beauty contests as used historically are slow to organize and not transparent, with difficult winner selection, while lotteries do not allow one to select the best providers, and are in this sense not optimal in terms of overall user satisfaction.

The principle of spectrum auctions is to sell to providers **licenses** that cover specific bands of the available radio spectrum over which they will be able to operate and to transmit communication data from and to their customers. Those licenses are for bands that are then reserved for the provider, and not used by competitors. An auction conducted properly allows one to allocate the resource to those who value it most, and can yield high revenues to governments. But the goals of regulators are also to ensure a fast deployment of services, good quality of service, and fair competition among providers, to the benefit of end-users.

Spectrum auctions have been conducted throughout the world. In the USA, between 1994 and 2012, the FCC conducted 93 auctions to allocate the spectrum in the country, permitting the acquisition of more than \$80 billion in revenues. Similar auction processes in the European Union in 2000–2001 for 3G licenses generated around \$100 billion. Auctions for 4G licenses on the 800-MHz band are now being held. In 2011, for example, in France, the three main operators Orange, SFR, and Bouygues Telecom were the winners, for a total gain of €2.639 billion. A lot of similar auctions are taking place worldwide, notably to use the spectrum freed thanks to the switch from analog to digital television.

Simultaneous multiple-round auctions (SMRAs), also called simultaneous ascending auctions, are often used to allocate the spectrum. This tool is, for example, what the FCC uses in the USA and what Sweden used in 2008 for 4G licenses. At each round, the bidders submit their bid in a sealed way (with no knowledge of other bids), and the results are then made public. For the next rounds, a minimal bid is imposed as the highest bid of the previous round, plus an increment. The process usually ends when no new bid is submitted.

However, there is a risk of tacit collusion among bidders in order to limit the final charges. Such suspicions were raised in Australia and New Zealand. This aspect has to be investigated with care. Also, a sufficient level of competition is required in order to keep prices high enough, otherwise the seller will not get much revenue. We furthermore notice that some auctions did have a disastrous outcome, such as in the USA in 1999, because the providers could not bear the costs of the bids. They took excessively high risks when bidding to ensure that they would be able to participate in the technological advances (and avoid losing their business). To make things worse, this was combined with a loss of confidence from stock markets in the telecommunications industry due to the dot com crisis, limiting their eventual investments. For this reason, the auctions had to be re-run. A similar outcome was observed in Germany and the UK for 3G licenses, illustrating that it is not a rare event.

Designing the auction process is therefore a key issue. There are many ways to select the winners and compute the charges, which will be described in the next chapter and in Chapter 5. A careful selection of the process is needed in order to end up with a spectrum allocation that is at the same time good for customers, allows providers to get a return on their investment, and lets governments earn money.

1.2.4 The network neutrality debate

We have described in the previous subsection the very active network neutrality debate on the relationships between content providers (CPs), i.e., organizations or individuals creating information, educational, or entertainment content, and Internet service providers (ISPs), i.e., the companies or organizations providing access to the network. Even though, in contrast to previous subsections, we do not aim here to highlight a potential paradox or problem that can be encountered without modeling, we want to stress that modeling can be helpful to ponder the arguments from both sides, especially from a regulator's point of view, in order to take the best possible decisions. Those decisions are critical because they will fashion the future of the Internet. The debate has indeed been very vivid, and the potential impact of regulation or non-regulation is so important that governments have held public consultations. This started in the USA, and more recently has been done in France (by ARCEP, the French regulator), in the UK (by Ofcom), and by the European Commission (EC) in 2010, after having been done in Norway (by the Post and Telecommunications Authority, NPT) and Sweden (by the Post and Telecom Agency, PTS) in 2009, among others.

The proponents of neutrality push forward several arguments. First, they point out that the Internet architecture is based on the Open Systems Interconnection (OSI) model, which is organized in layers managed by independent entities (see Table 1.2), the relations between layers being operated thanks to standardized protocols. Without neutrality, this successful and cheap model could disappear because ISPs may develop proprietary protocols that can harm the development of new applications by third parties. This structure also helps to preserve a level of anonymity and openness without which governments or political entities willing to limit information diffusion can control the network and harm *freedom of speech*. This point has been emphasized by the recent revolutions in

Table 1.2 The seven layers of the OSI model

Layer number	Name	Description
7	Application	Network process to applications, such as file transfer and e-mail
6	Presentation	Data encryption and decryption, makes the communications between two hosts possible
5	Session	Communication between hosts, management of sessions
4	Transport	End-to-end connections, decomposition in packets, reliability and flow control
3	Network	Routing and logical addressing
2	Data-link	Transfer between adjacent nodes
1	Physical	Media, signal, and binary transmission

Arabic-speaking countries, with information diffusion thanks to social networks, and the governments trying to cut down the traffic. Also, neutrality with cheap access to the network by CPs will encourage a high level of innovation from CPs that as a consequence will maintain demand from customers and revenues for ISPs. A last important argument for neutrality is *universality*: all consumers are entitled to reach meaningful content, whatever the technical limitations of their service, and to attach any type of device to the network. The basic idea behind that principle is to keep the whole population informed and able to communicate.

The opponents of neutrality, on the other hand, claim that ISPs have no incentive to invest in networking capacities because of the reduced revenues. They also claim, as demonstrated by the Chicago school of antitrust analysis, that **vertical integration**, meaning grouping access and content into a single service, is the economically most efficient paradigm. Vertical integration is actually a current trend, with big CPs such as Google investing in access, and ISPs offering portals with news and other services. The merger of AOL (an ISP) and Time Warner (a CP) in the USA is also a striking illustration. A third argument is that allowing service differentiation is often a way to increase not only the revenues of providers but also social welfare, by responding to the heterogeneity of users in terms of preferences. Another claim is that, even if it is true that the end-to-end architecture was a reason for the success of the Internet, there is a need for evolution because of economic changes. The universality principle is also questioned because it may harm investments in the network if access is granted. Finally, neutrality is argued to give more power to already powerful content providers such as Google.

We will analyze some of these arguments in the next chapters, but we believe here that the debate rather opposes a humanist (the proponents) to an economic (the opponents) view of the Internet.

1.3 A description of the actors

We have provided a few examples of the issues at stake, but, in order to get a full picture, we need to present all the actors, organized in different categories. A schematic view

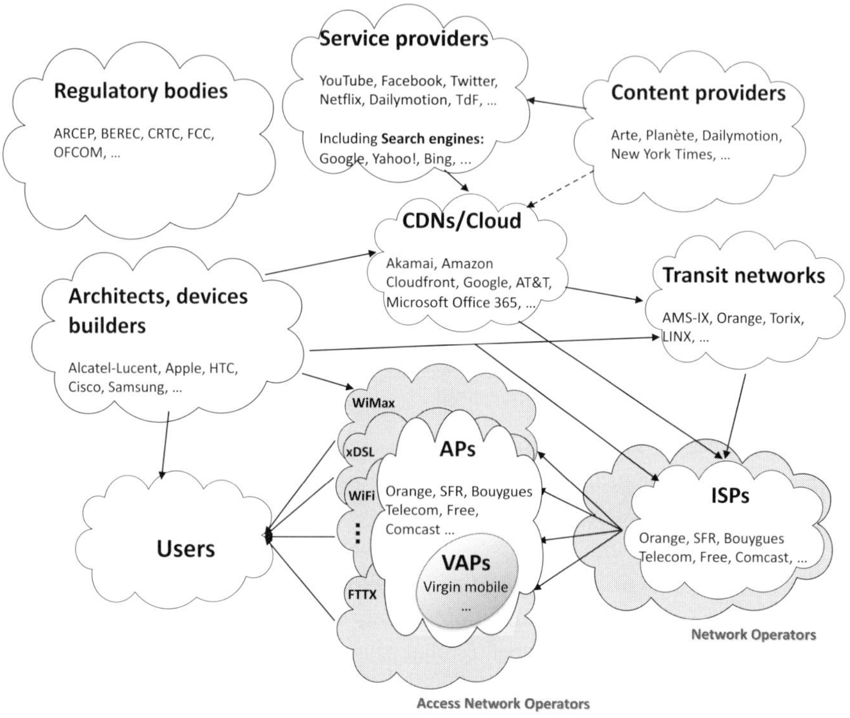

Figure 1.5 The actors of the Internet.

of the actors of the Internet is drawn in Figure 1.5. The various categories, and the presented examples of entities pertaining to those categories, are of course debatable, but we believe that it provides a good picture of the telecommunications ecosystem.

To be more specific, the actors can be categorized as follows.

1. *End-users.*

   The end-users are the actors to whom services are delivered, and whose interest is in getting those services if the prices are "reasonable." They are often represented through demand functions that describe their reactions to the offers of services, associated with their price and quality of service (QoS). Sometimes, for example for a wireless subscription, a device (a phone) is proposed at a discount. Internet boxes are similarly associated with Internet subscriptions. Those devices are there to make the subscription even more attractive, especially when the competition among providers is fierce. Modeling users' behavior and level of acceptance in terms of price for a given service has always been hard. This is even more difficult with the evolution of users' behaviors. To illustrate this, we can remark that users are less demanding in terms of voice quality: while they used not to accept a low-quality transmission on their fixed lines, this is less true now, because they have got used to it from their mobile phones and voice-over-IP services like Skype. Similarly, users are currently very flat-rate prone for Internet use. But may this change? There is huge pressure from providers to switch to usage-based pricing.

2. *Access network service providers.*

   Access network service providers, also called *eyeball providers*, are those in charge of the network resources at the access and offering access to the end-users. This access can be through a wired line at home, thanks to a modem or a box, or through radio links, thanks to WiFi, 2G, 3G, or 4G wireless networks. Wired access can be through the standard copper lines of traditional telephony, or fiber to the home (FTTH). Therefore several access technologies coexist, and not all providers operate on all technologies. They have to strategically decide on which ones to make investments in, in terms of infrastructure, or licenses for wireless technologies. As we highlighted in the previous subsection, spectrum licenses may be very expensive. Developing a wireless network with antennas covering an area sufficient to reach a large proportion of the population may also be costly, and even hardly accepted by people worried about the potential side effects of antennas.

   There are two kinds of access operator, those effectively operating a network, and virtual ones who do not own any infrastructure but can lease the network of a competitor to serve their customers. The development of virtual operators has been encouraged by regulators in the mobile industry to foster competition since very few operators actually own a license and an infrastructure. The interest for the primary or "real" operator to sublease its network is not obvious though, since this would mean losing customers to the virtual one. The issue must therefore be studied, as well as the design of rules to ensure that all operators would benefit from this. In Figure 1.5, we present the four main Internet network operators in France (Orange, SFR, Bouygues Telecom, and Free) and a virtual operator, Virgin Mobile. Note that Free is also partly a virtual operator for 3G telephony, since its architecture is not fully deployed yet and it therefore uses the Orange network.

   For access network providers in general, there is also the question of how to charge customers. The scheme has to be attractive for customers, but must also allow providers to make profits. There exist several potential pricing schemes, and there is a huge amount of research activity around this issue. Let us mention the two main pricing schemes in telecommunications: one is the above-mentioned flat-rate pricing, such that customers pay a subscription fee and can use the network resource as much as they want; and the other is usage-based pricing, such that customers pay for the amount of resource they have actually used. Flat-rate is the main scheme applied for Internet access, while usage-based is the traditional one for telephony. Interestingly, there has recently been a trend for changes: for instance volume-based pricing has been pushed forward for the Internet by providers (such as in Canada), and there are more and more flat-rate subscriptions for wireless telephony. We remark also that, thanks to voice-over-IP, telephony is more and more often included in the Internet subscription, hence a flat-rate-like pricing scheme.

   Note that pricing is not only a way for providers to make profits, but also allows them to control the network traffic, with higher prices leading to a smaller demand, and lower prices to a higher demand. Several innovative pricing schemes have been

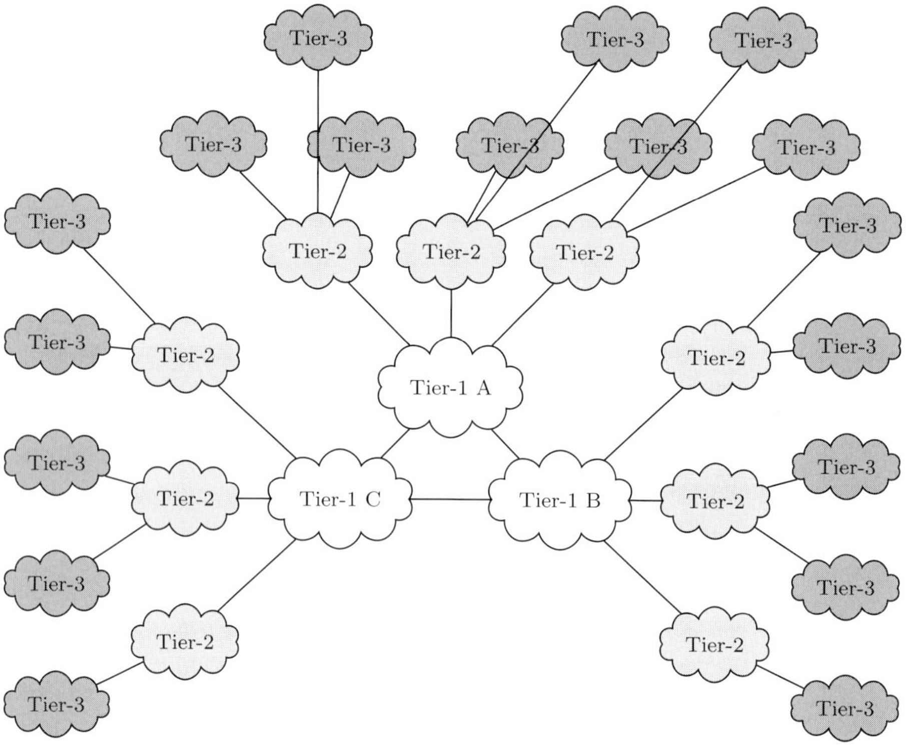

Figure 1.6 Hierarchy of transit providers.

imagined in the literature to statically or dynamically charge for network resources. This issue will be developed in Chapter 3.

3. *Transit providers.*

   The **transit providers** are those who run the network resources at the core. Their role is to take care of traffic transfer between access providers and/or other transit providers. As we have discussed, there is currently an organization of transit providers in a multi-tier hierarchy (even if the boundaries between levels tend to blur), with a clique of Tier-1 providers connected through peering links; more regional Tier-2 providers, namely customers of Tier-1 providers who can peer with some other providers but need to pay transit fees to reach some parts of the network; and smaller businesses known as Tier-3 providers, themselves customers of Tier-2 providers. The end-users are at the bottom of this hierarchy. Content providers used also to be customers of Tier-3s but tend to have an access point at a higher level of the hierarchy. This illustrates the necessity to understand the most profitable strategy for the actors, and the potential impact of a flatter (meshed) network topology. To get a somewhat (over-)simplified idea of the hierarchy, Tier-1 providers are supposed to guarantee global connectivity, and lower-level actors in the hierarchy are customers of providers one level up; see Figure 1.6, which describes the hierarchy and the ways to interconnect with other customers by going up and then down in the hierarchy.

A transit provider will sell upstream access to the Internet to a customer. Usually peering agreements are the economic relationships between Tier-1 providers, while a transit price per megabit per second per month is established between providers at different levels of the hierarchy.

We can remark that some access providers are also transit providers. This is typically the case of Orange in France, British Telecom in the UK, etc. This is part of the vertical integration paradigm that will be discussed throughout the book.

4. *Content providers (CPs).*

   From the usual definition, a content provider is an organization or individual that creates information, educational, or entertainment content for the Internet, CD-ROMs, or other software-based products. A content provider may, but need not, provide the software needed to access the material.

   In the Internet, content mainly used to be produced by users, and was located usually on their individual web pages, hosted by service providers, and delivered thanks to a content delivery network's computer systems (see below for descriptions of those actors). The content could be the user's web page on his access provider facilities, or could be made available on his social network account or through a photography service provider, etc. A user's content is now usually made available to other Internet users, or to authorized people only, depending on the context. This may include copyrighted material accessible through peer-to-peer. But we now also have "bigger" content providers, i.e., companies selling goods, providing music (Sony or Universal Studios), displaying news (CBS, BBC, newspapers such as the *New York Times*), providing videos (TV channels such as Arte and Planete in France). Some content providers make money by selling their content, while others provide the content for free but get their revenue by displaying advertisement banners.

   Content providers have to be differentiated from service providers in general.

5. *Service providers.*

   Service providers (or, more exactly, application service providers) are those providing facilities or doorways to/for content providers. We do not include here the ISPs, which have been assigned above to a specific class. Service providers host organizations'/companies'/users' content, so that it can be accessed worldwide. We can think of the portals of ISPs on which news and personal web pages can be published, YouTube and Dailymotion on which you can publish your videos or music, but also social networks such as Facebook and Twitter, to display your photos or personal news and opinions (we remark that Facebook could be seen as a content provider, since it owns the copyright of all published content). In most of those cases, the provider can make money thanks to (targeted) advertisements on the displayed pages. As another example, Netflix, in the USA, Canada, Latin America, the UK, and Ireland, is an Internet subscription-based service for watching movies and TV programs. TdF in Figure 1.5 is a TV broadcaster, hence offering its service to TV channels (content providers).

   Other specific categories of service providers we can think of are search engines and application stores. Search engines (such as Bing, Yahoo!, and Google) play a key role in the Internet, since in most cases end-users use them by inputting a keyword in order to reach the most relevant related content. They generate revenue by displaying commercial links in addition to the *organic* ones which are based only on relevance. An application store is an online store for downloading applications, mainly for mobile devices. The first noticeable one is Apple's online store (App Store) for downloading free and paid iPhone, iPod touch, and iPad applications, which was launched in 2008 with the iPhone 3G device. Its competitors now include Android Market (for Android OS), OVI Store by Nokia (for Symbian), BlackBerry Appworld, Windows Store by Microsoft (for Windows Mobile), etc.

6. *Content delivery networks (CDN)/clouds.*

   Content delivery networks are large distributed systems of servers deployed in the Internet. They are used by content and service providers for their infrastructures: while service providers provide the tools, the computer facilities are managed by CDNs. For example, Netflix's customer traffic is supported by Amazon's cloud services and Akamai. The role of CDNs is to produce the highest possible performance for content access by users. A CDN is a special class of service provider offering storage facilities among others that we prefer to single out. Basically, CDNs are paid by content providers and pay network providers for the traffic exchanged, but more specific economic relationships can exist. For example, Akamai has many contractual relationships with network operators: it usually pays nothing for the bandwidth when it has deployed servers within the network of the operator, or within an upstream provider, and has agreed to serve content only to the prefixes within the network and its downstreams.

   From the definition by the National Institute of Standards and Technology (NIST), cloud computing is "a model for enabling ubiquitous, convenient, on-demand network access to a shared pool of configurable computing resources (e.g., networks, servers, storage, applications, and services) that can be rapidly provisioned and released with minimal management effort or service provider interaction." The role of clouds is therefore to provide facilities for companies and individuals to perform operations without their having to buy their own costly resources. There is a lot of activity in cloud economics regarding the effort to define how cloud owners should be paid for the use of their resources.

7. *Network architects and device constructors.*

   Network equipment designers and manufacturers, such as Cisco and Alcatel-Lucent, are those providing network infrastructures (devices) to users, ISPs, content providers, etc. Their products are network infrastructures and software for performance-optimized transmissions with limited complexity, and *green* behavior, meaning selecting energy-efficient networking technologies and minimizing resource use whenever possible, which is a topic of current concern.

   Device constructors such as Apple, Nokia, Samsung, and Hewlett Packard are those providing the end-users with devices such as mobile phones and computers. They provide the "physical tools" for an efficient end-user use of applications provided by CDNs and CPs.

8. *Regulators.*

   A regulator is an agency established by a government. Its role is to control how the market operates. It intervenes if necessary, in order to ensure that some predefined criteria are respected. Typically, the criteria are based on the public interest, notions of fairness, etc. The regulatory bodies have an important role when some actors have a dominant position in the market. It can be a historic (national) network operator, which may prevent the entrance of competitors and restrict competition, or any other actor with a dominant position in such a specific role (Google for search engines, whose ranking policies have recently been accused of favoring YouTube content; and some time ago Microsoft's dominance on operating systems). Regulators also ensure that satisfying relationships pertain among the different sets of actors. Their role is, for example, central in the network neutrality debate, to define the rules that have to be implemented for the relations between content and network service providers.

1.4 Goals of the book

We aim in this book at helping the reader to better understand the (mainly economic) relations between the different actors in telecommunication networks, and would like to help those actors to design models for choosing their strategies more efficiently, taking into account the decisions of all other participants.

We remark that our goal is *not* to predict the evolution of the Internet. Indeed, as mentioned by Andrew Odlyzko in [239], "predicting the evolution of the Internet is an error-prone business." We again rather aim at providing tools for a better understanding of the telecommunications ecosystem and better decision-making for all actors, avoiding the pitfalls that could lead to costly errors.

We also want to give examples of how modeling and analysis should be used in telecommunication economics. We plan to describe the current behavior of the different players, how they get revenue, and how their economic relationships work. To that end, the business models for different settings are presented. This helps one to understand the behavior of the current Internet and some of the evolution that we are experiencing. For example, it allows one to understand the trends of vertical integration, with Google investing in access, service providers such as Netflix investing in CDNs, caching strategies by ISPs, Tier-2 networks trying to interconnect to bypass the transit cost through Tier-1 competitors, etc.

This book is part of the stream of works on *revenue management* [285], which tries to maximize the revenue of the actors. It is often admitted that without the application of such a theory and differentiation of products in a competitive context the providers would go soon out of business (as the outcome of *Bertrand competition* [106]).

1.5 Outline of the book

We hope that this introduction has highlighted the need for a better understanding of the relations between the different actors, for the economic good of those actors and of society itself. In particular, we have described why modeling those actors and analyzing their interactions mathematically might be helpful for taking better decisions and avoiding some (counter-intuitive) negative outputs. The rest of this book is organized as follows.

Chapter 2 describes the basic mathematical elements necessary in the rest of the book. It presents the basics of economic theory: how to represent the different actors, or more exactly their preferences and goals in a mathematical setting, and how their decisions impact the other actors. The goal of actors is to maximize their preferences, so we present, with examples, the main elements of optimization theory, as well as fixed-point theory to characterize stable outcomes. But, to understand the *interactions* between actors, with the decision of each one having potentially an impact on the outcome of others, we need the framework of *game theory*, which will be the key tool throughout the book. The main game theory results used in the book are provided there, as well as the explanation of numerous paradoxes such as the Braess paradox discussed earlier. We finally describe the elements of auction theory used in telecommunications.

Chapter 3 discusses the economics of ISPs with respect to end-users. While initially the telecommunication industry was a monopoly in most countries, increasing competition, and the congestion experienced, drives the need for careful definition of the charging scheme. One idea is to more closely align prices to real costs. However, taking care of user expectations and marketing is also of primary interest. The chapter describes the history of access pricing models and the expected trends of the future. It describes the flat-rate pricing scheme, volume-based pricing, and different congestion-based pricing alternatives, with propositions specific to each technology, as well as the desired properties from the perspectives of users and providers for the design of an efficient pricing scheme. Accounting and billing and the economics of bundling are also presented.

Chapter 4 deals with the economics of content providers, service providers, and content delivery networks/clouds. We start with some history on the related business models, and then describe how content/service providers (including search engines) get their revenue, in most cases through advertising. We present the basic principle, explain how ads are chosen, and describe the auction schemes between advertisers that are applied, along with their properties, as well as the existing tools to submit advertisers' bids in practice. The price paid by the advertiser can follow a large set of rules that we cover. We also explain in that chapter the business models of application stores, distinguishing free applications with advertisements from paid applications. We finish the chapter with the economics of content delivery networks and clouds: how those actors make money, the propositions made in the flourishing research literature, and the reasons for content owners to use those services are discussed.

Chapter 5 is about interactions among service providers. The first type of interaction we describe is the auctions for wireless spectrum. The huge amount of money required to buy 3G licenses a few years ago, and now 4G ones, forces providers to develop a serious understanding of the auction rules and of their own (and their competitors') best strategies. We then go to the pricing competition among access providers. Their actions concern the price (and QoS/capacity) definition in order to attract customers. But this has to be chosen with knowledge of what competitors can do. Indeed, pricing models that seem adequate in the monopolistic case have to be investigated in a competitive environment, because they may lead to results opposite to their initial goal in terms of behavior, efficiency, and survivability. We describe this competition as multi-level games, played on different time scales: on the largest ones, providers choose how much they invest, in terms of capacity, QoS provisioning, etc.; then they choose their price competitively; and, on the shortest time scale, users choose their provider. The games are usually solved by *backward induction*; that is, the games at a higher level are solved anticipating the solution of the later games on shorter time scales. Models of customer churn (i.e., switching of providers) are presented, as well as strategies of providers to prevent it, and the regulation rules that can be imagined for a more efficient economic model. The question of licensing the resource spectrum versus sharing it is also discussed, from the regulator's perspective. The last part of the chapter is about the economics of interconnection: it focuses on the economics of transit providers, the tendency toward a flatter Internet, and the links with wireless ad-hoc networks.

Chapter 6 discusses competition at the content and service level. Competition between content providers, between application stores, and between search engines, and its potential impact on the providers' decisions, are investigated.

Chapter 7 is about the interactions between content/service and network service providers. We describe the current and past economic relations, and then go to the sensitive network neutrality debate. The interest in *vertical integration*, that is the process of regrouping several steps in the production and distribution of a service under the control of a single company, is also discussed there. Typically, some service providers are investing in service delivery and getting their own networks. Similarly, ISPs can be both access and transit providers, and in addition can become (application) service providers.

2 Mathematical foundations: optimization, game theory, auctions

The goal of this chapter is to provide the economic vocabulary and the analytical tools that will be used throughout the book. We do not intend to be exhaustive here with regard to the problematics of optimization, game theory, or auctions; instead we focus on the minimal material that is needed to understand the derivations in the next chapters, and provide illustrations of those concepts and methods by presenting concrete examples. We will, however, indicate appropriate references for the reader interested in developing those aspects more deeply.

The chapter first introduces some key criteria used to quantify the economic performance of a situation from the point of view of an individual actor, of a group of actors, or of the system as a whole (Section 2.1). Then the emphasis is put on the mathematical tools and results that will be needed in order to characterize the specific outcomes obtained when optimizing such performance criteria (Section 2.2). When different stakeholders seek to optimize their own objectives through some decision, the resulting outcome can be suboptimal for all of them, due to their interactions. The analysis of those interactions and their consequences is the object of game theory; we provide the general framework of game theory and specify several types of games that will be of interest in this book (Section 2.3). Finally, we focus on the problem that is faced by a system planner dealing with self-interested actors: using the tools from game theory, the planner can anticipate the reactions of the actors to set some rules so as to drive the system in a desirable direction. This problem is known as the mechanism design problem, of which auctions are a specific case (Section 2.4).

In terms of the mathematical notation used throughout the book, we represent sets with calligraphic letters, vectors (matrices) with bold lower-case (upper-case) letters, and random variables with upper-case letters. When dealing with abstract concepts of no determined dimension, we use lower-case letters.

2.1 Basic economic theory

Telecommunication networks involve a lot of different stakeholders, all of which have their own preferences: while users mainly focus on the QoS they experience and the price they are charged (and sometimes other considerations such as the provider's reputation), service providers are interested in the revenue they get from subscribers, in the management costs incurred by user traffic, etc. Those discrepancies are the core reason for applying economic reasoning to telecommunication networks: since not all stakeholders are interested in the same outcomes, one has to consider the conflicts that appear – and the trade-offs that should be imposed – among the preferences of the actors in the whole ecosystem.

2.1.1 Representing actor preferences

The preferences of a stakeholder among the possible outcomes of a situation are generally represented through a **utility** function, which maps the set of outcomes to the set $\mathbb{R}$ of real values. The utility function can be used to determine, among several outcomes, the preferred one(s) of the considered actor: a stakeholder i with utility function U_i strictly prefers outcome o_1 over o_2 if and only if $U_i(o_1) > U_i(o_2)$. If $U_i(o_1) = U_i(o_2)$, then we say that the actor is indifferent between those two outcomes.

Note that utility functions are defined only up to a strictly increasing transformation: for a utility function U_i and any strictly increasing function $f : \mathbb{R} \mapsto \mathbb{R}$, the function $f \circ U_i$ represents exactly the same preferences of actor i among the outcomes. Let us denote by $\underline{U_i}$ the equivalence class of utility functions corresponding to the same preferences as U_i.

When monetary exchanges are involved, we often separate those exchanges from the other aspects of the outcome (resource allocation, quality of service, ...), and assume that each actor i is sensitive only to the non-monetary part q of an outcome and to the price p_i that he or she has to pay – which can be negative if actor i perceives a monetary reward. The prices paid are represented in a vector $\boldsymbol{p} = (p_i)_i$, so that an outcome can be decomposed into $o = (q, \boldsymbol{p})$. Additionally, it is often considered that the impact of price on utility is independent of non-monetary components, i.e., that, for two outcomes $o = (q, \boldsymbol{p})$ and $\tilde{o} = (\tilde{q}, \tilde{\boldsymbol{p}})$,

$$\forall x \in \mathbb{R}, \quad U_i(q, p_i + x) - U_i(q, p_i) = U_i(\tilde{q}, \tilde{p}_i + x) - U_i(\tilde{q}, \tilde{p}_i). \tag{2.1}$$

It is also (naturally) assumed that U_i is strictly decreasing in p_i, i.e., that, for a given non-monetary outcome, each actor i prefers to pay as little as possible. Finally, considering a reference outcome without monetary transfer q_0 (that classically corresponds to a situation where the actor refuses to enter the system), we assume that the price effect can always exceed the non-monetary part of an outcome: formally, we suppose that

$$\forall q, \quad \lim_{p_i \to -\infty} U_i(q, p_i) > U_i(q_0, 0) \quad \text{and} \quad \lim_{p_i \to +\infty} U_i(q, p_i) < U_i(q_0, 0).$$

Then we can define for each non-monetary outcome q the quantity

$$V_i(q) := \sup\{p_i : U_i(q, p_i) - U_i(q_0, 0) \geq 0\}. \tag{2.2}$$

That quantity, called the **valuation** of actor i for the outcome q, or his or her willingness to pay for that outcome, represents the maximum price he or she would be willing to pay to prefer the outcome over the reference situation $(q_0, 0)$. The function $V_i(\cdot)$ is then called the **valuation function** of the actor. Note that the properties above, and the value of V_i, are independent of the particular form of the utility function chosen to represent actor i's preferences, within the equivalence class $\underline{U}_i$.

Under those assumptions, the preferences of actor i can be represented by a **quasi-linear** form of the utility function, that is, by the difference

$$U_i(q, p_i) := V_i(q) - p_i. \tag{2.3}$$

We remark that this function conveys exactly the same preference information as does any representative of the equivalence class $\underline{U}_i$. In addition, it has the advantage of being interpretable in monetary-equivalent terms: since $V_i(q)$ is the value that the outcome has to actor i and p_i is the price he or she actually pays, $U_i(q, p_i)$ can be seen as the benefit (in monetary units) of actor i.

In most cases, we will use such quasi-linear utility functions in this book.

Example 1 Let i be a user wanting to download a file as fast as possible. Denoting by δ the download time and by p_i the total price paid by i for the service, the utility function $U_i(\delta, p_i)$ is non-increasing in δ. In addition, suppose that the user is willing to pay an extra unit price π per second of reduction in the download time, but only if that download time is less than some value $\delta_{\max}$: above that threshold, the user prefers not to use the service even if it is free. If $\delta = \delta_{\max}$ the user is indifferent between using the service (for free) and not using it.

In mathematical terms, $U_i(\delta_{\max}, 0)$ is the utility (for user i) of not using the download service. Additionally, the user being willing to pay π per second saved below $\delta_{\max}$ means that for all $0 \leq x \leq \delta \leq \delta_{\max}$ we have $U_i(\delta - x, p_i + \pi x) = U_i(\delta, p_i)$. Still assuming U_i strictly decreasing in p_i, we have from (2.2), applying our reasoning above with $q_0 = \delta_{\max}$ when $\delta \leq \delta_{\max}$,

$$\begin{aligned} V_i(\delta) &= \sup\{p_i : U_i(\delta, p_i) \geq U(\delta_{\max}, 0)\} \\ &= \sup\{p_i : U_i(\delta, p_i) \geq U(\delta, \pi(\delta_{\max} - \delta))\} \\ &= \pi(\delta_{\max} - \delta). \end{aligned}$$

When $\delta > \delta_{\max}$ the user is not willing to pay anything for a service improvement; hence he or she gets the same utility as if $\delta = \delta_{\max}$.

We therefore observe that the valuation function $V_i(\delta) = \pi[\delta_{\max} - \delta]^+$, where $x^+ := \max(0, x)$, reflects the preferences of the user: the user values one second at the unit price π, and is not willing to pay anything for the service if the delay exceeds $\delta_{\max}$. One of the interpretations that can be drawn from that expression is that the user is willing to pay up to $\pi\delta_{\max}$ for a "perfect" service (i.e., one with zero transfer delay).

Finally, the quasi-linear form of the utility function of the user is

$$U_i(\delta, p_i) = \pi[\delta_{\max} - \delta]^+ - p_i,$$

which can be used to compute the (monetary-equivalent) gain of a user paying p_i for a service with transfer delay δ.

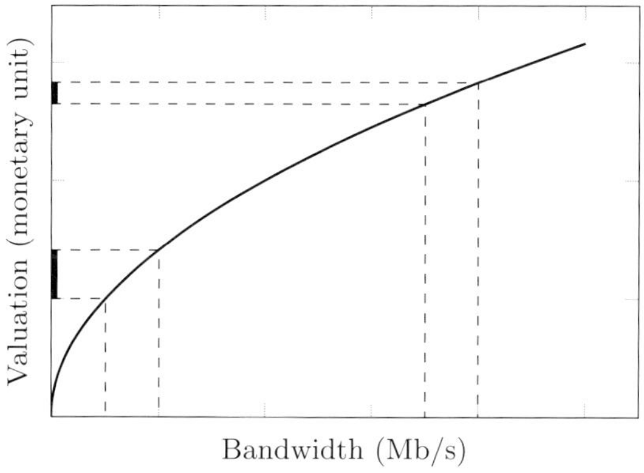

Figure 2.1 A concave valuation function.

2.1.2 Effect of prices on demand

Consider a user i who has the opportunity to buy a quantity q_i of a good (say, some communication capacity) at a *unit price* p (so that the total price paid is $p_i = q_i p$), and denote by $V_i(q_i)$ the willingness to pay of user i for that quantity. It is natural to consider V_i to be non-decreasing (the more bandwidth I get, the happier I am), but it is also often considered that V_i is *concave*, meaning that the additional benefit of an extra unit of good decreases with q_i. Indeed, obtaining an extra 1 Mb/s of capacity has a considerable impact on the transmission time of a file if the initial allocation was 1 Mb/s (the transmission time is then halved), whereas the impact is much smaller if the user already had $q_i = 1$ Gb/s, as illustrated in Figure 2.1. Assuming that V_i is differentiable, we therefore consider its derivative $V_i'(\cdot)$ to be non-negative and non-increasing. This function $V_i'(\cdot)$ is called the **marginal valuation** function of user i, representing the increase in the valuation per extra unit of good.

Now, when the user can choose the amount q_i^* of capacity to buy, he or she will do so in order to maximize his or her utility

$$U_i(q) = V_i(q_i) - q_i p,$$

which is a classical convex optimization problem when V_i is concave, with solution

$$\begin{cases} q_i^* = 0 & \text{if } V_i'(0) < p \\ q_i^* = +\infty & \text{if } \lim_{q\to\infty} V_i'(q) > p \\ q_i^* \in (V_i')^{-1}(p) & \text{otherwise,} \end{cases} \tag{2.4}$$

where $(V_i')^{-1}(\cdot)$ is the generalized inverse of V_i', i.e.,

$$(V_i')^{-1}(p) = \{x : \forall \epsilon > 0,\ V_i'(x - \varepsilon) \geq p \geq V_i'(x + \varepsilon)\}.$$

When V_i' is continuous and strictly decreasing, $(V_i')^{-1}(\cdot)$ is the classical inverse function of V_i'. The utility-maximizing quantity can also be seen as the one cancelling out the user's **marginal utility** $U_i'(q) = V_i'(q) - p$, which represents the utility increase per extra unit of good bought.

The application relating the quantity (2.4) that user i would buy at the unit price p is classically called the **demand function** of the user. In this book, we will most of the time denote this function by $D_i(\cdot)$.

Similarly, when considering the total quantity that would be bought by a set of users, we will use the term *demand function* to denote the aggregated demand (the sum of the demand functions of all users), which we will denote by $D(\cdot)$. There, the variations of D may come from the individual sensitivities of users to prices, but also from the heterogeneity among users, as illustrated in the next example.

Example 2 Consider a very large population of users, modeled as a continuum of total mass normalized to 1, who decide whether to subscribe or not to a service (say, the ADSL service) charged at a flat fee p. Assume that not all users value the service equally, and denote by $F(\cdot)$ the cumulative distribution function of the willingness to pay among the population; i.e., a proportion $F(p)$ of the population is willing to pay p or less for the service.

Then the aggregated demand function, which is also here the proportion of the population deciding to subscribe at price p, equals the proportion of users who are willing to pay more than p, that is

$$D(p) = 1 - F(p).$$

If the total mass of users is m instead of 1, then the demand function is simply $D(p) = m(1 - F(p))$. Note that users whose willingness to pay exactly equals p are indifferent between subscribing and not subscribing: their decision must therefore be specified in the model, especially if F is not continuous in p.

If, for example, the willingness to pay is uniformly distributed over an interval $[\theta_{\min}, \theta_{\max}]$, then the global demand function equals

$$\begin{aligned} D(p) = 1 - F(p) &= 1 - \frac{[\min(p, \theta_{\max}) - \theta_{\min}]^+}{\theta_{\max} - \theta_{\min}} \\ &= \min\left(1, \frac{[\theta_{\max} - p]^+}{\theta_{\max} - \theta_{\min}}\right). \end{aligned}$$

Note that, in that example, individual demands are discontinuous (users subscribe or not, in contrast to (2.4), where each user adapts his or her demand level to the unit price), but aggregated demand is continuous due to the assumptions of a large population and a continuous distribution of sensitivities.

An important characteristic of a demand function (be it individual or aggregated) is its variability with price. The measure classically used to quantify this variability is called **demand elasticity**, and is defined as the relative variation of demand with respect to a relative variation of price. Mathematically, the elasticity e_D when the unit price is p is given by

$$e_D(p) := \frac{\mathrm{d}D(p)/D(p)}{\mathrm{d}p/p}. \tag{2.5}$$

Note that the elasticity is negative in general (except for some extremely specific cases where demand can increase when the price increases), so economists often call the elasticity the absolute value of the term in (2.5), without confusion. In this book, we will not make that shortcut, and will manipulate e_D as a negative value. The importance of the elasticity concept comes from its strong influence on the behavior of a revenue-maximizing price setter: if demand does not decrease too much with price, then a price increase leads to a revenue increase, while on the other hand a high variability in response to price would not allow the price-setter to raise prices too much (since the diminution of demand would overcome the price increase in terms of total revenue). The threshold value of the elasticity is -1: if $e_D < -1$ ($e_D > -1$) then a revenue-maximizing price setter should decrease (increase) its price, since the derivative of the revenue $pD(p)$ can be rewritten as $D(p)(1 + e_D(p))$.

2.1.3 Global performance of an outcome

The interactions among all stakeholders (all types of service providers, users, regulators, possibly brokers) lead to an outcome, or possibly several different outcomes. Several metrics can be defined in order to compare them, so as to decide whether one outcome is preferable to another, and possibly design appropriate regulation rules leading to "better" outcomes.

All the criteria introduced below involve the utility or valuation functions of some (or all) stakeholders. Depending on which actors the regulator wants to favor, and what aspects are preponderant, the objective of a rule can be to maximize one of those metrics.

Pareto optimality

A minimal requirement for an outcome to be considered "good" is that, on deviating from that outcome, there is no way to make all actors happier at the same time. That condition is called **Pareto optimality**, and is formally defined as follows.

Definition 2.1 An outcome o is Pareto-optimal (or Pareto-efficient) if, for any feasible outcome $\bar{o} \in \mathcal{O}$,

$$\exists i : U_i(\bar{o}) > U_i(o) \Rightarrow \exists j : U_j(\bar{o}) < U_j(o), \tag{2.6}$$

where $\mathcal{O}$ represents the set of feasible outcomes. In words, from a Pareto-optimal situation it is not possible to increase the utility of an actor without harming any other actor.

Example 3 Let us consider a given amount Q of a resource (say, transmission capacity), that has to be shared among several users, without any price involved.

If the valuation functions of all users are strictly increasing in their allocation, and independent of the allocations of the others, then the Pareto-optimal outcomes are simply all outcomes such that all the capacity Q is allocated.

- If not all capacity is allocated, giving the remaining capacity (free of charge) to any user strictly increases that user's utility without harming any actor, thus violating (2.6),
- If all capacity is allocated, then it is not possible to strictly increase the utility (the share) of one user without strictly decreasing the utility of another user; i.e., (2.6) is satisfied.

Fairness measures

The **fairness** criterion addresses the question of the relative utilities that participants extract from an outcome. In general, that criterion is used when considering only the utilities (or allocations) of users, not the revenues of providers. Hence we will focus on that case here. Note, however, that, when several providers are involved and some revenues have to be split among them, some fairness considerations can be applied at the provider level as well. Also, in most cases those metrics are directly applied to the allocations instead of the utility functions, since their original aim was to provide objective guarantees of fairness of treatment among users, with no consideration for the possible heterogeneous perceptions that users have of their allocations. For simplicity, we present that case here, but the criteria defined below are also applicable to utilities.

Let us therefore consider an allocation $x = (x_i)_{\text{users } i}$ of some resource among users. Depending on how strict the equity constraints are (i.e., how much inequality is tolerable), different metrics can be used. The strictest case is called **max–min fairness**, and corresponds to the fact that no avoidable inequality is tolerated. Mathematically, an outcome is max–min fair if the smallest allocation among users is maximized first, then the second-smallest allocation is maximized, and so on. (If one is dealing with utilities, allocations should be such that the smallest utility is maximized, then the second-smallest one, etc.)

The **weighted fairness** criterion allows one to introduce preferences among users, while still avoiding extreme inequalities. The principle is the same as that of max–min fairness, except that a user-specific function of the allocation x_i is considered for each user i [202]: there the outcome should maximize the smallest value of $f_i(x_i)$ among all users, then the second-smallest value, and so on. The value $f_i(x_i)$ then represents the weight of user i when his allocation is x_i; note that the initial max–min fairness condition is a particular case of weighted fairness, with $f_i(x_i) = x_i$.

The **proportional fairness** criterion [148] is expressed using aggregate changes: an allocation x^* is proportionally fair if, for any other feasible allocation x, the sum of the relative changes is negative; i.e.,

$$\sum_{\text{users } i} \frac{x_i - x_i^*}{x_i^*} \leq 0.$$

Interestingly, it is proved in [148] that proportionally fair allocations are the feasible allocations maximizing the quantity $\sum_{\text{users } i} \log(x_i)$.

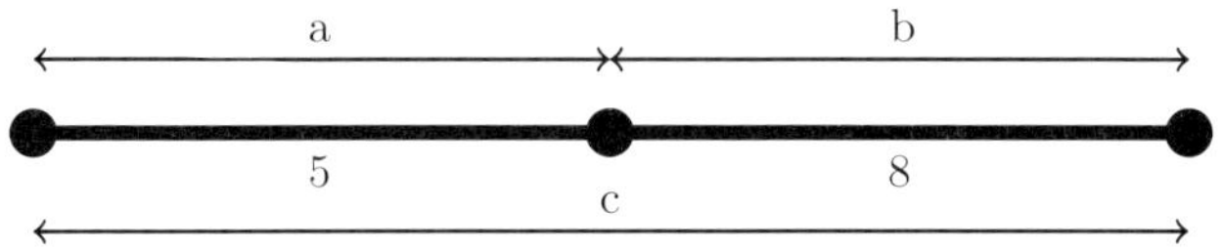

Figure 2.2 Topology of a simple network with three users a, b, and c. Arcs are labeled with their capacities.

Note that max–min and proportional fairness are two particular cases of **α-fairness**: an allocation x^* is α-fair for $\alpha \geq 0$, if, for any feasible allocation x, we have

$$\sum_{\text{users } i} \frac{x_i - x_i^*}{x_i^{*\alpha}} \leq 0.$$

The case $\alpha = 1$ corresponds to the proportional fairness condition, while $\alpha \to \infty$ leads to the max–min fairness condition. There also exist weighted versions of that criterion (see [218]), which we do not develop here.

A last fairness measure worth mentioning is **Jain's index**, which is the value $J(x)$ in the interval $[0, 1]$ defined by

$$J(x) = \frac{(\sum_i x_i)^2}{n \sum_i x_i^2},$$

where n is the total number of users. The index equals 1 if and only if all users have the same allocation.

Interestingly, the measures presented above can be grouped under a common mathematical framework, and can be seen as satisfying a given set of axioms, among which are continuity, homogeneity, and monotonicity (for details, see [164] and references therein).

Example 4 Consider the simple network topology in Figure 2.2, where three users (named a, b, and c) intend to obtain capacity on their path. The allocations corresponding to some different fairness criteria are given in Table 2.1. Note that, in that instance, all Pareto-optimal allocations are of the form $(x_a, x_b, x_c) = (5 - x_c, 8 - x_c, x_c)$, with $x_c \in [0, 5]$.

User welfare, social welfare, supply chain value

It is often interesting to quantify the impact that some mechanism (say, a regulation rule) has on the different types of stakeholders. To that end, it is tempting to *aggregate* the utility functions of those stakeholders through an addition. However, one has to recall from Subsection 2.1.1 that utility functions are defined only modulo a strictly increasing transformation, hence adding or comparing utilities has a meaning only if a common "utility unit" is considered. In general, the unit naturally chosen is the monetary unit, which is the one considered when utilities are expressed in a quasi-linear form as in (2.3): one can then compute the "aggregated value" (in monetary units) that an outcome

Table 2.1 The Pareto-efficient allocations satisfying different fairness conditions for Example 4

	x_a	x_b	x_c
Max–min fairness	2.5	5.5	2.5
Weighted fairness, with $f_1(x) = 4x$, $f_2(x) = x$, and $f_3(x) = x$	1	4	4
Proportional fairness	3	6	2
α-Fairness, with $\alpha = 0$	5	8	0
Maximizing Jain's index	1.923	4.923	3.077

has for some population, by adding the "benefits" (still in monetary units) that each individual in that population gets from the outcome.

That precaution being taken, the aggregated utility of the user population is often called the **user welfare**. On the other hand, when the focus is on the whole set of providers, the aggregated utility gives the global **supply chain value**, which corresponds to revenues minus costs.

Finally, when all actors are considered (users and providers), the sum of the utilities is called the **social welfare**. Quasi-linear utility functions of the form $U_i = V_i - p_i$ can represent the preferences of all actors:

- for users, V_i is the willingness to pay and p_i the price paid;
- for providers, both V_i and p_i are negative: $-V_i$ represents the costs of provider i, and $-p_i$ is the revenue for provider i.

If the money that is spent by some actors is earned by some others, the global amount of money within the system remains the same, i.e. $\sum_i p_i = 0$. Then the social welfare is actually

$$\sum_i U_i = \sum_i (V_i - p_i) = \sum_i V_i.$$

As a result, the social welfare can be computed independently of money transfers.

The notion of **efficiency** is in general associated with social welfare: an outcome is called *efficient* if the social welfare is maximized. By extension, the term can also be used with different measures representing global performance, such as, for telecommunication systems, the total amount of data exchanged or the proportion of calls accepted. In all cases, the objective function with respect to which a system or an outcome is efficient needs to be specified.

2.2 Mathematical tools

After introducing the notions and terms from economics necessary for this book, we now turn our attention to the few tools from applied mathematics that will be used. In the models developed in the next chapters, we will consider stakeholders trying to make strategic choices so as to maximize their own utility. We therefore devote a first part of this section to optimization tools, which will be useful to compute the best choices for actors.

Note that we focus on continuous optimization, i.e., optimization over a continuous set of decision variables. In contrast, discrete optimization involves other types of tools, divided into *combinatorial optimization* [61,158] and *integer programming* [302], which we do not develop here.

While the optimization results will be used to compute the best strategies that each particular actor should implement depending on the actions of the others, we will often focus on fixed points of those best-reply correspondences. Indeed, it is often expected that the interactions among stakeholders will lead to such fixed points, since they are such that no stakeholder has an incentive to change his or her action. Subsection 2.2.2 provides the main existence and uniqueness results that will be useful in the remainder of the book.

2.2.1 Continuous optimization methods

We keep this subsection short, limiting ourselves to the material necessary for the rest of the book. In particular, we do not provide proofs for the (classical) results given here, our goal being to clarify the vocabulary and tools applied in the economic models of the next chapters. For a more complete coverage of the topic of continuous optimization, the interested reader is referred to [18, 121].

In this subsection, we focus on optimization over a finite-dimensional set. Let us consider a natural number $d \in \mathbb{N}$, which will represent the dimension of our origin space. Our objective here will be to minimize or maximize some real function $f : \mathbb{R}^d \mapsto \mathbb{R}$, called the **objective function**, possibly subject to some constraints (non-negativity of prices, capacity limits, ...).

Gradient vector, Hessian matrix

In most cases, the function f we consider will be differentiable almost everywhere. For any $\boldsymbol{x} = (x_1, \ldots, x_d)$ such that f is differentiable, the **gradient** of f at $\boldsymbol{x}$ is the vector

$$\nabla f(\boldsymbol{x}) := \begin{bmatrix} (\partial f/\partial x_1)(\boldsymbol{x}) \\ \vdots \\ (\partial f/\partial x_d)(\boldsymbol{x}) \end{bmatrix}, \tag{2.7}$$

where $\partial f/\partial x_i$ is the partial derivative of f with respect to x_i. The gradient vector indicates the direction of steepest ascent of f at $\boldsymbol{x}$: it can therefore be used in numerical optimization algorithms – the so-called *gradient methods* – to iteratively decide the next candidate $\boldsymbol{x}$ maximizing (or minimizing, by following the direction of $-\nabla f$) the function f. See, for example, [38, 296] for descriptions of gradient methods. Note that the gradient can also be used to estimate changes in f when $\boldsymbol{x}$ slightly varies (as a first term of a Taylor series of f around $\boldsymbol{x}$); indeed, we have, for any vector $\Delta \in \mathbb{R}^d$,

$$f(\boldsymbol{x} + \Delta) = f(\boldsymbol{x}) + \Delta \cdot \nabla f(\boldsymbol{x}) + o(||\Delta||), \tag{2.8}$$

where $\cdot$ stands for the inner product in $\mathbb{R}^d$, and $||\ ||$ is any norm in $\mathbb{R}^d$. When $d = 1$ (i.e., $\boldsymbol{x} \in \mathbb{R}$), then the gradient is simply the derivative of the function.

Optimal values of f can be reached at a point where f is non-differentiable (if there are any), at a boundary of the domain delimited by the constraints, or at a point $\boldsymbol{x}^*$ with zero gradient (otherwise one can choose Δ small enough to strictly improve the objective). For the last case, some conditions involving the second derivatives of f can be checked to see whether $\boldsymbol{x}^*$ is indeed a local optimum, and (if so) whether it is a local maximum or a local minimum.

When the quantities $(\partial^2 f/(\partial x_i\, \partial x_j))(\boldsymbol{x})$ exist, one can define the **Hessian matrix** (or simply Hessian) of f at $\boldsymbol{x}$ as the matrix

$$\boldsymbol{H}_f(\boldsymbol{x}) := \begin{bmatrix} \frac{\partial^2 f}{\partial x_1^2}(\boldsymbol{x}) & \frac{\partial^2 f}{\partial x_1\, \partial x_2}(\boldsymbol{x}) & \cdots & \frac{\partial^2 f}{\partial x_1\, \partial x_d}(\boldsymbol{x}) \\ \frac{\partial^2 f}{\partial x_2\, \partial x_1}(\boldsymbol{x}) & \frac{\partial^2 f}{\partial x_2\, \partial x_2}(\boldsymbol{x}) & \cdots & \frac{\partial^2 f}{\partial x_2\, \partial x_d}(\boldsymbol{x}) \\ \vdots & \vdots & \ddots & \vdots \\ \frac{\partial^2 f}{\partial x_d\, \partial x_1}(\boldsymbol{x}) & \frac{\partial^2 f}{\partial x_d\, \partial x_2}(\boldsymbol{x}) & \cdots & \frac{\partial^2 f}{\partial x_d\, \partial x_d}(\boldsymbol{x}) \end{bmatrix}. \tag{2.9}$$

We remark that the Hessian matrix is symmetric, and can be used to determine whether a solution $\boldsymbol{x}^*$ of $\nabla f(\boldsymbol{x}^*) = 0$ is a local minimum or a local maximum:

- if $\boldsymbol{H}_f(\boldsymbol{x}^*)$ is positive definite (i.e., if for all vectors $\boldsymbol{v} \in \mathbb{R}^d \setminus \{0\}$ we have $\boldsymbol{v}^{\mathrm{T}} \boldsymbol{H}_f(\boldsymbol{x}^*)\boldsymbol{v} > 0$, with $\boldsymbol{v}^{\mathrm{T}}$ the transpose vector of $\boldsymbol{v}$), then $\boldsymbol{x}^*$ is a local minimum;
- if $\boldsymbol{H}_f(\boldsymbol{x}^*)$ is negative definite (i.e., $-\boldsymbol{H}_f(\boldsymbol{x}^*)$ is positive definite) then $\boldsymbol{x}^*$ is a local maximum;
- otherwise $\boldsymbol{x}^*$ may not be a local extremum.

We recall from [116] that the following assertions are equivalent, for a symmetric matrix $\boldsymbol{M}$:

(1) $\boldsymbol{M}$ is positive definite;
(2) all the eigenvalues of $\boldsymbol{M}$ are strictly positive; and
(3) all the upper-left-corner square submatrices of $\boldsymbol{M}$ (i.e., the upper-left i-by-i corner of $\boldsymbol{M}$, for $i = 1, \ldots, d$) have positive determinants (this characterization is called *Sylvester's criterion* [136]).

Also, the case when $\boldsymbol{M}$ is positive semidefinite (i.e., $\boldsymbol{v}^{\mathrm{T}} \boldsymbol{M} \boldsymbol{v} \geq 0$ for all $\boldsymbol{v} \in \mathbb{R}^d$) will be of interest. While $\boldsymbol{M}$ is positive semidefinite if and only if all its eigenvalues are non-negative, notice that Sylvester's criterion cannot be applied (all the upper-left-corner square submatrices of $\boldsymbol{M}$ may be non-negative, and $\boldsymbol{M}$ need not be positive semidefinite).

When it is defined, the Hessian matrix can be used to prove the concavity or convexity of f, as detailed below.

Convexity

A set $\mathcal{S} \subset \mathbb{R}^d$ is a **convex set** if, for any $\boldsymbol{a}, \boldsymbol{b} \in \mathcal{S}^2$ and any $\alpha \in [0, 1]$, the point $\alpha \boldsymbol{a} + (1 - \alpha)\boldsymbol{b}$ is in $\mathcal{S}$. The function f is a **convex function** on the convex set $\mathcal{S}$ if, for any

$\boldsymbol{a}, \boldsymbol{b} \in \mathcal{S}^2$ and any $\alpha \in [0, 1]$, we have

$$f(\alpha \boldsymbol{a} + (1-\alpha)\boldsymbol{b}) \leq \alpha f(\boldsymbol{a}) + (1-\alpha) f(\boldsymbol{b}).$$

Similarly, f is said to be strictly convex if, for any $\boldsymbol{a} \neq \boldsymbol{b}$ and $\alpha \in (0, 1)$, we have

$$f(\alpha \boldsymbol{a} + (1-\alpha)\boldsymbol{b}) < \alpha f(\boldsymbol{a}) + (1-\alpha) f(\boldsymbol{b}).$$

If the Hessian matrix $\boldsymbol{H}_f(\boldsymbol{x})$ of f is positive semidefinite (positive definite) for all $\boldsymbol{x} \in \mathcal{S}$, then f is convex (strictly convex) over $\mathcal{S}$. Then, the condition $\nabla f(\boldsymbol{x}^*) = 0$ is sufficient for $\boldsymbol{x}^*$ to be a (the unique) minimum of f over $\mathcal{S}$. Note, however, that this condition is not necessary, since the equation $\nabla f(\boldsymbol{x}) = 0$ may not have any solution in $\mathcal{S}$.

A function f is (strictly) concave if and only if $-f$ is (strictly) convex, hence all the results stated above can be adapted to that framework; for example, to show that $\nabla f(\boldsymbol{x}^*) = 0$ implies that $\boldsymbol{x}^*$ is a global maximum of f over $\mathcal{S}$ if $\boldsymbol{H}_f(\boldsymbol{x})$ is negative definite (i.e., if $-\boldsymbol{H}_f(\boldsymbol{x})$ is positive definite).

Unconstrained optimization

When the optimization is over the whole set $\mathbb{R}^d$, and the objective function f is differentiable, then the extrema (minima or maxima) are either for unbounded values of $\boldsymbol{x}$ (e.g., $f(x_1, x_2) = x_1 + x_2$), or at points $\boldsymbol{x}^*$ with zero gradient. Those characterizations then give candidates for an extremum, and the Hessian can be used to determine the nature of each candidate (local/global minimum or maximum, if any).

Optimization with equality constraints: Lagrange multipliers

Consider a differentiable function $f : \mathbb{R}^d \mapsto \mathbb{R}$, and a family $(g_j)_{i=1,\ldots,k}$ of k differentiable functions $g_j : \mathbb{R}^d \mapsto \mathbb{R}$. Assume we want to solve the problem

$$\begin{aligned} &\text{minimize} \quad f(\boldsymbol{x}) \\ &\text{subject to } g_j(\boldsymbol{x}) = 0, \qquad 1 \leq j \leq k. \end{aligned} \tag{2.10}$$

The method of Lagrange multipliers consists in finding all candidate solutions $\boldsymbol{x}$ as the solutions of a system of equations. Namely, one first defines the **Lagrangian** function L of the problem, as the function

$$\begin{aligned} L : \mathbb{R}^{d+k} &\mapsto \mathbb{R} \\ (x_1, \ldots, x_d, \lambda_1, \ldots, \lambda_k) &\mapsto f(x_1, \ldots, x_d) - \sum_{j=1}^{k} \lambda_j g_j(x_1, \ldots, x_d), \end{aligned}$$

where the scalar λ_j is called the **Lagrange multiplier** associated with constraint j. Then, a solution $\boldsymbol{x}^*$ of (2.10) is necessarily such that there exists $\boldsymbol{\lambda}^* = (\lambda_1^*, \ldots, \lambda_k^*) \in \mathbb{R}^k$ for which $\nabla L = 0$, i.e.,

$$\frac{\partial L}{\partial x_i}(\boldsymbol{x}^*, \boldsymbol{\lambda}^*) = 0, \qquad i = 1, \ldots, d, \tag{2.11}$$

$$\frac{\partial L}{\partial \lambda_j}(\boldsymbol{x}^*, \boldsymbol{\lambda}^*) = 0, \qquad j = 1, \ldots, k. \tag{2.12}$$

Note that (2.12) simply implies that $g_j(\boldsymbol{x}^*) = 0$ for all j, i.e., $\boldsymbol{x}^*$ satisfies all constraints. The conditions (2.11) are called **first-order necessary conditions**, and can be rewritten as

$$\nabla f(\boldsymbol{x}^*) = \sum_{j=1}^{k} \lambda_j^* \nabla g_j(\boldsymbol{x}^*), \tag{2.13}$$

i.e. the gradient of f at $\boldsymbol{x}^*$ is a linear combination of the gradients of the constraint functions at $\boldsymbol{x}^*$. The last point can be interpreted as follows: if there is a slight change in $\boldsymbol{x}$ such that f changes values, then this change is forbidden by the constraints (i.e., at least one constraint is violated). We now prove that mathematically: take a direction, represented by some non-zero vector $\Delta \in \mathbb{R}^d$, and consider an infinitesimal $\epsilon > 0$. An infinitesimal change in that direction (i.e., from $\boldsymbol{x}^*$ to $\boldsymbol{x}^* + \epsilon\Delta$) would give, from (2.8),

$$f(\boldsymbol{x}^* + \epsilon\Delta) = f(\boldsymbol{x}^*) + \epsilon\Delta \cdot \nabla f(\boldsymbol{x}^*) + o(\epsilon).$$

Therefore, f changes if and only if $\Delta \cdot \nabla f(\boldsymbol{x}^*) \neq 0$, which by (2.13) implies that $\Delta \cdot \nabla g_j(\boldsymbol{x}^*) \neq 0$ for at least one constraint j; i.e., constraint j is violated by a change of $\boldsymbol{x}$ in direction Δ.

Note that the conditions (2.11) and (2.12) are only necessary conditions for $\boldsymbol{x}^*$ to be a solution of the optimization problem (2.10). Those conditions therefore just provide solution candidates; they are not sufficient in general since they may actually yield local maxima, local minima, or just stationary points of f in the directions allowed by the constraints. When f and the constraint functions are twice differentiable, it is possible to narrow further the set of candidate solutions by using **second-order conditions**, which involve the Hessian matrices of f and of the constraint functions $(g_j)_{1 \leq j \leq k}$.

- **Second-order necessary condition.** In addition to solving the system $\nabla L(\boldsymbol{x}^*, \boldsymbol{\lambda}^*) = 0$, if $\boldsymbol{x}^*$ is a solution of (2.10) then the matrix

$$\boldsymbol{H} := \boldsymbol{H}_f(\boldsymbol{x}^*) + \sum_{j=1}^{k} \lambda_j \boldsymbol{H}_{g_i}(\boldsymbol{x}^*)$$

  is positive semidefinite on the space $\{\boldsymbol{y} \in \mathbb{R}^d : \forall 1 \leq j \leq k, \boldsymbol{y} \cdot \nabla g_j(\boldsymbol{x}^*) = 0\}$. Note that this space is the space of "authorized directions," i.e., directions not violating the constraints (from (2.8) applied to the constraints). The condition therefore is analogous to a convexity property in those directions.
- **Second-order sufficient condition.** If a solution $(\boldsymbol{x}^*, \boldsymbol{\lambda}^*)$ of the system $\nabla L(\boldsymbol{x}^*, \boldsymbol{\lambda}^*) = 0$ is such that the matrix $\boldsymbol{H}$ defined above is positive definite on the space $\{\boldsymbol{y} \in \mathbb{R}^d : \forall 1 \leq j \leq k, \boldsymbol{y} \cdot \nabla g_j(\boldsymbol{x}^*) = 0\}$, then $\boldsymbol{x}^*$ is a local minimum of f subject to the constraints $g_j(\boldsymbol{x}) = 0, 1 \leq j \leq k$.

It remains to compare the values of f at the points obtained, and possibly to use some global convexity properties of f, in order to draw conclusions about the optimization problem.

The method of Lagrange multipliers is summarized in Algorithm 1. Again, note that the algorithm returning a non-empty set does not imply that the problem has a bounded

Algorithm 1 The method of Lagrange multipliers.

Input: A differentiable function $f : \mathbb{R}^d \mapsto \mathbb{R}$, some differentiable constraint functions $g_j : \mathbb{R}^d \mapsto \mathbb{R}$ for $1 \leq j \leq k$.

Output: a set of candidate solutions $\boldsymbol{x}^*$ of the optimization problem

$$\begin{aligned} &\text{minimize} \quad f(\boldsymbol{x}) \\ &\text{subject to } g_j(\boldsymbol{x}) = 0, \qquad 1 \leq j \leq k. \end{aligned}$$

1. **Define the Lagrangian** over $\mathbb{R}^{d+k}$ as

$$L(x_1, \ldots, x_d, \lambda_1, \ldots, \lambda_k) = f(x_1, \ldots, x_d) - \sum_{j=1}^{k} \lambda_j g_j(x_1, \ldots, x_d).$$

2. **Find the set $\tilde{S}$ (if any) of solutions $(\tilde{\boldsymbol{x}}, \tilde{\boldsymbol{\lambda}})$ of** $\nabla L(\boldsymbol{x}, \boldsymbol{\lambda}) = 0$.
3. **Return** $\{\tilde{\boldsymbol{x}} : (\tilde{\boldsymbol{x}}, \tilde{\boldsymbol{\lambda}}) \in \tilde{S}\}$.

solution. Indeed, the method of Lagrange multipliers just narrows (greatly, in general) the search for optimal solutions, by stating that *if the problem has a bounded solution, it is in the set returned by Algorithm 1*. Then one has to carefully look at the specific problem to determine whether one of the candidates is really a solution. On the other hand, if the algorithm returns an empty set, then we know that there is no bounded solution of the problem: either the constraints are not jointly satisfiable, or f takes its extremal values for some unbounded values of $\boldsymbol{x}$.

Optimization with inequality constraints: the Karush–Kuhn–Tucker conditions

We consider that we have inequality constraints, of the form $g_j(\boldsymbol{x}) \leq 0$, for $1 \leq j \leq k$. We are still trying to minimize f, so the optimization problem is

$$\begin{aligned} &\text{minimize} \quad f(\boldsymbol{x}) \\ &\text{subject to } g_j(\boldsymbol{x}) \leq 0, \qquad 1 \leq j \leq k. \end{aligned} \tag{2.14}$$

Note that the case of inequality constraints is an extension of the case of equality constraints, since any equality constraint of the form $g(\boldsymbol{x}) = 0$ can be decomposed into two inequality constraints $g(\boldsymbol{x}) \leq 0, -g(\boldsymbol{x}) \leq 0$.

The method and intuition to narrow the search for potential solutions are quite similar to the case of the Lagrange method, the main difference being that the directions must now be distinguished by their sign: while, from a point $\boldsymbol{x}$, it may be that some vector Δ is a feasible direction with respect to the constraints, this need not be true for $-\Delta$. Indeed, consider an $\boldsymbol{x}$ satisfying the constraints, and an infinitesimal $\epsilon > 0$; if, for a constraint j, we have $g_j(\boldsymbol{x}) = 0$ then, from (2.8),

$$g_j(\boldsymbol{x} + \epsilon\Delta) = \epsilon\Delta \cdot \nabla g_j(\boldsymbol{x}) + o(\epsilon).$$

That is, Δ is a direction that is feasible with respect to constraint j if and only if $\Delta \cdot \nabla g_j(\boldsymbol{x}) \leq 0$.

We therefore find that distinction in the necessary conditions for a point $\boldsymbol{x}^*$ to be a local minimum of f under the constraints. Those conditions, called the **Karush–Kuhn–Tucker conditions** (or KKT conditions), can be formulated as follows.

Theorem 2.2 *If $\boldsymbol{x}^*$ is a local minimum such that the set $\{\nabla g_j(\boldsymbol{x}^*) : 1 \leq j \leq k, g_j(\boldsymbol{x}^*) = 0\}$ is a set of linearly independent vectors, then there exist non-negative constants $(\mu_j)_{1\leq j\leq k}$, called Karush–Kuhn–Tucker multipliers, such that*

$$\nabla f(\boldsymbol{x}^*) + \sum_{j=1}^{k} \mu_j \nabla g_j(\boldsymbol{x}^*) = 0, \tag{2.15}$$

$$g_j(\boldsymbol{x}^*) \leq 0, \qquad \forall j = 1, \ldots, k, \tag{2.16}$$

$$\mu_j g_j(\boldsymbol{x}^*) = 0, \qquad \forall j = 1, \ldots, k. \tag{2.17}$$

We remark that (2.16) simply states that $\boldsymbol{x}^*$ satisfies the constraints. The condition (2.15) can be interpreted, as in the case of the Lagrange multiplier method, in terms of the effects of a slight change from $\boldsymbol{x}^*$ onto a direction Δ: if there is a direction in which f strictly decreases, then at least one $\mu_j g_j$ strictly increases. Finally, the conditions (2.17), called **complementary slackness conditions**, state that only those constraints which are *active* at $\boldsymbol{x}^*$ (i.e., such that $g_j(\boldsymbol{x}^*) = 0$) must be considered: inactive constraints are such that $\mu_j = 0$, so that $\mu_j g_j$ cannot increase, while for active constraints $\mu_j g_j$ strictly increasing means that constraint j is violated.

The condition that the gradients of the active constraint functions be linearly independent is called *linear independence constraint qualification*. Note that some other conditions can be sufficient for (2.15)–(2.17) to be valid; we do not detail them, and refer the interested reader to [38].

Finally, it is worth mentioning that the KKT conditions are sufficient for $\boldsymbol{x}^*$ to be a global minimum when f and the constraint functions $(g_j)_{1\leq j\leq k}$ are continuously differentiable and convex.

Continuity properties of optimization results

Consider one optimization problem, where the objective function varies also with parameters not controlled by the decision maker. One can, for example, think of regulation policies, which may affect the revenues of a monopolistic provider (through taxes), but also reduce its allowed actions. The following result, called the **theorem of the maximum**, establishes some continuity properties of the actions and utility of the decision maker with regard to the variables it does not control. We remark that the optimization problems evoked in this subsection are formulated as maximization problems, in order to fit the framework of decision makers trying to maximize their utility; the counterparts of the results given here with minimization problems are also, of course, valid (one can observe that nothing is changed below when considering the minimization of $-f$ instead of the maximization of f).

We first recall that a **correspondence** between two sets $\mathcal{X}$ and $\mathcal{Y}$ is a map $h : \mathcal{X} \mapsto \mathcal{P}(\mathcal{Y})$, where $\mathcal{P}(\mathcal{Y})$ is the power set of $\mathcal{Y}$, i.e. the set of all subsets of $\mathcal{Y}$. The correspondence h is **upper hemicontinuous** if, for any sequence $(x_n, y_n)_{n\in\mathbb{N}} \in (\mathcal{X} \times \mathcal{Y})^{\mathbb{N}}$ – with $y_n \in h(x_n)$ for all n – converging to $(x, y) \in \mathcal{X} \times \mathcal{Y}$, we have $y \in h(x)$.

Theorem 2.3 (Theorem of the maximum [35]) *Consider two metric spaces, $\mathcal{X}$ and $\mathcal{Y}$, and a continuous function $f : \mathcal{X} \times \mathcal{Y} \mapsto \mathbb{R}$. Define the correspondence $x^* : \mathcal{Y} \mapsto \mathcal{P}(\mathcal{X})$ by*

$$x^*(y) := \arg\max_{x\in C(y)} f(x, y), \tag{2.18}$$

with $C : \mathcal{Y} \mapsto \mathcal{P}(\mathcal{X})$ a continuous and compact-valued correspondence. Then

(1) the function $f^* : \mathcal{Y} \mapsto \mathbb{R}$ defined by $f^*(y) := \max_{y\in C(x)} f(x, y)$ is continuous, and
(2) the correspondence x^* is upper hemicontinuous.

We remark that under the theorem conditions, $x^*(y)$ is trivially non-empty and compact for any $y \in \mathcal{Y}$.

In a regulation/monopoly interpretation, $\mathcal{X}$ can represent the space of all possible actions of the provider, $\mathcal{Y}$ the set of regulation policies, f the provider's revenue, and $C(y)$ the set of allowed actions of the provider under policy y. Under the continuity assumptions of the theorem, we can then say that the maximum revenue that the provider can attain varies continuously with y, and the set of revenue-maximizing actions x^* is upper hemicontinuous in y. In particular, if $x^*(y)$ is a singleton for all $y \in \mathcal{Y}$, then x^* is a continuous function of y.

The **envelope theorem** provides further insights regarding the evolution of the optimal value (in x) of f when the parameters y vary, in the unconstrained case as well as in the case with equality constraints of the form $g_j(x, y) = 0$ (instead of $x \in C(y)$ for the theorem of the maximum).

Theorem 2.4 (Envelope theorem [283]) *Let $k \in \mathbb{N}$, and consider $k + 1$ continuously differentiable functions $f : \mathbb{R}^d \times \mathbb{R}^p \mapsto \mathbb{R}$ and $g_j : \mathbb{R}^d \times \mathbb{R}^p \mapsto \mathbb{R}$, $1 \le j \le k$, and consider the maximization problem*

$$\max_{\boldsymbol{x}:g_j(\boldsymbol{x},\boldsymbol{y})=0,\forall j} f(\boldsymbol{x}, \boldsymbol{y}). \tag{2.19}$$

Denote by L the Lagrangian of the maximization problem; i.e., let λ_j be the Lagrange multiplier associated with constraint j for $1 \le j \le k$ so that

$$L(\boldsymbol{x}, \boldsymbol{y}, \boldsymbol{\lambda}) = f(\boldsymbol{x}, \boldsymbol{y}) - \sum_{j=1}^{k} \lambda_j g_j(\boldsymbol{x}, \boldsymbol{y}),$$

with $\boldsymbol{\lambda} := (\lambda_j)_{1\le j\le k}$. Assume that the problem (2.19) has a solution for each $\boldsymbol{y} \in \mathbb{R}^p$, which we denote by $\boldsymbol{x}^*(\boldsymbol{y})$. The corresponding Lagrange multipliers are denoted by $(\lambda_j^*(\boldsymbol{y}))_{1\le j\le k}$.

Let $f^*(\mathbf{y}) = f(\mathbf{x}^*(\mathbf{y}), \mathbf{y})$. We then have, for all $1 \le i \le p$,

$$\frac{\mathrm{d} f^*(\mathbf{y})}{\mathrm{d} y_i} = \left.\frac{\partial \mathcal{L}(\mathbf{x}, \mathbf{y}, \boldsymbol{\lambda})}{\partial y_i}\right|_{\mathbf{x}=\mathbf{x}^*(\mathbf{y}),\ \lambda_j=\lambda_j^*(\mathbf{y}), 1\le j\le k} \tag{2.20}$$

$$= \frac{\partial f}{\partial y_i}(\mathbf{x}^*(\mathbf{y}), \mathbf{y}) - \sum_{j=1}^{k} \lambda_j^*(\mathbf{y}) \frac{\partial g_j}{\partial y_i}(\mathbf{x}^*(\mathbf{y}), \mathbf{y}). \tag{2.21}$$

In the particular case with no constraints ($k = 0$), we have

$$\frac{\mathrm{d} f^*(\mathbf{y})}{\mathrm{d} y_i} = \left.\frac{\partial f(\mathbf{x}, \mathbf{y})}{\partial y_i}\right|_{\mathbf{x}=\mathbf{x}^*(\mathbf{y})}. \tag{2.22}$$

2.2.2 Fixed-point results

In the models we will study, optimization is performed by each stakeholder, taking into account her or his own objective (utility). In particular, we do not expect an actor whose current decision maximizes its utility to make any change. The situations when all actors are in that position are of particular interest, since they can be seen as stable outcomes – and, thus, possible long-term results – of the interaction. Mathematically, they can be characterized as fixed points of a function – or of a correspondence – describing the utility-maximizing action(s) for each decision maker, given the others' actions.

Consider a space $\mathcal{X}$ and a correspondence $F : \mathcal{X} \mapsto \mathcal{P}(\mathcal{X})$; a **fixed point** of F is an element $x_0 \in \mathcal{X}$ such that $x_0 \in F(x_0)$. When $F(x)$ is a singleton for all x, i.e., F is a function from $\mathcal{X}$ onto itself, a fixed point is an element $x_0 \in \mathcal{X}$ such that $x_0 = F(x_0)$. We provide here the main results regarding the existence and uniqueness of such points.

Existence results

The basic existence result, known as the **Brouwer fixed-point theorem**, was established in 1912 and deals with *functions*.

Theorem 2.5 (The Brouwer fixed-point theorem) *If f is a continuous function from a convex compact subset K of a Euclidean space (e.g., $\mathbb{R}^d$ with $d \in \mathbb{N}$) onto K, then f has a fixed point.*

Then the theorem was extended to *correspondences* in the **Kakutani fixed-point theorem**.

Theorem 2.6 (The Kakutani fixed-point theorem [146]) *If F is an upper hemicontinuous correspondence from a convex compact subset K of a Euclidean space onto $\mathcal{P}(K)$, then there exists an $x_0 \in K$ such that $x_0 \in F(x_0)$.*

Finally, it is worth mentioning the **Schauder fixed-point theorem**, which applies to functions but covers the case when the origin space is of infinite dimension; just the existence of a norm is needed.

Theorem 2.7 (The Schauder fixed-point theorem [262]) *Let K be a compact convex subset of a normed vector space, and f a continuous function mapping K into itself. Then f has a fixed point.*

Uniqueness results

We end this section with the **Banach fixed-point theorem**, also known as the **contraction mapping theorem**, which guarantees the existence and uniqueness of a fixed point for some specific types of functions, called contracting functions.

The theorem is valid on *complete metric spaces*, namely spaces $\mathcal{S}$ endowed with a *metric* – defining the distance $\text{dist}(x, y)$ between any two elements $x, y \in \mathcal{S}$ – such that any Cauchy sequence converges in $\mathcal{S}$. In other words, for any sequence $(x_n)_{n\in\mathbb{N}} \in \mathcal{S}^{\mathbb{N}}$ such that $\text{dist}(x_n, x_m)$ tends to 0 as both n and m tend to infinity, there exists $\bar{x} \in \mathcal{S}$ such that $\text{dist}(x_n, \bar{x}) \to_{n\to\infty} 0$.

Examples of complete metric spaces include Euclidean spaces, but also some infinite-dimensional spaces such as the space of continuous functions from $[0, 1]$ into $\mathbb{R}$, with the distance $\text{dist}(f_1, f_2) = \max_{x\in[0,1]} |f_2(x) - f_1(x)|$.

We can now state the theorem.

Theorem 2.8 (The Banach fixed-point theorem [156]) *Let $\mathcal{S}$ be a complete metric space, in which the distance between two elements x and y is denoted by* $\text{dist}(x, y)$. *Let $f : \mathcal{S} \mapsto \mathcal{S}$ be a* **contracting function** *from $\mathcal{S}$ into itself, i.e., there exists a real number $\rho < 1$ such that*

$$\forall x, y \in \mathcal{S}, \quad \text{dist}(f(x), f(y)) \leq \rho\, \text{dist}(x, y). \tag{2.23}$$

Then f has one and only one fixed point x_0. Moreover, that fixed point can be found as the limit of the sequence $(y_n)_{n\in\mathbb{N}}$ defined by y_0 taken arbitrarily in $\mathcal{S}$ and $y_{n+1} = f(y_n)$; and the speed of convergence can be described by the relation

$$\text{dist}(y_n, x_0) \leq \frac{\rho^n}{1-\rho}\, \text{dist}(y_1, y_0). \tag{2.24}$$

2.3 Game theory

Game theory is a branch of economics focusing on the interactions of self-interested decision makers. This section aims at providing the main concepts and results of game theory, as well as the associated vocabulary. We do not intend to be exhaustive here, but only to give the material that is used in the next chapters. Note that there exist several monographs offering a wider and more detailed coverage of the game theory field; we refer the reader interested in a more complete understanding of the field to [87, 107, 243]. With respect to those references, this section endeavors to systematically provide illustrations from the telecommunications world, some of which will be developed further in later chapters.

The added value of game theory in telecommunications

Before game theory concepts were applied to telecommunication systems, the main tools used for their design and study were coming from the optimization field. The objective could then be to optimize routing, resource provisioning for QoS, transmission power management, etc. Those tools are, of course, still applied extensively, but the multiplicity of the decision makers involved in networks has made it unrealistic to assume that one can control all actors. It has therefore become necessary to understand (and possibly anticipate) the behavior of those actors. But that involves taking into account the interactions of decision makers, each of which focuses selfishly on his own objective.

That framework can be seen as a kind of *distributed* optimization, with individual actors making their own – locally optimal – decisions. The *stable* points of the interactions will then be the situations where all local optimization problems are jointly solved. Those situations, called Nash equilibria, play a key role in game theory. Local problems being simpler to solve in general than global ones, it will be of particular interest to see whether the conjunction of local optimizations leads to an outcome that is globally satisfying. If that is not the case, then tools can be used to anticipate the impact of a change in the settings, like a modification of the regulation policy or the introduction of pricing.

The telecommunication ecosystem is full of situations with interacting agents, which can be analyzed through game theory. Of particular importance are the **externalities** in the system, which describe the costs or benefits resulting from the actions of the others. A few examples of interactions and (negative) externalities are given below.

- *Peer-to-peer file sharing systems.* In such systems, the benefit of each node directly comes from the resources (files and bandwidth) made available by the others. However, contributing to the system is costly in terms of resource allocation (and possibly legal risks in the case of copyrighted files); therefore, selfish nodes limit their offered resource. This behavior, called free-riding, diminishes the performance of actual contributors, and even threatens the existence of the system. The issue has led the designers of file-sharing software like BitTorrent to include some incentive mechanisms linking the receiving rate of a node to its contributions, but there still exist alternatives to get around this reciprocity obligation [173].
- *Grid/cloud computing.* The issue is extremely similar to the peer-to-peer case: each participant tries to benefit from others' computing power, while limiting its own contribution.
- *Routing (user level).* When routes taken by packets are decided by the sending node, each node will try to find the best route for its traffic (minimizing delay or energy, or maximizing throughput). But the intermediate links are shared with other flows: because of the negative externality coming from congestion, the choices of all nodes are interdependent.
- *Inter-domain routing (provider level).* One well-known routing behavior of providers, called *hot-potato routing*, consists in each ISP minimizing its transmission costs by giving the traffic to its successor ISP (identified by the Border Gateway Protocol) *as early as possible*. However this policy, driven by the self-interest of ISPs, is not optimal in general in terms of total transmission costs.

- *Ad-hoc networks.* The principle of ad-hoc networks is that intermediate nodes act as radio relays for any two other nodes willing to communicate. While this works well for military applications with nodes acting cooperatively, a prominent problem when nodes are owned by different self-interested individuals concerns the incentives to forward the traffic of neighbors. Just like in the case of peer-to-peer and grid computing systems, the service disappears if there are no contributors.
- *Congestion control.* The Internet's success is largely due to the TCP/IP stack, where the TCP congestion control scheme automatically adapts the transmission rates of flows to the network's capacities [141]. The mechanism relies on flow emitters to reduce their sending rate when a loss (interpreted as a congestion signal) is detected. However, each sender would improve its individual throughput by ignoring losses and maintaining a high rate, and the network performance will collapse if too many users adopt this selfish behavior.

  Exactly the same principle applies to congestion control at the link layer: in WiFi for example, nodes have to obtain the right to send data through a ready to send (RTS) message that they send on a shared channel. Nodes are supposed to wait a certain (randomly chosen) amount of time before sending their RTS message, in order to limit collisions (and, in the case of collisions, those RTS messages are backlogged). It is clearly better for each individual node to disobey that rule and always immediately send its RTS messages; however, as soon as two nodes simultaneously implement that strategy the system collapses (with infinitely repeated collisions).
- *Power control in wireless networks.* Increasing your transmission power will induce a better QoS, but at the expense of interference with others' (forming negative externalities).

In all those cases, selfish behavior of actors can lead to low-performance situations, and hence there is a need to better understand such behaviors and, if necessary, design some incentives/rules driving systems towards preferable outcomes.

2.3.1 Vocabulary and definitions

Studying *strategically interacting players*

In game theory, the decision makers are called **players**, and the way they choose their action is their **strategy**. The particularity of game theory is the presence of **interactions**: the benefit (i.e., utility) that an actor gets depends not only on his or her actions, but also on the actions of the other actors as illustrated just above. The basic assumption made in game theory is that players are **rational**, i.e., given the information available to them, they look for strategies that maximize their payoff. As a result of interactions, the choices that a player makes to maximize his or her utility depend on the choices of the others.

The set of players and their strategy sets, together with the definition of the utility functions, constitutes a **game**.

Definition 2.9 A strategic game Γ consists of

- a set $\mathcal{N}$ of players;

Table 2.2 A network association game

	3G	WiFi
3G	(3, 3)	(6, 4)
WiFi	(5, 6)	(1, 1)

- a set $\mathcal{A}_i$ of actions available to each player $i \in \mathcal{N}$; and
- for each player $i \in \mathcal{N}$ a utility function

$$U_i : \mathcal{A} \to \mathbb{R}$$

(with $\mathcal{A} := \prod_{i \in \mathcal{N}} \mathcal{A}_i$), characterizing the preferences of the player among all possible action configurations.

We then note that $\Gamma = (\mathcal{N}, (\mathcal{A}_i)_{i \in \mathcal{N}}, (U_i)_{i \in \mathcal{N}})$.

When the game has only two players and $\mathcal{A}_i$ is a finite set for each player, the game can be conveniently described via a table, where each row (column) corresponds to an action of player 1 (player 2). Inside the table, the utility values are represented in the form (U_1, U_2). Such a representation is called the **normal form** description of the game, and the matrix is sometimes referred to as the **payoff matrix**.

Example 5 (A network association game) Consider two users, having the choice to connect to the Internet through WiFi or 3G access. Since they have heterogeneous terminals and/or radio conditions, they may get different throughputs. In addition, due to congestion effects, that throughput is decreased if they both select the same technology. If we assume that the users are sensitive only to payoffs, then we can consider utilities equal to the throughput experienced in each case, which are represented in Table 2.2. The table reads as follows: if player (user) 1 chooses to connect to the WiFi access point while player 2 uses 3G access, then the former experiences a throughput of 5 Mb/s and the latter a throughput of 6 Mb/s.

Table 2.2 then contains all the information of the game, since from it we can deduce the number of players, their available strategies, and all the players' utilities for each configuration of actions.

What does each player know?

As will be further illustrated by the next chapters, the information available to each player can have a strong impact on the outcome. Even with only two players in a game, for any piece of information x there are infinitely many possibilities. Does each player know x? Does each player know whether the other one knows x? Does each player know whether the other one knows that he or she knows x?...

When the answers to those infinitely many questions are all positive, then x is said to be of **common knowledge**. Unless specified differently, it is considered that the set of players, their available strategies, and all utility functions are common knowledge. Similarly, the rationality of all players (i.e., the fact that they intend to maximize their utility) is supposed to be common knowledge.

On the other hand, players are assumed to make their decisions without any information about the choices made by the other players: each player then tries to anticipate the actions of the others, so as to determine his or her own action.

Example 6 Consider the game depicted in Example 5, and examine what happens under two different knowledge configurations regarding the utility values in Table 2.2.

- If both players know their own available strategies, but player 1 believes that player 2 has only the strategy "3G," then player 1 will play "WiFi." Should player 2 know this belief of player 1, then he or she plays "3G," that outcome being denoted by (WiFi, 3G).
- If both players know the full matrix of Table 2.2, but player 2 thinks that player 1 believes that player 2 only has the strategy "WiFi," then player 2 will anticipate that player 1 will play "3G" to maximize his or her throughput, to which player 2 will adapt by playing "WiFi." If player 1 is able to anticipate that action of player 2 (for example, if player 1 knows the beliefs of player 2 about his or her own beliefs), then the outcome is (3G, WiFi).

Quite counter-intuitively, having more information than another player does not necessarily offer an advantage, as will be illustrated later on (see Example 16).

The Nash equilibrium

The **Nash equilibrium**, which was introduced in 1951 [231], is the most important equilibrium concept in game theory. It represents an action profile such that no player can improve his or her utility through a unilateral deviation.

As pointed out above, the decision problem faced by each player $i \in \mathcal{N}$ is to choose an action $a_i \in \mathcal{A}_i$ so as to maximize his or her utility U_i. But that utility, and consequently the utility-maximizing actions, will depend on the actions of the other players, which we denote by $\boldsymbol{a}_{-i} := \{a_j, j \neq i\}$. An action profile $\boldsymbol{a} \in \mathcal{A}$ can thus be written $\boldsymbol{a} = (a_i, \boldsymbol{a}_{-i})$, and a Nash equilibrium is an action profile where each player maximizes his or her utility, given the actions of the others.

Definition 2.10 A Nash equilibrium of a strategic game $\Gamma = (\mathcal{N}, (\mathcal{A}_i)_{i\in\mathcal{N}}, (U_i)_{i\in\mathcal{N}})$ is a profile $\boldsymbol{a}^* \in \mathcal{A}$ such that, for every player $i \in \mathcal{N}$,

$$U_i(a_i^*, \boldsymbol{a}_{-i}^*) \geq U_i(a_i, \boldsymbol{a}_{-i}^*) \ \forall a_i \in \mathcal{A}_i. \tag{2.25}$$

Such a situation can be interpreted as sustainable: we can reasonably expect that, once it has been reached, no player will unilaterally change their actions, hence the term *equilibrium*. On the other hand, if (2.25) is not satisfied, then at least one player can obtain a strictly larger utility through an action change.

For those reasons, it is generally considered that the expected outcomes from a game are among the Nash equilibria (if any). In particular, when the game has a unique Nash equilibrium, the rationality of players and the fact that the game as a whole is common knowledge should lead all players to deduce that they should play their equilibrium strategy, and to anticipate that all other players will follow the same reasoning.

A Nash equilibrium is called a **strict Nash equilibrium** if the inequality in (2.25) is strict for any alternative $a_i \neq a_i^*$ for all players. Conversely, we have a **weak Nash equilibrium** if equality in (2.25) holds for one player i and one alternative $a_i \neq a_i^*$.

Note that we can also define, for some $\epsilon > 0$, the less restrictive notion of **ϵ-Nash equilibrium**, as an action profile $\boldsymbol{a}^*$ from which no player can improve his utility by more than ϵ, i.e.

$$U_i(a_i^*, \boldsymbol{a}_{-i}^*) \geq U_i(a_i, \boldsymbol{a}_{-i}^*) - \epsilon \;\; \forall i \in \mathcal{N}, \quad \forall a_i \in \mathcal{A}_i.$$

The stance consisting in treating Nash equilibria as good predictions of the game outcomes can be criticized for different reasons: limited rationality of players, uncertainty in the actions of the other players, lack of common knowledge of rationality, complexity of computing the utility-maximizing strategies, ... (see [28] and references therein). All those criticisms are valid, and the concept of Nash equilibrium therefore has to be considered with care. However, it remains of particular importance in that it helps explain a lot of observed phenomena. In addition, even when some assumptions (e.g., on the rationality of players) are relaxed, in several cases the Nash equilibrium can be the limiting outcome of some strategy-adaptation dynamics of players (see the references on evolutionary games in Subsection 2.3.8).

Finally, let us mention that some Nash equilibria are robust against the criticisms above. Such equilibria occur when each player has a **dominant strategy**, namely a strategy that maximizes his or her payoff *whatever the actions of the others*. It is therefore natural to assume that players will select such strategies, resulting in a **dominant-strategy Nash equilibrium**. An illustration of a game with dominant strategies will be provided in Example 8.

Finding the Nash equilibria

We describe here the general method to discover the set of Nash equilibria. The first step is to compute, for each player $i \in \mathcal{N}$, his or her **best-response correspondence**, denoted by BR_i, which identifies the utility-maximizing actions for any configuration of the actions of the other players:

$$\begin{aligned} \mathrm{BR}_i : \prod_{j \in \mathcal{N} \setminus \{i\}} \mathcal{A}_j &\mapsto \mathcal{P}(\mathcal{A}_i) \\ \boldsymbol{a}_{-i} &\mapsto \arg\max_{x \in \mathcal{A}_i} U_i(x, \boldsymbol{a}_{-i}). \end{aligned} \tag{2.26}$$

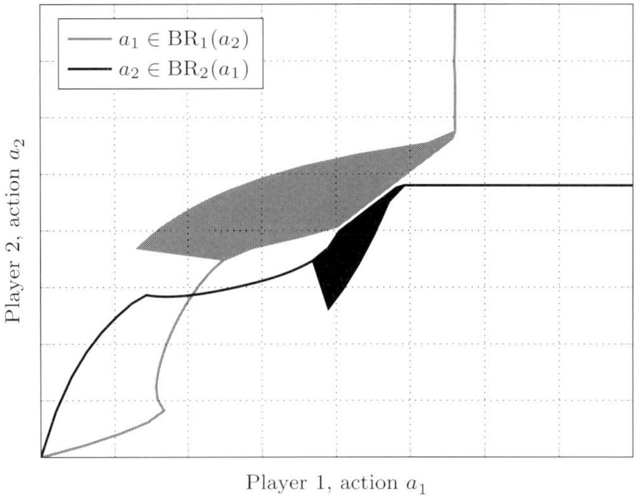

Figure 2.3 An example of best-response correspondences plotted on the same graph. We remark that, for a given action a_i, the best response of the other player can be an interval.

Then a Nash equilibrium is an action profile $\boldsymbol{a}^*$ such that, for all players $i \in \mathcal{N}$, $a_i^* \in \mathrm{BR}_i(\boldsymbol{a}_{-i}^*)$. Formulated differently, if we define the correspondence

$$\begin{aligned} \mathrm{BR} : \prod_{i \in \mathcal{N}} \mathcal{A}_i &\mapsto \mathcal{P}\left(\prod_{i \in \mathcal{N}} \mathcal{A}_i\right) \\ a &\mapsto \mathrm{BR}(a) = \left\{(b_1, \ldots, b_{|\mathcal{N}|}) : b_i \in \mathrm{BR}_i(\boldsymbol{a}_{-i}), \forall i \in \mathcal{N}\right\}, \end{aligned} \tag{2.27}$$

then a Nash equilibrium is exactly a fixed point of the correspondence BR. In several cases, the existence of such a fixed point derives from the theorems developed in Subsection 2.2.2.

When the game involves two players, each with a finite number of available actions, the Nash equilibria can simply be found by highlighting the best-response payoffs of each player in the payoff matrix itself. Then a Nash equilibrium is an action profile such that both players' payoffs are highlighted. (Note that the method can be generalized to the case of more players.)

When the game involves two players with a one-dimensional continuous set of actions, the Nash equilibria can be spotted graphically by plotting the best-response correspondences on the same graph, as in Figure 2.3. A Nash equilibrium is then an intersection point of the two curves.

Examples and general properties of the Nash equilibrium

This subsection provides a series of examples to illustrate the basic characteristics of the Nash-equilibrium concept. In the games we introduce, we apply the method given above to find the equilibria.

Table 2.3 The association game payoff matrix, with best-response payoffs in bold

	3G	WiFi
3G	(3, 3)	(**6**, **4**)
WiFi	(**5**, **6**)	(1, 1)

Table 2.4 The power game payoff matrix

	Normal	High
Normal	(Win, Win)	(Lose much, **Win much**)
High	(**Win much**, Lose much)	(**Lose**, **Lose**)

Example 7 For the association game of Example 5, we rewrite in Table 2.3 the payoff matrix, highlighting in bold the payoff of each player when his or her utility is maximized given the action of the opponent. The matrix then exhibits two Nash equilibria, namely the action profiles (WiFi, 3G) and (3G, WiFi).

Example 8 Consider another situation in wireless communications, taken from [134], where the focus is on transmission power: two users have to send information to the same base station, and have to choose between two power levels, referred to as "normal" and "high." Owing to interference, the throughput (and thus, the utility) of each user increases with his or her own transmission power, and decreases with the power of the other. If both users transmit with high power, a lot of energy is spent for poor performance, so the action profile (normal, normal) would be preferred by both users. Without entering into numerical values, the payoff situation is as described in Table 2.4, with best-response payoffs in bold. The game therefore has a unique Nash equilibrium, where both users transmit with high power. Note that this outcome is a dominant-strategy equilibrium, and yet it is not desirable for players, since both players would be better off with the action profile (normal, normal).

Example 8 is actually a version of the **prisoner's dilemma** problem (see, for example, [243], p. 16), which illustrates the fact that selfish players might not collaborate, even if this would be of benefit to all of them.

Example 9 (Jamming game) Consider two mobiles – a regular transmitter (player 1) and a jammer (player 2) – having to choose between two channels – c_1 and c_2 – to send data to a base station. If they choose the same channel, then a collision occurs and the transmission fails, which is perceived negatively by the regular transmitter and positively

Table 2.5 Payoff matrix for the **jamming game** (a, b, c, and d are strictly positive numbers)

	c_1	c_2
c_1	$(-a, \mathbf{b})$	$(\mathbf{c}, -d)$
c_2	$(\mathbf{c}, -d)$	$(-a, \mathbf{b})$

by the jammer. Inversely, the transmission succeeds if the mobiles select different channels, incurring a positive (negative) utility for the regular transmitter (the jammer), so that the payoff matrix is as in Table 2.5. We immediately observe that the game has no Nash equilibrium: for any action profile of the players, at least one player can improve his or her payoff through a unilateral move.

Let us summarize the first properties of the Nash equilibrium, which have respectively been evidenced by Examples 7, 8, and 9:

- a Nash equilibrium is not unique in general;
- a Nash equilibrium does not lead in general to a Pareto-efficient outcome; and
- a Nash equilibrium does not necessarily exist.

The last point can be dealt with by extending the strategy sets of players using probability distributions, as developed below.

Pure and mixed strategies, mixed-strategy equilibria

For a player $i \in \mathcal{N}$, a strategy consisting in selecting an action in $\mathcal{A}_i$ is called a **pure strategy**. As pointed out by Example 9, a game does not necessarily have a Nash equilibrium in pure strategies.

However, we can consider wider strategy sets, by allowing each player i to choose not only pure strategies, but any *probability distribution* over $\mathcal{A}_i$, forming a so-called **mixed strategy**. When $\mathcal{A}_i$ contains a finite number of actions, then a mixed strategy of player i is of the form $\pi_i = (\pi_i(a_1), \ldots, \pi_i(a_{|\mathcal{A}_i|}))$, with $\pi_i(a_j) \geq 0$ for all $j = 1, \ldots, |\mathcal{A}_i|$ and $\sum_{j=1}^{|\mathcal{A}_i|} \pi_i(a_j) = 1$. The value $\pi_i(a_j)$ then represents the probability that player i takes the action a_j when the game is played. Note that, when there is an action a_j such that $\pi_i(a_j) = 1$, the strategy π_i is actually a pure strategy.

Player utilities under mixed strategies

When one is dealing with mixed strategies, it is generally considered that players are interested only in their expected utility (i.e., they are insensitive to the uncertainty around their expected utility). As a result, denoting by $\bar{U}_i$ the utility of a player $i \in \mathcal{N}$ under the (mixed) strategy profile $\boldsymbol{\pi} := (\pi_1, \ldots, \pi_{|\mathcal{N}|})$, we have

$$\bar{U}_i(\boldsymbol{\pi}) := \mathbb{E}_{\boldsymbol{\pi}}[U_i] = \sum_{a \in \mathcal{A}} U_i(a) \underbrace{\prod_{k \in \mathcal{N}} \pi_k(a_k)}_{=\mathbb{P}(a)} . \tag{2.28}$$

Nash equilibria in mixed strategies

Allowing mixed strategies represents an extension of the strategy space of players: when $\mathcal{A}_i$ is finite for $i \in \mathcal{N}$, then the (mixed) strategy space is the convex hull $\mathcal{C}_i$ of all pure strategies, which is a (compact) polytope of dimension $|\mathcal{A}_i| - 1$ in $\mathbb{R}^{|\mathcal{A}_i|}$. The Nash-equilibrium concept can be expressed directly: a Nash equilibrium is a set of distribution functions $\boldsymbol{\pi}^* = (\pi_i^*)_i$ such that no user i can unilaterally improve his or her expected utility by changing alone his or her distribution π_i.

Formally,

$$\forall i, \forall \pi_i, \quad \bar{U}_i(\boldsymbol{\pi}^*) \geq \bar{U}_i(\pi_i, \boldsymbol{\pi}^*_{-i}),$$

where $\boldsymbol{\pi}_{-i} := (\pi_1, \ldots, \pi_{i-1}, \pi_{i+1}, \ldots, \pi_{|\mathcal{N}|})$.

Consider a **finite game**, i.e., a game with a finite set of players having a finite set of pure strategies each. Then, for each player $i \in \mathcal{N}$, since the utility of player i is continuous on $\mathcal{C}_i \times \prod_{j \neq i} \mathcal{C}_j$ (as can be seen from (2.28)), the theorem of the maximum implies that the best-response correspondence BR_i is upper hemicontinuous. Then, the Kakutani fixed-point theorem ensures that the correspondence

$$\begin{aligned} \mathrm{BR} : \prod_{i \in \mathcal{N}} \mathcal{C}_i &\mapsto \mathcal{P}\left(\prod_{i \in \mathcal{N}} \mathcal{C}_i\right) \\ \boldsymbol{\pi} = (\pi_1, \ldots, \pi_{|\mathcal{N}|}) &\mapsto \prod_{i \in \mathcal{N}} \mathrm{BR}_i(\boldsymbol{\pi}_{-i}) \end{aligned}$$

has a fixed point. We then reach the result obtained (with a different proof) by Nash [231].

Theorem 2.11 (Nash, 1950) *Any finite game has at least an equilibrium in mixed strategies.*

Let us go back to the jamming game described in Example 9. Since each player has only two pure strategies, the set of mixed strategies is one-dimensional for each player. If the transmitter (jammer) chooses a probability p_t (p_j) to transmit on channel c_1, then his or her respective utility U_t (U_j) is

$$\begin{aligned} U_\mathrm{t}(p_\mathrm{t}, p_\mathrm{j}) &= -a(p_\mathrm{t} p_\mathrm{j} + (1 - p_\mathrm{t})(1 - p_\mathrm{j})) + c(p_\mathrm{t}(1 - p_\mathrm{j}) + (1 - p_\mathrm{t}) p_\mathrm{j}), \\ U_\mathrm{j}(p_\mathrm{t}, p_\mathrm{j}) &= b(p_\mathrm{t} p_\mathrm{j} + (1 - p_\mathrm{t})(1 - p_\mathrm{j})) - d(p_\mathrm{t}(1 - p_\mathrm{j}) + (1 - p_\mathrm{t}) p_\mathrm{j}). \end{aligned}$$

The best-response correspondences can then easily be computed, since

$$\frac{\partial U_\mathrm{t}(p_\mathrm{t}, p_\mathrm{j})}{\partial p_\mathrm{t}} = (1 - 2p_\mathrm{j})(a + c)$$

and

$$\frac{\partial U_\mathrm{j}(p_\mathrm{t}, p_\mathrm{j})}{\partial p_\mathrm{j}} = (2p_\mathrm{t} - 1)(b + d),$$

leading to

$$\mathrm{BR_t}(p_\mathrm{j}) = \begin{cases} 1 & \text{if } p_\mathrm{j} < 1/2 \\ [0,1] & \text{if } p_\mathrm{j} = 1/2 \\ 0 & \text{if } p_\mathrm{j} > 1/2, \end{cases}$$

$$\mathrm{BR_j}(p_\mathrm{t}) = \begin{cases} 0 & \text{if } p_\mathrm{t} < 1/2 \\ [0,1] & \text{if } p_\mathrm{t} = 1/2 \\ 1 & \text{if } p_\mathrm{t} > 1/2. \end{cases}$$

Finally, the correspondence

$$\mathrm{BR}: \begin{cases} [0,1]^2 & \mapsto \mathcal{P}([0,1]^2) \\ (p_\mathrm{t}, p_\mathrm{j}) & \mapsto \{(x,y) : x \in \mathrm{BR_t}(p_\mathrm{j}), y \in \mathrm{BR_j}(p_\mathrm{t})\} \end{cases}$$

has a unique fixed point $(p_\mathrm{t}^*, p_\mathrm{j}^*) = (1/2, 1/2)$, which is the mixed-strategy Nash equilibrium of the game. Those values could be expected, due to the identical roles of the two channels in the game.

Interpreting mixed strategies

Introducing probabilities in player choices is recognized as difficult to interpret [27]. Several different interpretations can be given, whose relevance varies with the specific context considered.

The simplest and most direct view consists in players randomizing their action choice, selecting it from a "lottery" with appropriate probabilities.

Since it is not always satisfying to assume that players toss coins to select their actions, another interpretation can be preferred in the case of large populations of agents. In that view, each agent chooses a pure strategy, but does not know what particular agent he or she will be facing. In the jamming game, for example, a jammer does not know which particular user he or she will be trying to jam, but may know that a fraction $\pi_1(c_1)$ ($\pi_1(c_2)$) of regular transmitters will select the channel c_1 (c_2). The vector π_1 then represents the distribution of pure strategies among the population of transmitters, which is equivalent from the jammer's point of view to player 1 (the transmitter) randomizing his or her action choice.

For the case of individual players, maybe the most "intuitively satisfying" interpretation is the one proposed by Harsanyi [132], where the randomization comes from a lack of knowledge that players have about the other players' utilities. Each player bases his or her action choice on what he or she believes the others will play, which is perceived as random due to the uncertainty. This may lead to non-degenerate mixed strategies even when the uncertainty about other players' utilities tends to zero.

Example 10 Take the jamming game of Example 9: assume that playing c_1 incurs a *random* additional payoff gain of ϵ for player 1 (the transmitter) and η for player 2 (the jammer), leading to the payoff matrix of Table 2.6. Both players know the distribution of ϵ and η, but the realization value of ϵ (η) is known only to player 1 (player 2). As a result, each player selects his or her (pure) strategy depending on his or her random gain and

Table 2.6 Payoff matrix for the jamming game with uncertainty

	c_1	c_2
c_1	$(-a+\epsilon, b+\eta)$	$(c+\epsilon, -d)$
c_2	$(c, -d+\eta)$	$(-a, b)$

what he or she expects the other to play[1] – which he or she treats as random due to the unknown value of the random gain. If player 1 believes that player 2 will select channel c_1 with probability p_{j} (again, because of the value of η), then a (pure-strategy) best response for player 1 is to choose c_1 if the expected gain exceeds that of choosing c_2, that is, if $(-a+\epsilon)p_{\mathrm{j}}+(c+\epsilon)(1-p_{\mathrm{j}}) \geq cp_{\mathrm{j}} - a(1-p_{\mathrm{j}})$, i.e. $\epsilon \geq (c+a)(2p_{\mathrm{j}}-1)$. Otherwise, player 1 should prefer c_2. Similarly, denoting by p_{t} the probability with which player 2 thinks player 1 will select c_1, a jammer's best response is to choose c_1 if $\eta \geq (d+b)(1-2p_{\mathrm{t}})$. Take the simple case when both ϵ and η are uniformly distributed over $[-x, x]$, for $x > 0$: we can then check that the pure strategies

$$\text{for player 1:} \begin{cases} c_1 & \text{if } \epsilon \geq 0 \\ c_2 & \text{if } \epsilon < 0 \end{cases} \qquad \text{for player 2:} \begin{cases} c_1 & \text{if } \eta \geq 0 \\ c_2 & \text{if } \eta < 0 \end{cases}$$

lead to the probabilities $p_{\mathrm{t}} = p_{\mathrm{j}} = 1/2$ and constitute a Nash equilibrium for the game with uncertainty, for any value of x (the mixed strategies remain even if the uncertainty x tends to 0).

Price of anarchy, price of stability

Closely linked to the notion of efficiency or global performance (see Subsection 2.1.3) is the concept of the **price of anarchy**, a term coined by Koutsoupias and Papadimitriou [159]. That metric quantifies the loss of efficiency due to the selfishness of the actors involved, by comparing the outcome of the non-cooperative setting (the Nash equilibria) with the most efficient reachable outcome (that we call the social optimum).

The price of anarchy (PoA) is defined as the ratio between the performance measures of the worst equilibrium and the optimal outcome, so that we have PoA ≥ 1 in general. More precisely, denoting by $\mathcal{S}$ the set of all possible outcomes (that can be seen as the strategies implemented by the actors) and by $\mathcal{E}$ the set of Nash equilibria of the game, we have that

- if the global performance measure is of the form of a utility W (i.e., a metric that is *maximized* at the socially optimum outcome), then

$$\text{PoA} := \frac{\sup_{s\in\mathcal{S}} W(s)}{\inf_{s\in\mathcal{E}} W(s)}; \tag{2.29}$$

[1] Note that the game is then called a Bayesian game, a notion that will be defined properly in Subsection 2.3.3.

- if the global performance measure is of the form of a cost C (i.e., a metric that is *minimized* at the socially optimum outcome), then

$$\text{PoA} := \frac{\sup_{s \in \mathcal{E}} C(\boldsymbol{s})}{\inf_{s \in \mathcal{S}} C(\boldsymbol{s})}. \tag{2.30}$$

The PoA can be used to evaluate the need for an exterior intervention in the game (e.g., regulation). For example, if the PoA is close to 1, then introducing some coordination among players can yield only a limited improvement, which may be overcome by the costs of implementing the coordination: letting players behave selfishly appears to be the best option. In contrast, a large PoA suggests that the non-cooperative situation can be very inefficient, and that appropriate measures could significantly improve the global performance of the system; it is then up to the exterior controller (regulator) to properly design those measures so as to reduce the PoA.

Note that the PoA metric is pessimistic, in the sense that it considers the *worst* Nash equilibria and thus measures the worst loss of efficiency due to selfishness. The optimistic counterpart of the PoA is the so-called **price of stability** (PoS), which instead considers the *best* Nash equilibria, and thus measures the minimum cost – in terms of the global performance measure – of having a stable outcome. Mathematically,

- if the global performance measure is of the form of a utility W, then

$$\text{PoS} := \frac{\sup_{s \in \mathcal{S}} W(\boldsymbol{s})}{\sup_{s \in \mathcal{E}} W(\boldsymbol{s})}; \tag{2.31}$$

- if the global performance measure is of the form of a cost C, then

$$\text{PoS} := \frac{\inf_{s \in \mathcal{E}} C(\boldsymbol{s})}{\inf_{s \in \mathcal{S}} C(\boldsymbol{s})}. \tag{2.32}$$

Example 11 Take the association game of Example 5, with the payoffs given in Table 2.2, and consider social welfare as the global performance measure. The optimal outcome is the action profile (WiFi, 3G), and the game has two equilibria, where both players select different technologies. We then have $\text{PoA} = 11/10$ and $\text{PoS} = 1$ for that game.

Example 12 The jamming game of Example 9 has a unique Nash equilibrium (in mixed strategies), such that

$$\text{PoA} = \text{PoS} = \frac{\max(b - a, c - d)}{(b - a)/2 + (c - d)/2}.$$

Note that PoA is unbounded if we allow any values a, b, c, and d in Table 2.5, and that $\text{PoA} = 1$ if and only if $b - a = c - d$.

2.3.2 Non-atomic games

As the number of players in a game grows, the number of action profiles explodes, and in most cases the complexity of finding the Nash equilibria soars accordingly. However, when a game involves a very large number of players so that the individual effect of each player on the perceived payoffs becomes negligible, some simplifications occur. Such interaction situations are called **non-atomic games**: no individual player is *atomic*, i.e., impacts the utilities of the players through his or her own actions. Rather, the interactions among players occur via *aggregated* action changes: a proportion of the players switching choices does have an effect on the utilities.

Non-atomic games were originally used to study transportation networks [33, 249, 297], the players being commuters choosing what route to take towards their destination. In telecommunications, some large-scale networks involve a lot of users whose individual decisions do not significantly affect the perceived congestion, which corresponds to the non-atomic game framework.

The most simple instance of a non-atomic game was introduced as early as 1920 in Pigou's book [249]; we provide here an interpretation of that instance in the context of wireless communications.

Example 13 (Pigou's instance) Let us consider a large population of wireless users in a common WiFi hotspot, willing to send data on the Internet. The aggregated amount of data to send is normalized to 1 (say, a gigabyte), and is made of many individually negligible flows. Each user has the choice of whether to connect to the WiFi hotspot, or to use the 3G interface of his or her mobile to connect to a 3G cellular network. Assume that wireless users focus exclusively on minimizing their energy consumption (to maximize the battery lifetime of their mobiles). While the energy cost to reach the 3G base station is quite high (normalized to one energy unit per bit sent), the WiFi cost can be much lower (due to the smaller distance to the access point) but is sensitive to congestion: because of the WiFi collision-avoidance mechanisms, the energy cost increases with the load of the hotspot. Making the simplifying assumption that the energy cost on WiFi is proportional to the hotspot load, the non-atomic game played among users can be represented as in Figure 2.4, where

- the left-hand node represents the origin of the flows (i.e., the position of the users, considered as common);
- the right-hand node is the destination (the Internet); and
- each arc represents one strategy of a user: the top arc corresponds to the WiFi, whose cost for each user depends on the WiFi load, and the bottom arc corresponds to the 3G access, with a load-independent energy cost.

Such a representation therefore contains all the game elements: the players (with total weight 1), the strategies, and the utilities (here, the costs) of each player depending on all players' actions.

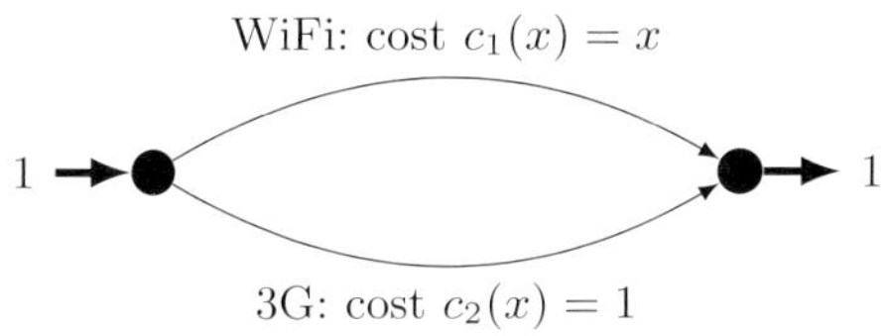

Figure 2.4 Pigou's instance. Arcs are labeled with their cost functions.

It is then easy to check that this game has an equilibrium, where all users choose the WiFi access and experience a cost of one energy unit per bit sent. Interestingly, that equilibrium is not Pareto-efficient: if any proportion $y \in (0, 1)$ of the total flow were to be sent through the 3G, then the affected users would perceive the same energy cost, but the other users would be better off, with a cost of $1 - y$ energy units per bit. However, such outcomes are not equilibria since 3G-connected users would then have an interest to switch to the WiFi network to reduce their cost.

The game described in Example 13 falls within the framework of **routing games**, where players try to select a cost-minimizing route between some origin and some destination on a network, whose links have load-sensitive costs. The typical applications are in transportation networks; however, those types of game can also be found in other domains, such as supply chain management [228] and telecommunication networks.

Non-atomic routing games in general may involve complex network topologies and multiple origin–destination pairs (and hence different types of players). There, the description of the game consists in

- a graph (nodes and arcs) describing the network topology,
- a set of origin–destination pairs and their respective demands (cost-independent or elastic), and
- a cost function for each arc.

Owing to the non-atomicity of players, the conditions for a situation to be an equilibrium can be expressed in a simple way known as **Wardrop's first principle**, and stating that each player selects one of the cheapest paths:

> The journey times [*the costs*] on all the routes actually used are equal, and less than those which would be experienced by a single vehicle [*a single player*] on any unused route. (Wardrop, 1952 [297])

The key intuition is again that, if there were an available route with a strictly lower cost, then part of the traffic would have an interest to switch to that one. That principle characterizes the outcome of non-cooperative behavior in non-atomic routing games, which is called a **Wardrop equilibrium**. The socially optimal outcome, on the other hand, is expressed in terms of aggregated (or equivalently, average over the population) cost, through **Wardrop's second principle**:

> The average journey time [*the total cost*] is a minimum. (Wardrop, 1952 [297])

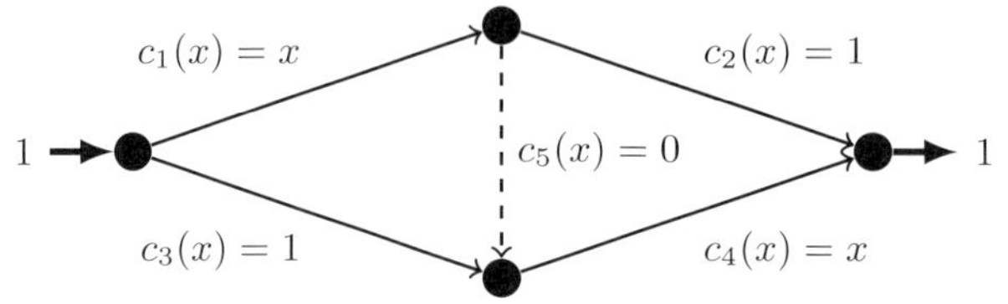

Figure 2.5 The Braess instance. Arcs are labeled with their cost functions.

Example 14 Take Pigou's instance, as represented in Figure 2.4. Denoting by $y \in [0, 1]$ the proportion of the demand on the 3G, the average cost among users is $1 \times y + (1 - y) \times (1 - y) = 1 - y + y^2$, which is minimum for $y = 1/2$ and equals $3/4$. But Wardrop's first principle can only be satisfied with $y = 0$; hence there is a unique equilibrium, with average cost $1 \times 1 = 1$.

From (2.30), the resulting PoA is $4/3$, meaning that user selfishness leads to a 33% increase in the average cost with respect to a centralized cost-minimizing situation.

Routing games have been the subject of great interest in the last decade, which witnessed several important new results regarding the properties of the PoA with the total user cost as the performance measure. For example, it has been discovered that, when the cost function of each link is affine in its load, the PoA is always below $4/3$ for non-atomic routing games, independently of the network topology and the demands (number and location of origin–destination pairs, and associated flow). We remark from Pigou's instance that this bound is tight, and can be attained with extremely simple games. Tight upper bounds have also been proved for other families of cost functions and more and more general settings (see [53, 62, 248, 260]).

As pointed out previously, selfishness can drive systems towards inefficient outcomes. More interestingly, it may happen that, because of selfishness, a situation that seems to have been improved (because of some extra resource) finally turns out to be worse for all players. Such a phenomenon was first evidenced by Braess [44], and is called a **Braess paradox**; practical situations have been highlighted in Chapter 1.

Example 15 (The Braess instance [44]) Consider the non-atomic routing game of Figure 2.5, initially ignoring the dashed link. One unit of demand has to reach the right-end node from the left-end one. Let x_t be the proportion of demand on the top path (links 1 and 2), and let x_b be the proportion of demand using the bottom path (links 3 and 4). Then the cost on the top (bottom) path equals $1 + x_t$ $(1 + x_b)$. By Wardrop's first principle, $x_t = x_b = 1/2$ (because $x_t + x_b = 1$ here); hence there is a common cost of $3/2$ for all players.

Then consider what happens if a link with no cost is added between the north and south nodes: a third (zig-zag) path is available (using links 1, 5, and 4), and will be used by some demand $x_z = 1 - x_t - x_b$. Then the respective costs of the top, bottom, and zig-zag paths are $x_t + x_z + 1$, $x_b + x_z + 1$, and $x_t + x_b + 2x_z$. The last cost is always lower than the costs of the two previously existing paths (since $x_t + x_b + x_z = 1$). The zig-zag path is actually strictly cheaper than the two others unless it is chosen by all players, in which case all paths have equal cost and Wardrop's first principle is satisfied. The last outcome is therefore the unique Wardrop equilibrium of the new game, and each user experiences a cost of 2. The total amount of demand being 1, the aggregated cost is also 2, which is strictly above the cost at equilibrium when link 5 was absent.

The Braess paradox corresponds to a very counter-intuitive phenomenon, since one could expect that all users would be better off (or at least that the aggregated payoff of players would be higher) when some resource is added. Its formal definition, which can be applied to payoffs in the form of utilities or of costs (negative utilities), follows.

Definition 2.12 (The Braess paradox) A situation where adding a resource reduces the aggregated payoff of all players is called a *Braess paradox*.

2.3.3 Bayesian games

We use the term **Bayesian game** to refer to a game where the information of at least one player is incomplete, but each player has some prior probability distribution he or she believes the lacking information to follow. The information in question concerns in general

- the payoffs of one or more players in the different possible outcomes (possibly including one's own payoff, as in Example 16 below),
- the beliefs of the other players regarding those payoffs, and/or
- the available strategies of the other players.

Example 10 did actually describe a Bayesian game. The incomplete information was with respect to the exact value of the other player's payoff, each player knowing his or her own payoff function but having only prior knowledge of the distribution of the other's.

A possible interpretation of the incomplete information, which also has the advantage of solving the problem of the infinite hierarchy of beliefs – "what does each player believe that the other's beliefs are?" and so on – was proposed by Harsanyi [131]. It consists in representing the beliefs of each player regarding

- the actual game played – called the *state of nature*, describing the available strategies and the payoffs of all players, and
- the beliefs of the other players,

using the notion of *type*. The type of a player can then be thought of as his or her *state of mind*, and allows one to answer all the belief-related questions regarding that player (like "what does he or she believe that the other players believe regarding his or her beliefs?"). Then all players have the same prior distribution (that is common knowledge) regarding the distribution of the *state of the world*, namely the joint distribution of the states of nature and of the players' types. Finally, each player knows only his or her own type. The reader interested in a thorough analysis of the belief structure is referred to [307].

Bayesian games are called such because they involve **Bayesian players**, i.e., players who use their priors to maximize their payoffs. Moreover, if the game consists in sequential actions, then players also update their beliefs (about the state of nature and the types of the other players) by using Bayes' rule.

In most cases, as in the study of mixed strategies it is assumed that players focus exclusively on maximizing their expected payoffs given their available information. Such players are called **risk-neutral**: they do not consider the variability (the risk) of their payoff around its expected value. That assumption can be questioned when the stakes of the game are of large importance to the player: in general a person would prefer being given \$1M for sure, rather than having a 1% chance of winning \$101M. That person is then **risk-averse**, while a multinational bank managing billions of dollars would be more likely to be risk-neutral, and hence would prefer the second alternative.

Risk-neutral players should base their action choices on the information they have, that is, their own type and the prior on the others' types. Within that framework, the counterpart of the Nash equilibrium can be formulated as follows. A strategy profile (mapping types to action decisions for each player) forms a **Bayesian–Nash equilibrium** if, for each player and each type, the *expected* payoff of that user is maximized with that strategy given that all players follow it.

Note that the Bayesian–Nash equilibrium must be thought of as an **ex-ante** equilibrium: given the information available *before* playing the game, following the Bayesian–Nash equilibrium strategy is the best choice. In contrast, in general this does not describe an **ex-post** equilibrium (i.e., an equilibrium in the complete-information game), since it may turn out that some players would have been better off taking another action, had they known the state of nature and the exact actions of all players.

We provide here an example of a (sequential) Bayesian game, with two risk-neutral players.

Example 16 (Negative value of information) Consider two investors (players), having to decide which of two systems/services, A and B, to develop (say, WiFi or HiperLAN in the wireless technology area at the beginning of the 2000s). The decisions are taken sequentially, i.e., first player 1 decides where to invest, and then player 2 observes that move and makes his or her investment decision.

Only one system will eventually be adopted by customers. From the point of view of both players, we assume that each system can be successful (or a failure) with an *a priori* probability of $1/2$, and this probability is assumed to be common knowledge.

Of course, for each player, making the right choice is better than investing in the unsuccessful system, but also being the only one making the good investment offers an additional competitive advantage. The payoffs of the form (Payoff_1, Payoff_2) given in Figure 2.6 reflect those phenomena. Given the time difference between the decisions made by the two players, the game can be represented as a tree, where player 2 makes his decision, knowing the action previously taken by player 1.

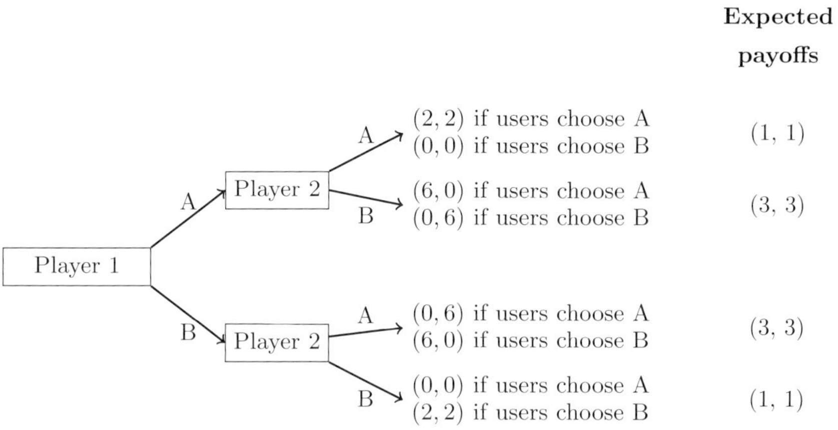

Figure 2.6 A decision tree, with payoffs under each possibility for the successful system. The expected payoffs (given the uncertainty about the successful system) appear on the right.

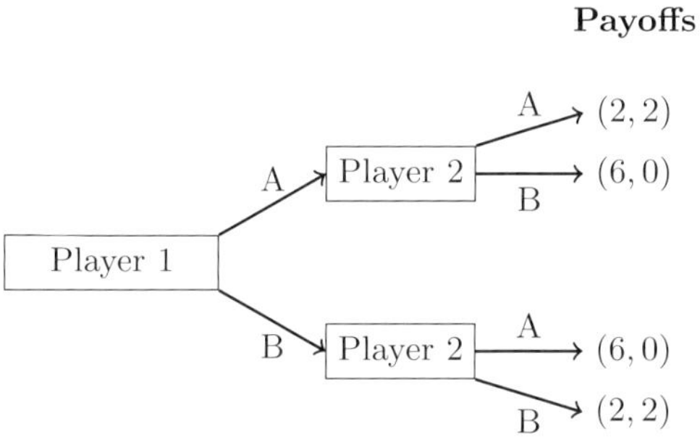

Figure 2.7 The decision tree when player 1 knows which system will be successful.

We first investigate the Bayesian game where no player knows which system is successful. The knowledge pattern here can be called, as in [307, Example 2, page 433], the *commonly known uncertainty about the state of nature*. Considering that each player maximizes his or her expected payoff, we immediately see that player 2 will always make a choice different from player 1, resulting in an expected payoff of 3 for each player (versus 1 for each player if player 2 follows player 1's action).

Now, consider a slight variation of that game, where player 1 knows which system will be successful. Player 2 does not have that information, but knows that player 1 has it; a formalization of the players' beliefs in that case is provided in [307, page 433] under the name *two players with incomplete information on one side*. The decision tree of Figure 2.6 reduces to the one in Figure 2.7. It is now in the best interest of player 2 to follow player 1's choice (since that choice is the correct one). As a result, both players end up with an expected (actually, deterministic here) payoff of 2.

In the above example, having some extra information actually reduced the expected payoff of both players.

2.3.4 Congestion games

Congestion games refer to situations first studied by Rosenthal in 1973 [259], where the strategy sets and the player payoffs satisfy the following structure:

- there exists a finite set $\mathcal{T}$ of "primary factors" (that can be interpreted as elementary resources);
- the strategy set of each player i is made of some subsets of $\mathcal{T}$: his or her action is the choice of such a subset $\mathcal{T}_i$;
- with each primary factor $k \in \mathcal{T}$ there is associated a cost function $c_k : \mathbb{N} \mapsto \mathbb{R}$, whose value under a given strategy profile depends only on the *number* n_k of players whose strategies include k; and
- the cost (negative payoff) of player i under a given strategy profile $(\mathcal{T}_i)_i$ is the sum of the costs of the primary factors included in his or her strategy, i.e., $\sum_{k \in \mathcal{T}_i} c_k(n_k)$.

Routing games where the contribution of each player to the cost of taking a road is the same are simple examples of congestion games. There, $\mathcal{T}$ is the set of roads, the strategy of each player is made of the paths (sets of roads) linking their position to their target destination, and players seek to minimize the total travel cost. In the context of wireless networks, one just has to recall that transmission rates (times to send/receive some amount of data) depend in general on the number of nodes connected to the same channel (access point plus frequency), where congestion occurs. The strategy set of each player is then the set of channels to which he or she has access; an action is the selection of just one primary factor – a channel.

Rosenthal established that congestion games always have a Nash equilibrium in pure strategies [259], a result extended by Monderer and Shapley, who showed that they are a particular case of *potential games* [220] (see Subsection 2.3.5 for definitions and main results).

While the set of players was assumed finite in [259], the definition can be extended to non-atomic games, where for computing the cost of a primary factor we consider the *proportion of the whole player population* (instead of the number of players) selecting that factor. In that case we talk about **continuous congestion games**; the non-atomic routing games presented in Examples 13 and 15 fall within that framework.

2.3.5 Potential games

Potential games [220] are a family of games in which utility functions have a particular structure: the individual incentives corresponding to an action change can be expressed through a common function, called the *potential*. Such games exhibit interesting properties.

More precisely, a game $\Gamma = (\mathcal{N}, (\mathcal{A}_i)_{i \in \mathcal{N}}, (U_i)_{i \in \mathcal{N}})$ is an **exact potential game** if there exists a function $\Phi : \prod_{i \in \mathcal{N}} \mathcal{A}_i \mapsto \mathbb{R}$ such that for all players $i \in \mathcal{N}$ and action profiles $\boldsymbol{a}_{-i} \in \prod_{j \in \mathcal{N} \setminus \{i\}} \mathcal{A}_j$, we have, for any $a_i, \tilde{a}_i \in \mathcal{A}_i$,

$$U_i(a_i, \boldsymbol{a}_{-i}) - U_i(\tilde{a}_i, \boldsymbol{a}_{-i}) = \Phi(a_i, \boldsymbol{a}_{-i}) - \Phi(\tilde{a}_i, \boldsymbol{a}_{-i}). \quad (2.33)$$

Weighted potential games correspond to a less restrictive definition, where payoff variations need only be *proportional* to the potential variations. The game $\Gamma = (\mathcal{N}, (\mathcal{A}_i)_{i\in\mathcal{N}}, (U_i)_{i\in\mathcal{N}})$ is a weighted potential game if there exists a vector $\boldsymbol{w} \in (\mathbb{R}^{++})^{|\mathcal{N}|}$ of strictly positive weights such that

$$\forall i \in \mathcal{N}, \forall \boldsymbol{a}_{-i} \in \prod_{j\in\mathcal{N}\setminus\{i\}} \mathcal{A}_j, \forall a_i, \tilde{a}_i \in \mathcal{A}_i,$$
$$U_i(a_i, \boldsymbol{a}_{-i}) - U_i(\tilde{a}_i, \boldsymbol{a}_{-i}) = w_i(\Phi(a_i, \boldsymbol{a}_{-i}) - \Phi(\tilde{a}_i, \boldsymbol{a}_{-i})). \tag{2.34}$$

Such a game is also called a $\boldsymbol{w}$**-potential game**.

Both exact and weighted potential games are **ordinal potential games**, for which only an ordinal relation is needed between the utility variations and the potential, i.e.,

$$\forall i \in \mathcal{N}, \forall \boldsymbol{a}_{-i} \in \prod_{j\in\mathcal{N}\setminus\{i\}} \mathcal{A}_j, \forall a_i, \tilde{a}_i \in \mathcal{A}_i,$$
$$U_i(a_i, \boldsymbol{a}_{-i}) > U_i(\tilde{a}_i, \boldsymbol{a}_{-i}) \Leftrightarrow \Phi(a_i, \boldsymbol{a}_{-i}) > \Phi(\tilde{a}_i, \boldsymbol{a}_{-i}). \tag{2.35}$$

Finally, we have a **generalized ordinal potential game** if the ordinal relation is only one-directional:

$$\forall i \in \mathcal{N}, \forall \boldsymbol{a}_{-i} \in \prod_{j\in\mathcal{N}\setminus\{i\}} \mathcal{A}_j, \forall a_i, \tilde{a}_i \in \mathcal{A}_i,$$
$$U_i(a_i, \boldsymbol{a}_{-i}) > U_i(\tilde{a}_i, \boldsymbol{a}_{-i}) \Rightarrow \Phi(a_i, \boldsymbol{a}_{-i}) > \Phi(\tilde{a}_i, \boldsymbol{a}_{-i}). \tag{2.36}$$

Consider a game, and a sequence of action profiles $(\boldsymbol{a}_{(0)}, \boldsymbol{a}_{(1)}, \ldots)$ such that, between two consecutive profiles, only one player changes his or her actions. Such a sequence is called a **path**. The existence of a potential can be verified by focusing on paths of length 4, and summing the payoff variations of the successive players changing their actions.

Theorem 2.13 (Existence of a potential [220]) *A game Γ with a finite number of players is a potential game if and only if, for any path of length 4 of the form $(\boldsymbol{a}_{(0)}, \boldsymbol{a}_{(1)}, \boldsymbol{a}_{(2)}, \boldsymbol{a}_{(3)}, \boldsymbol{a}_{(4)})$ such that $\boldsymbol{a}_{(0)} = \boldsymbol{a}_{(4)}$ and all profiles $\boldsymbol{a}_{(0)}, \boldsymbol{a}_{(1)}, \boldsymbol{a}_{(2)}$, and $\boldsymbol{a}_{(3)}$ are distinct, we have*

$$\sum_{j=1}^{4} U_{i_j}(\boldsymbol{a}_j) - U_{i_j}(\boldsymbol{a}_{j-1}) = 0,$$

where i_j is the index of the player whose action differs between $\boldsymbol{a}_{(j-1)}$ and $\boldsymbol{a}_{(j)}$.

The most important properties of potential games are summarized below: they establish results regarding the existence of pure-strategy equilibria, and the convergence of some action-change policies.

Theorem 2.14 ([220]) *If the game $\Gamma = (\mathcal{N}, (\mathcal{A}_i)_{i\in\mathcal{N}}, (U_i)_{i\in\mathcal{N}})$ has an ordinal potential Φ, then the set of its Nash equilibria is the set of the Nash equilibria of the game $\Gamma_\Phi = (\mathcal{N}, (\mathcal{A}_i)_{i\in\mathcal{N}}, (\Phi)_{i\in\mathcal{N}})$. In particular, if Φ admits a maximum over $\prod_{i\in\mathcal{N}} \mathcal{A}_i$ (which is always true if $\mathcal{N}$ and $\mathcal{A}_i$ are finite sets) then the game has a Nash equilibrium in pure strategies, at this potential-maximizing strategy profile.*

Let us consider again paths of action profiles as in Theorem 2.13, and add the constraint that the user changing strategies between two consecutive steps strictly improves his or her utility. Such a sequence is called an **improvement path**. The game is said to have the **finite improvement property** if any improvement path is of finite length.

Theorem 2.15 (Finite improvement property of potential games [220]) *A finite game has the finite improvement property if and only if it has a generalized ordinal potential.*

An interesting consequence of Theorem 2.15 is that, if we allow only one player at a time to change his or her actions, in order to strictly improve his or her payoff, then the process stops after a finite number of moves, after which there are no more payoff-improving action changes for any player: in other terms, the resulting action profile is a Nash equilibrium. A particular type of such an action-change policy is the case of best replies: in a game with a generalized ordinal potential, having players successively (possibly using a round-robin policy) selecting a best response (if their current action is not already a best response) to the current actions of the others converges to a Nash equilibrium.

A more elaborate process of action adaptation, when actors play the same game repeatedly over time, consists in considering for each player the empirical distribution of his or her previous actions as a prediction of the (mixed) strategy he or she will play, and selecting the next action from among the best responses to that profile of strategies. Such a process is called **fictitious play**.

Theorem 2.16 (Fictitious play property of potential games [219]) *Any finite weighted potential game has the fictitious play property, i.e., for any first moves of players and any tie-breaking rules for players, the fictitious play process converges to a Nash equilibrium.*

Note that the fictitious play considered here consists in all players making their (myopic) moves *simultaneously*. As pointed out in [36], some more results can be established when dealing with *alternating* player moves.

Let us finally mention the link between congestion games and potential games.

Theorem 2.17 ([220]) *Every congestion game is a potential game. Every potential game is isomorphic to a congestion game, i.e., there exists a bijection from the strategy sets of the initial game towards some strategy sets of a congestion game, leaving the utilities unchanged.*

Example 17 (A variation of the battle of the sexes) Consider two companies building some complementary goods (say, mobile terminals and base stations, for example), and having to decide between two technologies to develop, for example WiMAX and LTE. The technology can be successful, yielding revenues, only if there are both terminals and base stations implementing it. Hence the two companies should make the same choice. However, because of the current development state of both firms and the possible technical differences between the technologies, each company may prefer one technology or the other to be the common choice.

Such a model was initially called the **battle of the sexes** (see [243], p. 15), since it modeled the decisions of a couple about where to go out, with each spouse having some different preferences (but overall wanting to be together).

Table 2.7 The payoff matrix for the battle of the sexes

	WiMAX	LTE
WiMAX	(**9**, **7**)	(−1, −5)
LTE	(−2, −3)	(**8**, **15**)

Table 2.8 A $(1, 3/2)$ potential for the battle of the sexes of Table 2.7

	WiMAX	LTE
WiMAX	11	3
LTE	0	12

Table 2.9 A prisoner's dilemma payoff matrix

	Normal	High
Normal	(6, 6)	(0, **10**)
High	(**10**, 0)	(**1**, **1**)

Assume that the respective investment costs on WiMAX for the two firms equal $(1, 3)$, while they are $(2, 5)$ on LTE. No revenues arise if the firms make different choices, and the respective revenues in a WiMAX world are assumed to be $(10, 10)$, while they are $(10, 20)$ in an LTE world. The resulting payoffs appear in Table 2.7.

The game exhibits two Nash equilibria in pure strategies, which are the situations when both firms make the same choice; player 1 prefers the (WiMAX, WiMAX) strategy outcome and player 2 the (LTE, LTE) one.

It can be observed that this game is a weighted potential game, with weights 1 for player 1 and $3/2$ for player 2, if we define the potential function Φ by the values in Table 2.8. We trivially observe that the game has the finite improvement property, as is implied by Theorem 2.15: no improvement path can be of length larger than 2.

Example 18 (Potential for the prisoner's dilemma) Consider the prisoner's dilemma game, corresponding to two wireless users fixing their transmission power, as presented in Example 8. Let us assign some numerical values to the payoffs of Table 2.4: assume that, when just one user selects a high transmission power, he or she gets the largest rate and a payoff of 10 while the other one gets 0; and that both players choosing high (normal) power yields a payoff of 1 (6) each. We then have an exact potential game, with the potential values given in Tables 2.9 and 2.10. Here again, we can easily check that the game has the finite improvement property, and that the maximum of the potential corresponds to a Nash-equilibrium strategy profile.

strategy i when player choices are made according to the repartition $\boldsymbol{x}$. Those utility functions are assumed to be extended to a vicinity $\bar{\mathcal{X}}$ of the feasible strategy set $\mathcal{X}$ (delimited by the conditions $\sum_{i\in\mathcal{S}^{(p)}} x_i^{(p)} = m^{(p)},\ p \in \mathcal{P}$) in a continuous manner, so that partial derivatives can be computed.

A non-atomic game is then said to be a **full potential game**, if there exists a function Φ (the potential), continuously differentiable on $\bar{\mathcal{X}}$, such that

$$\frac{\partial \Phi}{\partial x_i^{(p)}}(\boldsymbol{x}) = U_i^{(p)}(\boldsymbol{x}). \tag{2.39}$$

The following useful results were established in [263].

Theorem 2.21 *If all utility functions are continuously differentiable, a necessary and sufficient condition for the non-atomic game to be a full potential game is that*

$$\frac{\partial U_i^{(p)}(\boldsymbol{x})}{\partial x_j^{(q)}} = \frac{\partial U_j^{(q)}(\boldsymbol{x})}{x_i^{(p)}}, \qquad \forall p, q, i \in \mathcal{S}^{(p)}, j \in \mathcal{S}^{(q)}. \tag{2.40}$$

In a full potential game, a repartition $\mathbf{x}$ *of the populations among the strategies is an equilibrium if and only if* $\mathbf{x}$ *satisfies the first-order Karush–Kuhn–Tucker conditions of the problem of maximizing the potential* Φ *under the feasibility constraints* ($x_i^{(p)} \geq 0, \forall p \in \mathcal{P}, i \in \mathcal{S}^{(p)}$ *and* $\sum_{i\in\mathcal{S}^{(p)}} x_i^{(p)} = m^{(p)},\ p \in \mathcal{P}$).

Full potential games also have interesting properties in terms of convergence of strategy dynamics (i.e., evolution of the strategy repartitions among the population); see [263, 265] for details.

Example 20 The non-atomic routing games introduced in Subsection 2.3.2 are full potential games. Each origin–destination pair in the network corresponds to a population p, whose total mass $m^{(p)}$ equals the demand level on that pair, and whose strategy set $\mathcal{S}^{(p)}$ is made of all possible routes in the network linking that origin to that destination, indexed by i (i.e., a route $\mathcal{R}_i$ is a subset of the total set $\mathcal{L}$ of links in the network). Then the payoff $U_i^{(p)}$ can be taken as the opposite of the total cost on route $\mathcal{R}_i$, for $i \in \mathcal{S}^{(p)}$, that is, the sum of the costs of all links used in that route.

For a repartition of player choices $(x_i^{(p)})_{p\in\mathcal{P},i\in\mathcal{S}^{(p)}}$, one can easily compute the total load of each link ℓ in the network as $x_\ell := \sum_{p\in\mathcal{P},i\in\mathcal{S}^{(p)},\ell\in\mathcal{R}_i} x_i^{(p)}$. Those link loads affect the costs on routes $-U_i^{(p)} = \sum_{\ell\in\mathcal{R}_i} c_\ell(x_\ell)$, with c_ℓ the cost function of link $\ell \in \mathcal{L}$. From Theorem 2.21, it is then easy to check that the function $\Phi = -\sum_{\ell\in\mathcal{L}} \int_{t=0}^{x_\ell} c_\ell(t)\mathrm{d}t$ is a full potential of the routing game.

Consider Pigou's instance, as described in Example 13. That routing game counts only one population with total mass 1 and two action choices (route 1 with a congestion-sensitive cost, or route 2 with a constant cost). Denoting by x_i the load on route $i = 1, 2$, one can observe that $\Phi(x_1, x_2) := -x_1^2/2 - x_2$ satisfies (2.39) and therefore is a potential for the game. Maximizing that potential under the constraints $x_i \geq 0$ and $x_1 + x_2 = 1$ yields $(x_1, x_2) = (1, 0)$, which is the equilibrium of the routing game.

2.3.6 Stackelberg games

A **Stackelberg game** (or leader–follower game) [293] is a type of game where a player (or a set of players) can commit to an action and assume that the others will adapt to that action. Such a player is then called the **leader**, while the others are **followers**. In practice, a leader position arises when there is some sequentiality in the way actions are taken and some players observe the moves of their opponents before making their own choices. In such settings, the actor who plays first can anticipate the reaction of the player(s) after him or her, and take the best decision with that constraint. The resulting action profile is called a **Stackelberg equilibrium**; it is not in general a Nash equilibrium since the leader's action need not be a best response to the follower's choice. When different decision times are involved, this comes from the fact that the leader can no longer change his or her action.

Note that having sequential decisions is not necessary in order to have a Stackelberg situation; actions can take place simultaneously, but in that case the leader has to inform and convince the follower about the strategy he or she will follow, the difficulty for the leader being to be credible about not switching to a best response.

Example 21 In the association game of Example 5, imagine that player 1 chooses his or her access technology first, and therefore acts as the leader. He or she anticipates that player 2 will make a selfish choice by selecting the technology not used by player 1 (from payoffs in Table 2.2). Player 1 then knows that the payoff vector will be (6, 4) if he or she chooses to use WiFi, and (5, 6) if he or she selects 3G. The Stackelberg strategy for player 1 is then to choose WiFi.

A similar reasoning applied to the battle of the sexes (Example 17) shows that being a leader would allow player 1 to guarantee a payoff of 9, and that both players invest in WiMAX (see the payoffs in Table 2.7). Conversely, should player 2 be the leader in that game, the Stackelberg equilibrium would consist in both firms investing in LTE.

We remark that, in both cases, the Stackelberg equilibrium corresponds to a Nash equilibrium of the initial game (although this is not the case in general), with the advantage of yielding a unique solution.

The previous example illustrates situations where being a leader offers an advantage over the follower; however this is not always the case.

Example 22 Consider the jamming game, whose payoffs are in Table 2.5, where player 1 is a leader. Then, for any channel he or she selects, player 2 (the jammer) will choose the same channel, yielding a payoff vector $(-a, b)$: the leader obtains his or her lowest possible payoff, and the jammer his or her highest possible one. The situation is inverted if player 2 becomes the leader: player 2 is then disadvantaged.

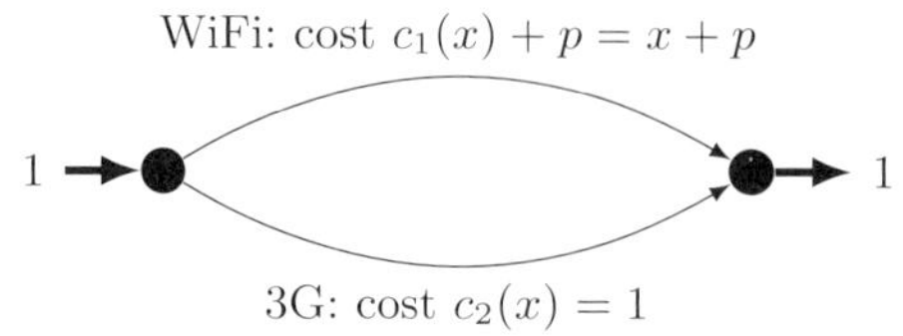

Figure 2.8 Pigou's instance with pricing on the top arc. Arcs are labeled with their cost functions.

Stackelberg games naturally arise when a decision maker (network administrator, designer, service provider, ...) wants to optimize a utility function (e.g., revenue or social welfare) by playing on parameters such as price or resource management, and this utility depends on the reaction of users who have their own objectives. The leader then has to find out the parameters driving to the best outcome for him or her, knowing that users will adapt to those parameters (and, maybe, also play a non-cooperative game among themselves).

Among typical situations, one can think of a provider playing on prices and capacities and users reacting through traffic rates, possibly with congestion effects involved in the user game.

Example 23 (Pricing in a non-atomic routing game) Take the routing game of Example 13, but where the WiFi operator can charge the service at a unit price p (say, per bit). Then, in addition to the energy cost of each technology, this monetary cost has to be taken into account by users when choosing which technology to activate. The operator is assumed to advertise the price level $p \geq 0$ before users make their decision, so that users play the game drawn in Figure 2.8. The users' reaction to the price level p results in a demand repartition among the two technologies satisfying the Wardrop equilibrium condition, i.e., where only cheapest routes are used. In this example, for a given value of p that repartition is of the form $(x_1(p), 1 - x_1(p))$, with the load $x_1(p)$ on the WiFi being $x_1(p) = [1 - p]^+$, where $y^+ := \max(y, 0)$. Indeed, for $p \geq 1$ WiFi cannot be used without being strictly more costly than 3G, while for $p < 1$ we get that both technologies are used and have equal cost 1.

Depending on the operator's objective, the pricing decision can differ. For example, if the operator aims at maximizing revenue $px_1(p)$, then the optimal price is the solution of the maximization problem $\max_p p[1 - p]^+$, which gives $p^{\text{revenue}} = 1/2$.

Backward induction and subgame-perfect equilibrium

For simplicity, consider first a unique leader. Once the leader has taken an action, the follower(s) then play(s) a game parameterized by the leader's choice. For each value of the leader's action, that parameterized game (that is a trivial game – an optimization problem to solve – if there is only one follower) is called a **subgame** of the original game.

From the point of view of the leader, one way to determine the best strategy to follow is to "solve" (i.e., find the equilibria of) all the possible subgames and compute the corresponding utility as a *function* of his or her own action, so that the action to take is the solution of an optimization problem.

Note that the subgames may also involve players in another leader–follower interaction, so that other subgames can be defined. Also, there can be several leaders (i.e., players making decisions at the same time), whose conjunct decisions affect the follower's or followers' actions.

The global method to analyze those multi-level games consists in computing the equilibria of all possible subgames, which are taken then as inputs for the games played at a higher level (i.e., at the leader level). When for each subgame the actions are maximizing the utility of the players involved at that level, given the actions of the other players of the same and higher levels, it is said that we have a **subgame-perfect equilibrium**. We remark that the analysis method proceeds in the opposite direction to the one in which decisions are taken: we first consider the players with no followers, which can be seen as the players who play last, and then take their reaction as an input for the leaders, repeating that process until the level where players are only leaders. For that reason, the resolution method is called **backward induction**.

Example 24 (Regulation and competition) Let us consider the context of wireless networks. Regulators decide the rules which operators have to follow: they can take the form of taxes, spectrum allocations, constraints on the commercial practices, etc. Once the rules have been fixed, operators have to take decisions regarding investments in technologies, sites at which to locate access points, and offers to propose to users. Finally, users make their decisions about which offers to subscribe to (if any).

That interaction setting corresponds to a (Stackelberg) game with three decision levels, where the regulators are the leaders, the operators are followers (with respect to regulators) but also leaders (with respect to users), and the users are followers. The backward induction method would then translate into the following procedure.

1. Determine the user situation resulting from any fixed value of the operators' and regulators' decisions, and define that outcome as a function of those decisions. (Users possibly interact through a game, in which case an equilibrium has to be found.)
2. Determine the operators' equilibrium decisions for any fixed value of the regulators' action. For that, consider that the operators play a non-cooperative game, considering the imposed rules as fixed and the user behavior as a function of their own actions. Define then the operators' behavior as a function of the regulators' action.
3. Finally, consider the regulators, who have to fix rules in order to maximize their own objective, being expressed as a function of the rules only.

We remark that we have not raised the questions of the existence and uniqueness at each game level. The cases of non-existence at some followers' level are difficult to treat, since the leader then cannot anticipate how the followers will react; in some cases, one can reason on ϵ-Nash equilibria to deal with that problem. On the other hand, if the

Table 2.11 The prisoner's dilemma payoff matrix

	Normal	High
Normal	(6, 6)	(0, **10**)
High	(**10**, 0)	(**1**, **1**)

followers' game has several equilibria, one possible (pessimistic) approach is for each leader to consider the least favorable equilibrium (for him or her), i.e., to optimize the worst case.

The notion of a Stackelberg game developed here will be used in various contexts throughout the book.

2.3.7 Repeated games

Let us consider again the interferences between two wireless users choosing their transmission power (Example 18), with the payoff values recalled in Table 2.11.

The novelty here is that a time dimension is added: we now assume that players can change their transmission power (say, each second) during their connection. As a result, both users are playing the same game (called the *simple game* or the *one-shot game*) *repeatedly* over time; but, during each time period when they interact, they can each use the information from the previous moves of the other to make their decision. Such a situation is called a **repeated game**.

A player's strategy in the repeated game is now a function that at each time period associates an action (or a distribution over the possible actions) with the one-shot game strategies that have been played so far.[3] That set of previously played strategies is called the **history of the game**. Moreover, during each time period the players could make their action choice also dependent on some public signal, i.e., they can play *correlated* strategies [25]. We will not explicitly use such correlated strategies, so we do not enter into the details here.

For example, player 1 may discover that player 2 systematically uses the same transmission power as that chosen by player 1 in the previous period. Such a strategy, called *tit-for-tat*, is then likely to incentivize player 1 to keep his transmission power to a normal level: while a high power improves player 1's immediate payoff by 4 (if player 2 chooses normal power) or 1 (if player 2 chooses high power), ensuring that player 2 plays "normal" rather than "high" represents a minimal improvement of 6 for player 1. So player 1 may give up the immediate gain, to guarantee the larger future gains yielded by player 2 also playing "normal."

The overall payoff for a player will depend on the whole sequence of actions played throughout the total interaction duration, through the payoffs obtained during each period. The classical measure considered for player i's overall payoff for the repeated

[3] We remark that, in general, mixed strategies are not directly observable, since only actions are observed. Nevertheless, the general results given here hold also when only actions are observed [105].

game is the sum of the (one-shot) utilities that he obtained at each time period, where a per-period discount factor $\delta \leq 1$ is used to compare gains obtained at different times. Consequently, if we denote by $\boldsymbol{s}_t$ the set of strategies for the one-shot game that are played at time t, the normalized[4] expected utility of provider i is

$$V_i = (1 - \delta) \sum_{t=0}^{+\infty} \delta^t U_i(\boldsymbol{s}_t). \tag{2.41}$$

The discount factor represents the "patience" of the players: the smaller it is, the more players value present gains against future ones. That factor δ can have different interpretations.

- It can represent the price in the current period of a monetary unit in the next period: if a per-period interest rate $r > 0$ is applied, then this gives $\delta = 1/(1 + r)$.
- Independently of interest rates, the factor δ can also stand for the probability of the game continuing during the next period. For our example with wireless users, the interaction ends when one communication finishes or one user moves and connects to another access point.
- The most realistic interpretation is to consider that δ represents both of those aspects, and could for example be expressed as

$$\delta = \frac{1}{1 + r} \mathbb{P}(\text{game continues during next period}).$$

The possible outcomes of a repeated game are investigated through different versions of the so-called folk theorem, which we summarize here.

As before in this chapter, we denote by n the number of players and by $\mathcal{A}_i$ the action set of player i in the one-shot game, and we represent the actions of all players as a vector $\boldsymbol{a} := (a_1, \ldots, a_n) \in \prod_{i=1}^n \mathcal{A}_i$, where $a_i \in \mathcal{A}_i$ is the action of player i.

Let us first focus on the one-shot game, and denote by $U_i^{\min}$ the **minimax utility** of player i, i.e. the minimal utility that he or she could ensure if the other players tried to minimize his or her utility:

$$U_i^{\min} := \min_{\boldsymbol{a}_{-i} \in \mathcal{A}_{-i}} \max_{a_i \in \mathcal{A}_i} U_i(a_i, \boldsymbol{a}_{-i}),$$

where $\boldsymbol{a}_{-i}$ ($\mathcal{A}_{-i}$) stands for the actions (set of possible actions) of players different from i, and we write the overall action vector as $\boldsymbol{a} = (a_i, \boldsymbol{a}_{-i})$. Consequently, $U_i^{\min}$ is the worst utility that i can ensure if he or she knows the actions of his or her opponents. It can also be interpreted as the worst utility that the other players could impose on player i, and the corresponding strategies that they play to do so are called the *minimax strategies* against player i.

For the simple one-shot game of Table 2.11, the minimax strategy of each player is "high," and his or her minimax utility equals 1.

[4] The normalization is through the multiplicative constant $(1 - \delta)$, which is introduced so that the sum in (2.41) gives the weighted mean of the player utilities over time periods, with weight δ^t for period t.

The set of "reachable" utility vectors of the simple game is also of interest when studying their repeated version. Let us denote it by $\mathcal{U}$, so that formally we have

$$\mathcal{U} := \{(U_1(\boldsymbol{a}), \ldots, U_n(\boldsymbol{a})) : \boldsymbol{a} \in \prod_{i=1}^{n} \mathcal{A}_i\}.$$

The basic version of the **folk theorem** is then as follows.

Theorem 2.22 (The folk theorem) *For any* $(v_1, \ldots, v_n)$ *in the convex hull of* $\mathcal{U}$*, such that* $v_i > U_i^{\min}$*, if* δ *is sufficiently close to* 1 *then there exists a Nash equilibrium of the infinitely repeated game where, for all* i*,* $V_i = v_i$.

The proof is quite simple: first, the utility vector $(v_1, \ldots, v_n)$ is reachable via a (correlated, possibly mixed) strategy vector $(s_1, \ldots, s_n)$ (see, for example, [243] for a proof). The Nash strategies consist in each player i playing s_i while everybody does, and if a player j deviates then all other players should minimize his or her utility forever, playing the minimax strategy against j. Therefore, the immediate gain for a deviator i being able to get more than v_i on one period would be discouraged (when δ is large enough) by the future sanctions imposing a utility $U_i^{\min}$ on all future periods; hence the equilibrium.

However, in repeated games the notion of Nash equilibrium might not be strong enough: for example the previously described Nash strategies constrain each player to sanction the first deviator forever, although this might be costly to them. Therefore, there might be no incentive to sanction the deviators. As a result, for a potential deviator the threat of being sanctioned might not be credible, and the deviator can reasonably expect not to be sanctioned.

For that reason, we use the stronger equilibrium concept of *perfect* (or subgame-perfect) Nash equilibrium, which corresponds to the one introduced in Subsection 2.3.6. A **perfect Nash equilibrium** strategy should be a Nash equilibrium strategy for any subgame of the game, i.e., in all possible situations, even those which are not supposed to be attained when players follow the Nash equilibrium (see Example 25 below). Therefore, if there exists a perfect Nash equilibrium implying sanctions, then it includes incentives for the sanctioning players to actually perform the sanctions.

We give below three versions of the folk theorem involving perfect equilibria, with different hypotheses.

Theorem 2.23 (Aumann–Shapley [26]/Rubinstein [261]) *If* $\delta = 1$ *then the folk theorem holds with perfect equilibrium.*

Theorem 2.24 (Friedman [104]) *For all* $\boldsymbol{v} = (v_1, \ldots, v_n)$ *in the convex hull of* $\mathcal{U}$*, if* $v_i > U_i(\boldsymbol{a}^{\mathrm{NE}})$*, where* $\boldsymbol{a}^{\mathrm{NE}}$ *is a Nash equilibrium of the simple game, if* δ *is sufficiently close to* 1*, there exists a perfect Nash equilibrium of the infinitely repeated game where* $V_i = v_i$ *for all* i.

Theorem 2.25 (Fudenberg and Maskin [105]) *With two players, the folk theorem holds with perfect equilibrium.*

For more than three players, assume that the convex hull $\mathcal{V}^*$ of the reachable utility vectors dominating the minimax utilities is of dimensionality n. Then for any $\boldsymbol{v} = (v_1, \ldots, v_n) \in \mathcal{V}^*$ *there exists a perfect equilibrium of the infinitely repeated game in which player i's average payoff is* v_i *for* δ *sufficiently close to* 1.

The proofs for those results are built in the same way: the perfect equilibrium strategy consists in playing during each period the same strategy (that gives the utility vector $\boldsymbol{v}$) until someone deviates, and then following a sanction procedure against the first deviator. The constraint of the equilibrium being perfect makes the sanction procedure more complex than for the classical version of the folk theorem. We do not enter into the details of those strategies in this chapter, and instead direct the interested reader to [26, 104, 105, 261], but we do rather develop an example.

Example 25 (Sustainable cooperation among wireless nodes) Studying the repeated version of the interference game (with the payoffs of the one-shot game given in Table 2.11), we investigate the conditions to have a perfect equilibrium where both users use a normal transmission power and get a payoff of 6 each. Indeed, on applying [105], one can define the following strategy, made of two phases:

- phase A; play "normal" while the other does, but if a deviation is observed switch to phase B; and
- phase B; play "high" a number of times $\nu(\delta)$, and re-start phase B each time the other player plays "high," then go back to phase A.

Let us compute $\nu(\delta)$ in terms of the constraints for a perfect equilibrium.

- The sanction must be sufficient for the deviation to be non-beneficial: a deviation from "normal" to "high" yields an immediate payoff gain of 4 to the deviator (with respect to the (normal, normal) action profile), who then has to undergo a sanction due to the opponent switching to phase B, which represents for the deviator a loss of 6 for $\nu(\delta)$ periods (if he or she goes back to a normal transmission power) or of 5 for all future periods (if he or she sticks to the high power). Those future sanctions overcome the immediate gain if $4 < \min(\sum_{i=1}^{\nu(\delta)} 6\delta^i, \sum_{i=1}^{\infty} 5\delta^i)$, which can be only satisfied when $\delta > 4/9$, and for
$$\nu(\delta) \geq \frac{\log(6) - \log(10\delta - 4)}{-\log(\delta)} - 1.$$
- The sanction duration $\nu(\delta)$ must also incentivize a deviator to accept the sanction and show his or her willingness to go back to the initial situation once he or she has deviated, i.e. he or she should prefer being sanctioned and getting a payoff 0 for $\nu(\delta)$ periods before obtaining 6, to sticking to a high transmission power and getting a payoff of 1 in all remaining periods (since the other player remains in phase B). This translates into the condition $\sum_{i=\nu(\delta)}^{\infty} 6\delta^i > 1/(1-\delta)$, which gives
$$\nu(\delta) < \frac{\log(6)}{-\log(\delta)}.$$

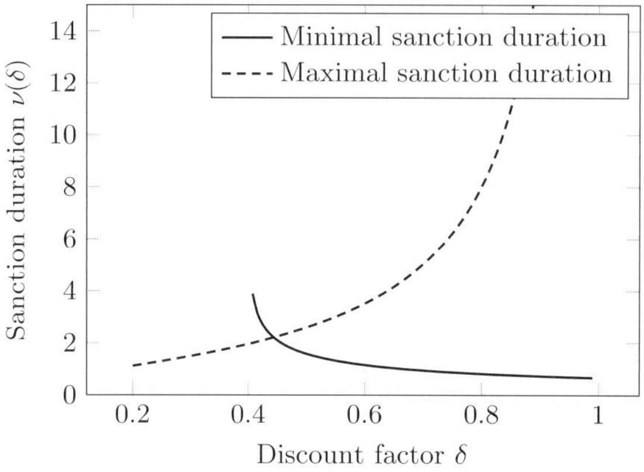

Figure 2.9 The minimal (maximal) sanction duration for the sanction to be dissuasive (acceptable) for a deviator.

When $\nu(\delta)$ satisfies those two constraints, the two-phase strategy described before forms a perfect equilibrium: in phase A there is no incentive to deviate due to the sanction threat, whereas in phase B the deviator is better off accepting the sanction and playing "normal" while the other player gets the highest possible payoff when sanctioning him or her (hence there is no disincentive to sanctioning). Plotting in Figure 2.9 the maximal and minimal values of $\nu(\delta)$ from the two constraints, we conclude that cooperation (i.e., both players using normal transmission power) is sustainable through a perfect Nash equilibrium when $\delta > 0.5$, with a sanction duration of $\nu(\delta) = 2$ periods.

2.3.8 Further reading

The previous subsections present a necessarily limited view of the extremely rich set of situations that can be modeled through game theory. We intended to cover the main notions and results necessary for the rest of the book; however, we would like to mention a few other types of games that we consider of interest in the future of network economics, even if they are not directly applied in this book.

Coalitions/coalitional games

The term **coalitional games** refers to situations where the players have to decide whether to form a coalition (i.e., a team trying to maximize their aggregated payoff), and which players to cooperate with. In the context of telecommunication network economics, such situations can, for example, model mergers or agreements between providers (of the same good, or of complementary goods such as network access and content). In the game-theory vocabulary, such games are also called **cooperative games**. Coalitional games are the subject of a whole part of the book [243], to which we refer the interested reader. In that reference, the authors assume that the global coalition (i.e., involving all players) can reach a total payoff at least equal to the cumulative payoffs of any partition of players into several coalitions.

Forming a coalition allows one to avoid the effects of competition among its members, which enables the coalition to obtain larger payoffs. As an example, if the competing access providers decide to collude, the coalition can then behave as a monopoly and optimize revenue, avoiding price competition such as that studied in Example 19, but at the expense of users in this case.

The difficulty in coalitional games is to ensure sustainability. In other terms, payoffs should be such that no player – or group of players – would be better off leaving the coalition. Assuming that payoff is transferable among players, as is the case with monetary revenues, the problem is then that of how to share the total payoff of the coalition among its members. When the sustainability property holds, the payoff repartition is said to belong to the **core** of the game.

Among the most famous revenue-sharing rules, the **Shapley value** (which was invented by the recent winner of the Nobel prize for economics, Lloyd Shapley) consists in sharing the total payoff of the coalition, in proportion to the average (over all possible coalitions) marginal contribution of each player (see [243], p. 291). The Shapley value has some desirable properties, such as ensuring that the solution will be in the core in special cases where the aggregated payoff function v of coalitions is convex, i.e., when, for any coalitions $S, T \subset \mathcal{N}$, we have $v(S) + v(T) \leq v(S \cup T) + v(S \cap T)$.

Evolutionary games

We finally would like to mention **evolutionary game theory**, which was initially introduced in evolutionary biology [206]. The idea is to investigate the evolution of some characteristics (genes) of species, more precisely of the distribution of those characteristics within an ecosystem, as a result of the interactions. Interactions within the ecosystem affect the reproduction rates of the individuals differently according to their characteristics, so that, over generations, the best-fitted genes should gain prevalence and the less-fitted should vanish. This raises an analogy with the previous sections, where the most payoff-rewarding strategies are assumed to be selected by each player, given the actions of the others. Note that, in the context of species evolution, we do not have *strategic* interactions *per se*, since individuals cannot control their genes here. However, evolutionary games appear to be of interest in the context of this book, mainly for two reasons.

First, the equilibrium concepts in evolutionary game theory are closely linked to the Nash-equilibrium concept, as we explain now. A (possibly mixed) strategy s in a population is said to be an **evolutionarily stable strategy** if, when a small proportion of the population deviates from it to play some other strategy s', then s is strictly a better fit (i.e., provides a larger payoff or reproduction rate) than s'. In particular, s is a best response to the current strategy distribution, i.e., the concept of evolutionarily stable strategy is a refinement of that of Nash equilibrium. Also, if the game described by the characteristics (as strategies) and the reproduction rates (as payoffs) has a strict Nash equilibrium, then those equilibrium strategies are evolutionarily stable. Hence there is a close relation among those concepts, with the notion of evolutionarily stable strategy lying between those of Nash equilibrium and strict Nash equilibrium (see [88]).

Second, some evolutionary-like behaviors arise in game theory, when rational players have a very partial knowledge of the interactions, observing only their own payoff after the interaction has occurred. There, no strategical reasoning is possible since players do not know the strategies employed by others or their payoff functions. However, when the game is repeatedly played over time each player can try to *learn* what action to take, given the information he has. It is then possible to define a **reinforcement learning algorithm** such that, at each iteration of the game, players try an action under their current (mixed) strategy, updating that strategy by raising (reducing) the weight of the action if it yields a large (small) payoff. The strategies considered by a player then evolve similarly to the distribution of characteristics in species over generations, possibly converging to Nash equilibria of the interaction game [269, 300].

A non-technical presentation of the concepts of evolutionary game theory can be found in [264], while the reader interested in a deeper approach is referred to [266].

2.4 Mechanism design and auctions

In this section, we consider the problem of a decision maker (e.g., a network owner or a regulator) who is willing to optimize some global objective, but needs some information that is privately owned by other agents (typically, users) in order to do so. For example, for resource allocation in a network to maximize social welfare, a regulator needs to know the users' payoff functions, information that users would reveal only if it were in their interest. A **mechanism** is a set of rules chosen by the decision maker to that end: it consists in

- a set of available strategies for each agent and
- an **outcome rule**, which maps the strategy profiles of agents to an outcome (say, an allocation of resources, possibly together with payments).

The decision maker is interested in the outcome of the game, and therefore intends to design rules in order to attain his or her objective(s), given the fact that the game will be played by selfish agents; hence the name **mechanism design** for that field of game theory.

We are typically here in a leader–follower or Stackelberg situation (see Subsection 2.3.6), where the leader is the *designer* of the game played by the followers. The mechanism-design problem thus corresponds to a very specific type of Stackelberg game, where the focus is on *information revealing*: the followers will disclose information strategically according to the rules set by the leader, which he or she has to take into account.

Because of the asymmetry of information, we are also in the framework of Bayesian games (see Subsection 2.3.3): all agents (including the mechanism designer) reason on a common prior distribution of the *types* of the others, where the type θ_i of an agent i characterizes his or her privately known information (including the payoff function).

Hence the notion of Bayesian–Nash equilibrium – with agents maximizing their *expected* payoff – plays a central role in mechanism design.

2.4.1 General principles and desirable properties

The most important results of mechanism design were developed in the 1970s, when several possibility and impossibility results were established. We provide some of them in this section, and refer to Chapter 2 in [244] for a synthetic and more exhaustive presentation. In the telecommunication world, mechanism design sees its most important applications in the allocation and pricing of the radio spectrum, which is commonly performed through auctions, a specific kind of mechanism. In the USA, such auctions have been organized since 1994 by the Federal Communications Commission (See http://wireless.fcc.gov/auctions for the latest US spectrum auctions news). This will be described in Chapter 5.

A mechanism designer may want to choose rules so that the resulting situation satisfies different kinds of properties, defined below. All properties refer to the game built by the designer, and consider that game to have some equilibrium.

- **Efficiency**. A mechanism is *efficient* if there is an equilibrium where social welfare is maximized.
- **Incentive compatibility**. A mechanism is called a **revelation mechanism** if each player is asked to reveal his or her type, i.e., if the strategy set S_i of each player i equals the set of possible types Θ_i of that player. A revelation mechanism is **incentive-compatible** if each player revealing truthfully his or her type is an equilibrium.
- **Individual rationality**. This property refers to the incentive for players to actually participate in the mechanism, and therefore assumes that players have an outside option. A mechanism is **individually rational** if, at equilibrium, their payoff is non-negative.
- **Budget balance**. When a mechanism involves monetary transfers (payments), it is desirable that all money exchanges within the system clear each other at equilibrium. When that is the case, the mechanism is said to be budget-balanced. A less restrictive property, called **weak budget balance**, requires that the aggregated amount paid by users be non-negative, i.e., there is no need to inject money into the system for the mechanism to work, but the system can generate financial gains.
- **Robustness against collusion**. While mechanisms are designed for self-interested agents with private information, it could be in the interest of several players to coordinate their actions towards the mechanism. This property states that, at equilibrium, no association can become better off through a joint move.
- **Scalability**. This last property refers to the implementation complexity of the mechanism: a mechanism is scalable if the computational complexity and/or the storage space needed grow reasonably with the number of players.

Very importantly, note that two additional dimensions must be appended to those properties.

1. The **type of equilibrium** considered is of paramount importance: a property holding under a dominant-strategy equilibrium should be preferred to a (weaker) Bayesian–Nash equilibrium. Ultimately, considering only (classical) Nash equilibria must be questioned because of the privately owned type information.
2. The **time status** of the property, i.e., whether the property holds *ex-ante* or *ex-post*. Ex-ante individual rationality – players do not lose *in expectation* by participating in the mechanism – is indeed different from ex-post individual rationality – players never lose, whatever the others' types are. Similarly, the budget balance is also particularly sensitive to that distinction, with ex-ante budget balance being much weaker, and much less desirable, than ex-post budget balance.

The most powerful results of mechanism design concern what is possible and what is not, among the combinations of the aforementioned properties. We recall some of them in the next subsections.

2.4.2 The revelation principle

The **revelation principle** is a key result of mechanism design. It basically establishes that any mechanism can be changed into an incentive-compatible revelation mechanism with exactly the same outcome. To understand the revelation principle, let us first summarize how a mechanism works, recalling that each player (agent) i is of type $\theta_i \in \Theta_i$, and knows his or her own type but has only a prior distribution knowledge of the other players' types.

1. The designer selects strategy sets (that can be different from Θ_i: we do not limit ourselves to revelation mechanisms) for all players, and commits to an outcome rule derived from the observed actions taken by players.
2. Players behave strategically to choose their actions, depending on the rules of the mechanism and on their own type. Let us denote by $s_i(\theta_i)$ the strategy selected by player i if he or she is of type θ_i.
3. The outcome rule is finally applied.

When the (Bayesian) game defined by the mechanism has an equilibrium $\boldsymbol{s}^*(\boldsymbol{\theta}) = (s_1^*(\theta_1), \ldots, s_{|\mathcal{N}|}^*(\theta_{|\mathcal{N}|}))$, the final outcome of the mechanism *at equilibrium* can be seen as a function $f(\theta_1, \ldots, \theta_{|\mathcal{N}|})$ of the players' types only, called the **social choice function** of the mechanism. The revelation principle then states that such a mechanism has an equivalent incentive-compatible revelation version.

Theorem 2.26 (Revelation principle [111, 225]) *Consider a mechanism whose outcome in Bayesian–Nash (dominant-strategy, Nash) equilibrium equals $f(\boldsymbol{\theta})$ for all vectors $\boldsymbol{\theta} = (\theta_1, \ldots, \theta_{|\mathcal{N}|})$ of player types. Then there exists a revelation mechanism that is incentive-compatible in Bayesian–Nash (dominant-strategy, Nash) equilibrium, with the same outcome $f(\boldsymbol{\theta})$ for all $\boldsymbol{\theta}$.*

The principle of the proof is quite simple: the revelation mechanism just needs to mimic the initial situation (including the strategic behavior of the players). More precisely, each player i is asked to reveal his or her type θ_i, and the new mechanism commits to apply the outcome rule of the initial mechanism, precisely to the equilibrium vector $(s_1^*(\theta_1), \ldots, s_{|\mathcal{N}|}^*(\theta_{|\mathcal{N}|}))$. The difference is that this equilibrium vector is no longer computed by the players, but by the mechanism. As a result, since the profile $\boldsymbol{s}^*(\boldsymbol{\theta})$ is a Bayesian–Nash (or dominant-strategy) equilibrium of the initial mechanism, no player can do better than declaring truthfully his or her type. For a detailed proof, the reader is referred to [243].

As a consequence of the revelation principle, in order to study the outcomes that can be reached through a mechanism, it is sufficient to consider only incentive-compatible revelation mechanisms. However, notice that, on the practical side, revelation mechanisms are not always implementable (e.g., when the type spaces are of large dimension), leading to the necessity to investigate other schemes.

2.4.3 Auctions: a specific type of mechanism

An **auction** is defined as follows:

> An auction is a market institution with an explicit set of rules determining resource allocation and prices on the basis of bids from the market participants. (McAfee and McMillan [208])

In telecommunications, mechanism design can mainly be used to allocate a resource (radio spectrum to be shared among operators, radio channels among users, bandwidth over the links of a network among concurrent flows, . . .). Resource consumption (or production) is then sanctioned through *prices*; hence auctions appear as an appropriate kind of mechanism.

Note that the definition above contains the three necessary ingredients of an auction: the designer has to determine

- under what form the participants will be able to submit **bids** (some information is required by the designer, in most cases the amount of money that the participant is willing to pay for some good);
- what **allocation rule** to apply, i.e., how to distribute the resource among the participants depending on their bids;
- and finally what **pricing rule** to implement, again as a function of the submitted bids.

The literature on auction mechanisms in the telecommunication world covers several aspects: auctions can indeed be envisaged

- to be run at each router to decide which packets to discard [177] (see Chapter 3);
- to be used to share the bandwidth of some communication link [55, 67, 184, 271, 273] or the computational capacity of a machine [45] among users (see Chapter 3);
- to allocate transmission time slots among users in wireless networks [281, 282];
- at a higher level, to sell transmission licenses to wireless operators [71], a method widely used in most countries (see Chapter 5);
- to allocate advertisement spots on web pages and search-engine result pages (see Chapter 4); and
- to allocate storage and CPU capacity in grid/cloud systems (see Chapter 4).

In this whole section, we focus on cases where players have quasi-linear utility functions, i.e., the utility of a player i is the difference between his or her valuation for the allocation (the price he or she is willing to pay, that is his or her *type* θ_i) and the price he or she actually has to pay. Also, unless specified otherwise, the valuation function of each player depends only on his or her allocation, i.e., it is independent of the allocations of the other bidders.

2.4.4 First-price auctions

Consider a single *indivisible* item (say, a license to transmit on a specific radio-frequency band) that is sold through an auction to a set $\mathcal{N}$ of potential buyers. In a **first-price auction**, each buyer $i \in \mathcal{N}$ is asked to submit a (sealed) bid with a price b_i. Then the item is awarded to the highest bidder, who then has to pay his or her bid, while losing bidders do not pay anything.

Let us consider the bidding strategy choice from the buyer's point of view. In a Bayesian context, each buyer $i \in \mathcal{N}$ has a valuation (type) θ_i for the item (and the value of not getting it is 0), but can only reason probabilistically about the other bidders' valuations. Clearly, no buyer would bid strictly above his or her valuation, since this would bring utility 0 in the best case (when the buyer loses the auction) and a strictly negative utility otherwise. Similarly, bidding truthfully ($b_i = \theta_i$) always yields a null utility to player i, i.e., he or she is indifferent between participating in the auction and not doing so. It is therefore in the players' interest to set $b_i < \theta_i$, since this ensures a non-negative utility in all cases, and a strictly positive utility $U_i = \theta_i - b_i$ when player i wins the auction. However, the optimal value to bid depends on the expected bids of the opponents, which depend on their types; hence there is a possibly complex problem to solve (see [208] for the equilibrium bidding strategies under some different settings).

2.4.5 Iterative open auctions

Still for a single indivisible item to allocate, one can use other types of auctions with *unsealed* (open) bids that each bidder can update over time. Two patterns can be highlighted.

- **Ascending auctions** (also called English auctions), where potential buyers iteratively increase their bid until only one interested buyer remains, the item being allocated to that bidder at his or her final bid price. This is the most commonly used auction scheme, being applied by auctioneers such as Sotheby's, Christie's, etc.
- **Descending auctions** (also called Dutch auctions), where a large starting price is advertised, and is progressively decreased until one buyer claims to be interested, and gets the item at the currently advertised price. They have been used for perishable products such as flowers (in the Netherlands), fruit (in Germany), and oak in the wine industry.

We remark that there is a dominant bidding strategy in ascending auctions that consists in following the bid increase as long as the highest bid is below one's valuation for the item. When all bidders follow that strategy, the auction ends with the highest-valuation bidder getting the item, and being charged just above the second-highest valuation among bidders (the difference arising from the discreteness of bid updates will be neglected). Interestingly, with that scheme, the highest-valuation bidder avoids revealing his true valuation, since the auction stops before the price reaches that level.

Computing one's bidding strategy in descending auctions is more difficult, and is actually exactly equivalent to what happens in first-price auctions. Indeed, in a descending auction bidders should decide at which price to declare their interest, which leads to the same allocation and price as in the first-price auction: the bidder selecting the highest value gets the item and pays that value.

2.4.6 Second-price auctions

A **second-price auction** (or Vickrey auction) [290] to sell a single item works as follows: potential buyers submit sealed bids and the item goes to the highest bidder, but the price charged equals the *second-highest* bid value.

Some straightforward reasoning shows that this auction mechanism is (dominant-strategy) incentive-compatible, i.e., bidding one's real valuation is a dominant strategy: over-bidding brings no benefit and instead creates the risk of paying more than one's valuation when a truthful bid would have led to losing the auction, while under-bidding leads to the risk of losing the auction and getting utility 0 in some cases when bidding truthfully would have led to a strictly positive utility. Interestingly, notice that, when bidders play their dominant strategy (i.e., the truthful one), the second-price auction leads to exactly the same outcome as the ascending auction (that also has dominant strategies): the item goes to the bidder with the largest valuation, who pays the second-highest valuation.

2.4.7 Revenue-equivalence results

The **revenue-equivalence theorem** holds for situations when a single seller has an indivisible item to auction. It compares the outcomes between two different auction mechanisms, for the seller and the bidders.

Theorem 2.27 (Revenue-equivalence theorem [226, 255, 290]) *Consider two auction mechanisms such that*

- *bidders are risk-neutral;*
- *bidder valuations for the item are independently distributed over a given interval, with a finite and strictly positive density;*
- *the bidder with the lowest possible valuation expects a null utility; and*
- *the bidder with the highest valuation always wins the item.*

Then, for Bayesian–Nash equilibria of the bidding games, both schemes yield the same expected revenue to the seller, and each bidder gets the same utility in both cases.

Note that the framework of the theorem is quite restrictive: for example, we expect to study auctions for spectrum channels, where more than one item is to be sold. However, it clearly illustrates some of the limitations in mechanism design: once some conditions have been set (here, the four assumptions in the theorem), there is no degree of freedom left in terms of expected utility of the participants. In particular, the auction designer cannot get more revenue than he or she gets with second-price auctions (i.e., the expected second-highest valuation among all bidders).

A generalized formulation, known as the **revenue-equivalence principle**, can also be proved.

Theorem 2.28 ([232], p. 236) *For two mechanisms implementing the same social choice function (defined in Subsection 2.4.2) in Bayesian–Nash equilibrium, we have that, if for each player i there is a type θ_i^0 such that both mechanisms have the same expected payment for player i, then the two mechanisms have the same expected payments from each player and the expected revenues of the sellers are the same.*

2.4.8 Vickrey–Clarke–Groves auctions

Vickrey–Clarke–Groves auctions [58, 123, 290], or **VCG auctions**, refer to the following auction mechanism among agents with quasi-linear preferences.

1. Each player i is asked to declare his or her whole valuation function V_i, or equivalently his or her type (i.e., we have a direct mechanism) – let us denote by $\hat{V}_i$ the declared valuation function of player i (not necessarily the true one; it may be beneficial to lie).
2. The allocation $\boldsymbol{a}^{\mathrm{VCG}}$ is chosen from among those that maximize social welfare computed with declared valuations:
$$\boldsymbol{a}^{\mathrm{VCG}} \in \arg\max_{\boldsymbol{a}} \sum_{i \in \mathcal{N}} \hat{V}_i(\boldsymbol{a}). \tag{2.42}$$
3. The price p_i^{VCG} that each bidder i is charged equals his or her **social opportunity cost**, i.e., the total loss of (declared) value that he or she imposes on the others through his or her presence:
$$p_i^{\mathrm{VCG}} = \underbrace{\max_{\boldsymbol{a}_{-i}} \sum_{j \in \mathcal{N}\setminus\{i\}} \hat{V}_j(\boldsymbol{a}_{-i})}_{:=h_i(\hat{\boldsymbol{V}}_{-i})} - \sum_{j \in \mathcal{N}\setminus\{i\}} \hat{V}_j(\boldsymbol{a}^{\mathrm{VCG}}), \tag{2.43}$$
where $\boldsymbol{a}_{-i}$ is the allocation vector that would have been computed, had player i not participated in the auction.

Note that the function h_i in (2.43) does not depend on the action of player i (his or her declared valuation function $\hat{V}_i$). Therefore, that function can be changed without affecting the incentive properties (and thus the outcome) of the mechanism. With a general function h_i in (2.43), the auction scheme is called a **Groves mechanism** (VCG auctions then being a special case of the Groves mechanism).

It can quite easily be proved that Groves mechanisms are dominant-strategy incentive-compatible (that is, bidding $\hat{V}_i = V_i$ is always among the best strategies), and thus dominant-strategy-efficient because of the allocation rule (2.42). Moreover, as established in [122], Groves mechanisms are the only direct mechanisms that are jointly dominant-strategy incentive-compatible and efficient. The degree of freedom left by the choice of the function h_i can then be used to reach other properties, the specific one in (2.43) leading to individual rationality (which can easily be checked) and being the smallest one with that property; in other terms, among Groves mechanisms, the VCG rules yield the largest revenue (sum of payments) to the system under the constraint of ex-post individual rationality.

In summary, VCG auctions turn out to hold several of the desirable properties mentioned earlier, namely *dominant-strategy incentive-compatibility*, *dominant-strategy efficiency*, and *ex-post individual rationality*. Moreover, when only potential buyers participate in the auction, all prices charged are non-negative, i.e., the scheme is *ex-post weakly budget-balanced.*

2.4.9 Combinatorial auctions

A **combinatorial auction** is a specific type of auction aiming at allocating several items, where participants can bid on bundles (combinations) of those items. Applying the results of the previous subsection, one can use the VCG allocation and pricing rules to guarantee incentive-compatibility and efficiency, as well as individual rationality.

The challenges in combinatorial auctions are mostly computational: since the number of combinations grows exponentially with the number of items, finding a welfare-maximizing allocation is often computationally difficult. We do not enter into those considerations here, but the interested reader is referred to [81] and to Chapter 11 in [232].

In terms of economic properties, Groves mechanisms (like the VCG scheme) seem to be among the best candidates, since they are the only mechanisms guaranteeing incentive-compatibility and efficiency. However, they also have some disadvantages, as illustrated below.

Example 26 (VCG prices in combinatorial auctions) Let us consider two items (say, spectrum licenses) to be auctioned, and three bidders (say, wireless operators).[5] Bidder 1 is interested in item 1 only, with valuation 10 (monetary units). Similarly, bidder 2 is willing to pay 12 to obtain item 2, and is not interested in item 1. Finally, bidder 3 is interested only in getting both items, with valuation 16, and is not willing to pay anything to get just one item.

In VCG auctions, each bidder would be asked to reveal his or her valuation vector, for all combinations of goods. Considering that each bidder is willing to pay 0 to get no item, a bid should be a three-dimensional vector specifying the bidder's valuation for

[5] This example has been extracted from the presentation of [2].

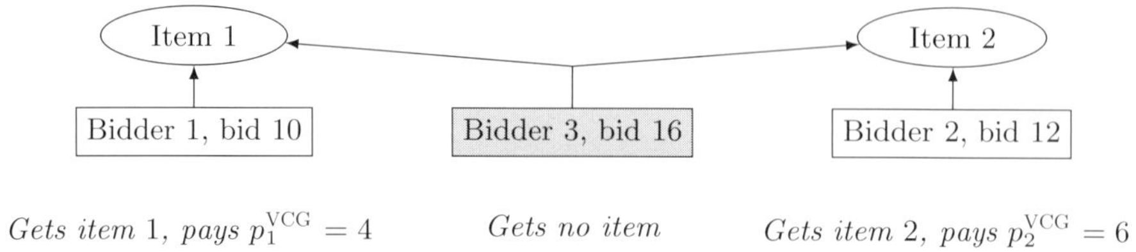

Figure 2.10 Bids and corresponding VCG allocations and prices.

the respective combinations of items $\{1\}$, $\{2\}$, and $\{1, 2\}$. With the values given above, we would have bids $b_1 = (10, 0, 10)$, $b_2 = (0, 12, 12)$, and $b_3 = (0, 0, 16)$.

The bids and results are displayed in Figure 2.10: (declared) social welfare is maximized with value 22 when item 1 (item 2) is allocated to bidder 1 (bidder 2). Had one of those bidders been out of the game, both items would have gone to bidder 3 for a total declared welfare of 16. The VCG price charged to bidder 1 (bidder 2) therefore equals $16 - 12 = 4$ $(16 - 10 = 6)$.

Those prices then raise a problem: bidder 3 does not get the items, whereas the total price paid by the winning bidders for both items equals 10, a value strictly below his or her bid. This situation can seem unfair: why should bidder 3's request be denied while his or her bid is above the total price paid?

The problem raised in Example 26 can also be interpreted as a coalition issue: bidder 3 could agree with the seller to buy both items for some price in $(10, 16)$, making both of them better off. In that sense, the VCG prices are said not to be in the core, as evoked in Subsection 2.3.8. That issue is considered important by UK policymakers in the context of spectrum auctions [71].

An auction scheme not having that default, i.e., such that no group of losing bidders could collude with the seller in a mutually beneficial way, is called a **core-selecting auction**. Nevertheless, it can be shown that it is in general impossible to define an incentive-compatible core-selecting auction, since when the core exists VCG auctions are already core-selecting: if VCG auctions do not lead to core outcomes, no mechanism does [118]. Hence there has been extensive recent research on the trade-offs between the desirable properties (incentive-compatibility, efficiency) for such types of auctions [79, 80].

2.4.10 Double-sided auctions

We end this section with auction schemes involving both sellers and buyers as participants. In other words, the mechanism is used to jointly determine

- how much resource – and for what price – each seller will provide to the market, and
- how those resources – and for what price – are allocated among buyers.

The auction concerns both the supply and demand sides; hence the name **double-sided auction**.

In a direct double-sided auction, sellers declare the minimum price they are willing to be paid to provide (possibly different quantities of) a resource, while buyers reveal the maximum price they are willing to pay (possibly for different quantities). Like in the previous cases, VCG auctions can be applied, and guarantee incentive-compatibility, efficiency, and individual rationality. However, a major issue with VCG payments is that they do not provide (even weak) budget balance; i.e., in many cases the total amount paid by buyers does not cover the amount to be paid to sellers. Since VCG auctions yield the largest revenue among the incentive-compatible, efficient, and individually rational schemes, it appears that it is not possible to add budget balance to those properties: for the mechanism to keep its incentive and efficiency properties, one has to inject money into the system, a clearly undesirable situation [187].

Because of this impossibility result, the mechanism designer has to decide which properties to relax (among efficiency and individual rationality) in order to build a budget-balanced scheme.

Example 27 (Double-sided VCG auction for a single good) Consider a single indivisible good that can be produced by a seller (player 0) for a cost $C > 0$ (or, equivalently, a valuation $V_0 = -C$), with n potential buyers interested in that good. Buyer i is willing to pay up to V_i for it, and we assume that $V_1 > V_2 > \ldots > V_n$. We suppose here that $V_1 > C$, so that the socially efficient outcome is to produce the good and allocate it to buyer 1.

If we apply a double-sided VCG auction, all actors will bid truthfully, so that from the allocation rule (2.42) that efficient outcome is selected, and the social welfare equals $V_1 - C$. Now let us consider the prices derived from the pricing rule (2.43).

- *Seller revenue.* Only buyer 1 would be affected by the absence of the seller, and would lose V_1 in terms of value (recall that prices paid are not considered in (2.43)); therefore, the seller should receive a payment of $-p_0^{\text{VCG}} = V_1$.
- *Buyer* 1 *payment.* In the absence of buyer 1, the good would be allocated to buyer 2 if $V_2 \geq C$ (thus buyer 2 would win V_2 in value, and the other actors are not affected), and not produced if $C > V_2$ (the seller then saving C, with no buyer being affected). As a result, bidder 1 should pay $p_1^{\text{VCG}} = \max(V_2, C)$.

We then immediately observe that the scheme is not (even weakly) budget-balanced, since $\sum_{i=0}^{n} p_i = p_0 + p_1 = \max(V_2, C) - V_1 < 0$.

There are, however, cases where VCG-like results can be obtained together with budget balance, namely when the players are non-atomic.

Example 28 (Double-sided uniform auction) Consider a divisible good to be produced and consumed by a large number of stakeholders. The typical example is that of electricity spot markets (i.e., the market for immediate usage), where the sets of buyers and sellers are large enough, and the individual impact of each buyer or seller is negligible. Then applying the VCG approach leads to prices to pay and rewards for producers that are extremely close to the same marginal value, with equality if players are non-atomic: each participant purchasing (producing) a quantity q_i of good pays (receives) $q_i u$, where u is called the **market clearing price** (or market price) and is the unit price such that total declared demand exceeds total declared supply.

More precisely, each potential buyer i is asked to declare his or her demand function $d_i(p)$, specifying the quantity he or she would buy at price p. Similarly, each potential producer j declares his or her supply function $s_j(p)$, specifying the quantity he or she would produce and sell at price p. Then, assuming all demand functions to be continuous, decreasing, and tending to 0 for large prices, and all supply functions to be continuous, strictly increasing, and null at 0, there exists a unique market clearing price $u = \min\{p : \sum_{\text{buyers } i} d_i(p) = \sum_{\text{sellers } j} s_j(p)\}$. This is precisely the unit price used in the scheme: the total quantity $\sum_{\text{buyers } i} d_i(u) = \sum_{\text{sellers } j} s_j(u)$ is exchanged at that unit price, ensuring budget balance. Here no stakeholder can play the system by declaring an untruthful supply or demand function, because of his or her negligible individual impact on the price u computed. That mechanism is called a **uniform auction**, due to the fact that the same unit price is applied to all sellers and buyers. Here we have a two-sided uniform auction, but uniform auctions can also be applied on one-sided markets.

2.4.11 Towards *computational* mechanism design

In the previous subsections, we have focused on economic properties, deliberately ignoring computational constraints. However, in practice direct mechanisms are often difficult to implement due to the size of the type sets, which can be of large dimension and incur a heavy communication burden. Similarly, applying some allocation and pricing rules (such as the VCG ones) may involve solving problems that are computationally hard. Finally, if a mechanism is not dominant-strategy incentive-compatible, then computing one's optimal bidding strategy can also be computationally difficult.

All those types of new constraints, which are often binding in the telecommunication context, are considered in the specific field of **computational mechanism design**, which we do not develop in this book. Wide coverage of that area of mechanism design can be found in Part II of [232].

2.5 Conclusion

We close this chapter with the few basic highlights to keep in mind when reading the remainder of the book. The first one is that, since telecommunications networks involve several types of actors with individual interests, selfishness does need to be taken into account. Hence the use of game-theoretic models to analyze the interactions in telecommunication contexts. Optimization tools allow one to determine the best strategies from the point of view of stakeholders (considered as separated optimization problems), and the outcomes that are desirable for a regulator (thanks to global optimization tools).

The second important fact is related to the main solution concept from game theory, namely the Nash equilibrium (and its non-atomic counterpart, the Wardrop equilibrium): a Nash equilibrium can help predict the rational outcomes of a game, but it is non-optimal in general (in the sense that all actors may be better off in a non-equilibrium situation). Also, the information available to the players is of the utmost importance, and leads to different kinds of equilibrium notions (Nash equilibrium in complete information, Bayesian–Nash equilibrium when only distributional information is available).

Finally, the outcome of a game can be modified by changing the game rules (through mechanism design), or by repeating the game over time. However, designing a mechanism with a given set of desirable properties is not always doable; therefore, one often has to relax those properties and find some satisfying trade-offs.

3 Economics of access service providers

This chapter aims at describing the history and practice of Internet and telecommunications charging, their mathematical foundations, and the research activity and propositions in this domain.

3.1 History and evolution of access pricing models

Internet access has historically been and often still is charged through a *flat-rate* pricing scheme, meaning that users pay a subscription fee to their provider, and then can use the service as much as they want at no additional cost. It corresponds to an "all-you-can-eat" buffet. We can remark, and are going to develop here, that flat-rate pricing exists in many other contexts, but also that it has not been and still is not always the case in telecommunications, and that there is a trend to move from a flat rate to a more general **usage-based** scheme, meaning here that the price you will pay will depend on your consumption pattern. For an extensive description of flat-rate versus usage-based pricing, we advise the reader to refer to [241].

To discuss a bit more the evolution and comparison of pricing schemes, we can first remark that flat-rate pricing is commonly admitted in other areas. It is typically the case for television (cable TV, but not only), for which there is a subscription fee, and the user can watch as many shows as he wants. In telephony, flat-rate pricing has always been the traditional charging scheme for local calls in the USA, even before the emergence of voice-over-IP. In public transportation, it is customary to have a flat-rate fee (monthly, yearly, weekly, for several days for tourists) for tickets such that you can travel as much as you want during a certain period, in order to limit congestion due to personal cars or to promote tourism.

While flat-rate is now considered the standard pricing scheme and a kind of social benefit by customers in (at least wired) Internet usage, this pricing process has not always been in use. The first Internet access providers appeared at the end of the 1980s, using the regular telephone network (with rate 56 kb/s), before the advent of ADSL (and the expected future FTTH, *fiber to the home*). There was an amount of *time* under which users could use the Internet for a given fixed cost, but any additional use would be charged with a fee per unit of time (cap-based pricing). In this case, what was charged was the connection time or the use of the telephone line, but not the Internet use itself. This type of pricing has not disappeared: we can still find it at some hotspots, in hotels for example, with a number of hours of connection to a wireless network for a given price, the use of which can be stopped and resumed. But note that it does not have a clear link with operating cost in this last case.

The flat rate for the Internet therefore mainly arrived with ADSL technology. Its application took place in 1996 in the USA when AOL, the main access provider for residential users, put it in place after years of resistance. In France and the UK, similarly to other countries, ADSL arrived in 1999 and flat-rate pricing was immediately applied, but sometimes stopped for a while because the business was (already) seen as not viable. Some providers even proposed free Internet access in exchange for the display of advertisement banners on the screen.

Flat-rate pricing has always been questioned though. As early as in 1905, for the phone service in New York City, an unlimited service was considered unfair for light users, and was thought to lead to an inefficient service. Flat-rate pricing is now being heavily attacked by ISPs, claiming that (i) it does not solve congestion, because there is no incentive to reduce usage during consumption peaks, and (ii) it is unfair since very few users contribute a large portion of the Internet traffic. For example, it has recently been claimed by a Time Warner spokesman that 50% of network traffic actually comes from only 5% of subscribers. It is argued that flat-rate pricing is no longer viable given the massive acceleration in Internet traffic and the load it puts on carrier networks. A cause of worry for access providers comes from the fact that interconnection is generally made through volume-based charges, which makes it difficult to combine with a flat rate at the end-user access, because, when an end-user increases his data traffic at no cost, the access provider has to pay for the extra traffic. There has been a lot of pressure exerted by ISPs on governments/regulators to be allowed to implement usage-based pricing. It was recently allowed, at the beginning of 2011, by the regulatory body in Canada (the Canadian Radio-television and Telecommunications Commission (CRTC)), with a switch from a flat rate to **data caps** as requested by major providers (Bell being the main one), and smaller competitors have been forced to follow. This type of scheme is expanding worldwide.

Data caps consist in allowing each user a given fixed maximal amount of data volume each month for a given fee. Usually, each gigabyte (GB) above will be charged at a fixed unitary rate (i.e., volume-based pricing). Bell representatives claim that data-cap billing methods and rates are satisfactory for the vast majority of customers, while those who gorge themselves on bandwidth must simply pay the price for it. Arguing that half of their customers use less than 5 GB a month, they set a data cap of 25 GB per month. A customer going above that limit (about one in ten) pays overage fees of CAD$1 to CAD$2.50 per gigabyte, depending on province. This policy has been followed by main access providers in the USA. Comcast implements a 250-GB data cap that was tested in 2008. Comcast also increased the cap to 300 GB per month in 2012; 300 GB is approximately equivalent to downloading 50 high-definition movies. For customers exceeding this cap, the company will charge an extra $10 for every 50 GB of data that a user uses each month. Time Warner Cable and AT&T also tested in 2008 different data caps in small areas. Will this trend be extended to the rest of the world? In any case, this type of usage-based pricing needed the support of regulators (the American

Federal Communications Commission (FCC) in the USA and CRTC in Canada), and was a condition for receiving adherence by access providers to network neutrality rules (see Chapter 7).

It is often asserted that usage-based charging can have positive effects, but also very negative outcomes [241], needing therefore to be analyzed attentively before application. On the positive side, usage-based pricing can lower the end-users' bill and lead to a better use of bandwidth. But users are often reluctant to be subjected to such pricing schemes, resulting in less broadband adoption. Also, the introduction of usage-based pricing creates economic difficulties for other actors, for instance content delivery networks or content providers such as Netflix. Those aspects will be discussed more in this chapter.

On the other hand, usage-based pricing is commonly admitted in wireless. It is common to offer contracts for a given number of minutes of call duration, with every minute in excess being additionally charged. Here too, a cap is therefore applied. But flat rates for wireless telephony are also proposed, at quite a high charge. Customers do not complain about usage-based schemes, probably because resource scarcity is more willingly endured and congestion better tolerated. Talking about wireless data, because of the traffic increase, there is also a move towards data caps. In the USA in 2011, all major operators switched to this pricing scheme: Verizon (new) customers can choose a monthly volume of 2 GB for \$30, 5 GB for \$50, or 10 GB for \$80. Each gigabyte consumed above the defined cap would cost \$10. AT&T implements a similar strategy. T-Mobile, on the other hand, even if also implementing data caps (\$10 for 200 MB, \$20 for 2 GB, \$30 for 5 GB or \$60 for 10 GB), does not charge for consumption in excess, but chooses to slow down to 100 kbps the communication speed of connections when the cap is reached. Among the four main providers in the USA, Sprint as of 2012 was the only one still offering unlimited data plans to its new subscribers. In addition, providers have been applying a policy to "incentivize" (or force) old users with unlimited data plans to move to the new data caps. This trend of data caps is observed worldwide. In many cases, an unlimited plan is or was advertised, but a single cap is actually implemented, with a slowed connection speed when the cap is exceeded. This was typically the case in France, where the policy was applied, even if most customers were not aware of the imposition of speed reductions (speed reductions are made more explicit now in contracts and advertisements). Data charges are often bundled with phone call caps (and potentially SMS/MMS caps), making it difficult to compute the real charges for data. Though, in 2012, the French operator Bouygues Telecom offered a specific data cap of 500 MB with a limited speed if the consumption goes above, for 9.90 Euros, and 5 GB for 19.90 Euros monthly.

Another way to attract customers for long-term contracts (one or two years) is to offer a big discount on a terminal. This way, subscribers have the feeling that they are saving a huge amount of money while operators ensure the "loyalty" of their customers.

Finally, cable providers also implement usage-based pricing for Internet access while there is a flat rate for TV usage. This is due to historical reasons, but does not stick to real operating or bandwidth costs given that video is consuming more than data.

To summarize and conclude this introduction, it is considered by many that it is now time to switch from flat-rate to innovative pricing schemes to better match the real cost and value that end-users get from the service, a bit similarly to the revenue-management techniques which have been adopted in the airline industry since the 1980s.

3.2 Expectations of users and ISPs, impact on other actors

In order to define the most appropriate pricing schemes and values, access providers need to take into account the various constraints imposed on each of them and their own requirements and usual expectations. But any change or new decision can have an important impact on other actors, not only end-users and access providers: content providers, content delivery networks, grids, etc., can be affected. This may therefore lead to changes in the network structure, which have to be carefully investigated, especially by regulatory bodies.

But we first focus on the directly impacted actors: users and access providers. The goals and requirements of customers (i.e., end-users) and sellers (i.e., access providers) are in general different (see also [279], where this is described with sometimes different arguments).

From the customers' side; the main requirements are the following.

- An important requirement is the **predictability** of the bill. End-users want to know how much they will pay beforehand. This is the reason for their clear preference for a flat rate. In a report published in 2011, the OECD revealed that the rapid development of mobile broadband was due to inexpensive flat-rate mobile data plans.
- Users also want to *understand* the way they are charged and *control* their bill. The pricing scheme has to be as simple as possible, even if this is at the expense of economic efficiency.
- A third aspect is *traceability*. The bills are expected to be easy to verify. *Transparency* is a key word here.

As a consequence, the preference for a flat rate is often explained by three main factors.

- The **insurance effect** of a flat rate. Users prefer to pay a higher cost in order to avoid the risk of having to pay a huge bill at some point, even if on average they will pay less with usage-based pricing, illustrating significant risk aversion. They do not want to worry about the consumption plan.
- The **mental accounting cost**: users are fine with paying more if this simplifies the evaluation of the price they are going to pay. This can be referred to as the *peace of mind* that avoiding computations brings.
- Finally, users have a tendency to *over-estimate* their real usage, hence their readiness to pay an additional flat-rate charge.

This explains why it has been observed that, when possible, most users on metered plans choose to move to unlimited plans even if they will pay more then. This was the case for AT&T's Internet customers who switched to $19.95 flat rate while they were on average paying $11 or $12 per month on metered plans [241], without increasing their time spent online.

From the sellers' side, the goals and preferences are as follows.

- They wish to *maximize their revenue*. Revenue generation is achieved through subscription fees from customers, and/or usage fees if it is not a flat-rate scheme that is implemented. We often get a combination of both, like in traditional telephony or electricity charging plans. However, the choices have to be made considering the potential choices of competitors; this competitive context, as well as its impact on decisions, is the topic of Chapter 5.
- *Maximizing demand* and market share is also of interest. Providers seek to attract as many customers as possible and to incentivize them to use (and therefore pay for) more resources.
- They also wish to *limit costs*. This can be done through simplified operations and billing, if such simplification is not at the expense of economic efficiency. The real costs have to be clearly identified and measured to ensure benefits. Charging in a cost-oriented way is often observed in highly competitive contexts, and strongly pushed by regulators.
- They prefer a pricing scheme with some *flexibility*, to better adapt to network and economic conditions, and better control demand, QoS, and revenue. Above all, flexibility allows changes in the pricing strategies without additional cost from the seller's point of view, and it allows one to deal with price discrimination, segmentation, and different types of contract and services.
- Finally, the *interoperability* of the pricing schemes of the various providers, in the case of exchanges of traffic, is another issue: providers have to agree on exchanges, and this has to be done in a way that is transparent or understandable from the customers' point of view.

A provider has to choose between (i) controlling traffic in the case of limited resources, to avoid congestion, offering a better quality to (some) services, and (ii) over-provisioning its network resources, resulting in limited needs for control and potentially more users, but incurring more infrastructure costs. This induces an important difference of strategy between wireless and wireline access network providers. As said by Craig Moffett from Bernstein Research and quoted in [241]:

> In the calculus of revenue per megabyte, wireline operators are attempting to manage the numerator – that is, they are trying to use price as a mechanism to increase average revenue per user and align their business with a new delivery model as video moves to the web – while wireless operators are focusing on the denominator, and are trying to reduce runaway growth to prevent being crushed by capital spending.

From the point of view of a mechanism designer (the one who elaborates a model; this can also be a regulator), the properties to be satisfied are the ones introduced in Subsection 2.4.1 and summarized below.

1. *Individual rationality*. This ensures that participating in the game/business will yield a non-negative utility. It therefore encourages people to participate, since no user will lose anything.
2. *Incentive compatibility*. The players' best interest is to declare their real valuation, i.e., to be truthful. Lying or cheating does not bring any improvement. This is of interest from an "ethical" point of view, but also for the mechanism designer who can, thanks to that, more easily understand/forecast the output of the game.
3. *Efficiency*. The mechanism results in a maximized sum of utilities of all players. Revenue maximization is another option.
4. *Budget balance*. The sum of money exchanged is null. This may seem an obvious property with buyers just paying the amount due to sellers, but, with some mechanisms (such as double-sided VCG auctions), this is not necessarily the case.
5. *Decentralization*. Decentralized implementation can lead to large-scale application of the mechanism.
6. *Robustness against collusion*. There is no incentive for players to collude.

But it has to be remarked that the definition of a pricing scheme cannot be separated from the related issues of accounting and billing (which are also related to e-commerce). *Accounting* is the process of collecting data usage for the resources that are considered. It is necessary to generate the *bill* sent to customers. Accounting (see [279]) is made of three sub-phases: *usage metering*, consisting in recording usage according to accountable events, *rating and discounting*, consisting in computing the charge according to prespecified rules, and *billing*, namely sending the bills and the details of the consumption to the customers. Metering depends on the type of pricing considered (time-of-day, duration-based, distance-based, volume-based, QoS-based, etc., and all possible combinations thereof). Metering and billing costs increase with the complexity of the pricing scheme. Billing can follow several rules: we can have pre-paid bills, periodic bills, or on-demand bills. There are requirements associated with accounting and billing. The first one is *accuracy*: measurements (which could be from statistics) of real consumption have to be as close as possible to what is stated in the bill. Another is *time delivery*: the bill is to be provided to customers at the expected moment, as defined in their service contract. A third is *security*, with its three components that are authentication, data integrity, and confidentiality, to ensure that nobody can have access to and/or corrupt the data.

Finally, as mentioned briefly at the beginning of this section, changing a pricing scheme can have an important, even catastrophic, impact on third parties. Indeed, a switch from flat-rate to usage-based pricing dramatically influences the use of cloud-based storage services, content delivery networks, and volume-consuming content providers such as Netflix: users can be deterred from using such services if they are charged according to volume. This is the reason why an ISP such as Comcast states that a data cap of 300 GB is approximately equivalent to downloading 50 high-definition movies, and therefore that it should not be a factor, in order to avoid complaints and legal actions. It can be noted that in Canada Netflix has already changed its strategy by offering a streaming option with reduced quality that uses two-thirds less bandwidth than the higher-quality option. It also brings forth questions about competition, since most Internet service providers are also offering pay-television services: bundling without a fee for TV biases the competition with other video content providers like Netflix. Usage-based pricing can thus drastically limit other businesses and services innovation; hence there is a need for scrutiny from regulatory bodies. This pertains to the *network neutrality* debate because packets are differentiated according to their origin. A similar issue has been highlighted in Asia, with a wireless provider not counting Facebook traffic in its usage-based plan in order to attract some customers [241], with this application alone being authorized unlimited use.

3.3 Flat-rate pricing

Flat rates consist in requiring a fixed subscription fee, and then letting the customer consume as much as he or she wants at no additional cost.

One of the concerns about flat rates, quite apart from the unfairness of equal charges for light and heavy users, is that they might lead to increased usage, and hence more congestion. During the transition from metered to flat-rate fees, access providers have experienced a traffic growth from users ranging from 0% to 200%, depending on the provider. On average, the data volume has doubled because of flat-rate pricing [241]. Those traffic increases may create congestion, but on the other hand mean larger benefits for users, society, and developers, fostering business, innovation, and access to culture and knowledge. In other words, without this flat-rate mechanism, it is believed that many users would have been reluctant to adopt the Internet; moreover, it has been a boost for education (thanks to live streaming, video, and access to data for people in general and students in particular), employment (thanks to teleworking), and innovation (with the appearance of services such as clouds and content delivery networks that have introduced a great amount of added value to the network).

The use of flat-rate pricing in broadband networks has some theoretical foundations. It is indeed often admitted in economic studies that charges should mimic actual costs, especially in competitive contexts; and it is known that consuming an additional gigabyte in that case has hardly any marginal cost (but this is not true for wireless usage involving scarcer resources). Even though determining marginal costs is not easy, it was, for instance, established by the CTRC in Canada in 2011, as an indication of real cost, that the wholesale usage-based billing for transit between ISPs was ranging from CAD$0.07 to CAD$0.195 per GB, and that this price was constantly decreasing.

To describe a first and simple mathematical model, consider an access broadband provider without (visible) congestion problems. Its goal is to define the flat-rate access price p to charge each customer. The higher the price, the smaller the number of subscribers. This can be represented by a demand function $D(p)$ decreasing with price, which can be defined as the "volume" of subscribers (which is abusively seen here as a continuous value to simplify) willing to pay p for Internet access. Typically, $D(0)$ is the volume of subscribers interested in access, while $D(\infty)$ is generally 0. Again, in this model, there is no notion of QoS; only price counts. The revenue of the provider is then

$$R(p) = pD(p),$$

with each user paying the access fee p. The provider generally sets this price in order to maximize revenue. Assuming that $pD(p) \to 0$ as $p \to \infty$, which is a reasonable assumption, the maximum necessarily exists. If $R(p)$ is a concave function it can be easily determined.

Example 29 Assume for example that $D(p) = e^{-p}$. Then $R(p) = pe^{-p}$, and its derivative is $R'(p) = e^{-p}(1 - p)$. The revenue is then maximized at $p = 1$, yielding a revenue $R(1) = 1/e$.

In order to apply this type of maximization, the access provider will need to properly estimate the demand function.

As an equivalent type of modeling, we can look at the utility of users instead of aggregating all of them in the demand function. As in [217], consider a total mass m of users (the equivalent of $D(0)$ above), and the set of users indexed by a parameter θ. A type-θ user has a valuation function $V(\theta, x)$ when consuming a volume of data x, and a quasi-linear utility function

$$U(\theta, x) = V(\theta, x) - p \tag{3.1}$$

when a flat-rate subscription p is applied, assuming here that no congestion cost is felt by users. But, with the charging scheme being here independent of the volume x, and assuming $V(\theta, x)$ to be increasing in x (i.e., the more data you use, the better it is for you), each customer of type θ can use the maximum volume $\bar{x}$ possible ($\bar{x} < \infty$ because of the operator's finite capacity) because there is no need for him or her to restrain usage. We can then write

$$\tilde{V}(\theta) = V(\theta, \bar{x}) \text{ and } \tilde{U}(\theta) = \tilde{V}(\theta) - p.$$

Users subscribe to the service as soon as they get a positive utility, i.e., $\tilde{V}(\theta) \geq p$. If there is a distribution over the population for the parameter θ with density f, the mass of users actually subscribing when the price is p, i.e., exactly the above demand function $D(p)$, can be obtained by evaluating

$$D(p) = m \int \mathbb{1}_{\{\tilde{V}(\theta) \geq p\}} f(\theta) \mathrm{d}\theta.$$

It is easy to check that this is a decreasing function of p.

Example 30 Following [217], we assume that $V(\theta, x) = \theta\sqrt{x}$, so that there is a decreasing marginal utility associated with an increasing volume, leading here to $\tilde{V}(\theta) = \theta\sqrt{\bar{x}}$. If the parameter θ comes from a uniform distribution over [0, 1], we get

$$\begin{aligned} D(p) &= m \int 1\!\!1_{\{\tilde{V}(\theta)\geq p\}} f(\theta)\mathrm{d}\theta \\ &= m \int_{\min(1, p/\sqrt{\bar{x}})}^{1} \mathrm{d}\theta \\ &= m\left[1 - \frac{p}{\sqrt{\bar{x}}}\right]^{+}, \end{aligned}$$

still with our notation $[\cdot]^{+} = \max(0, \cdot)$.

The maximization of $R(p) = pD(p)$ can also be solved easily, leading to a price $p = \sqrt{\bar{x}}/2$ and a revenue $m\sqrt{\bar{x}}/4$.

In the above description, we have assumed that congestion has no effect (no negative externality) on users' behavior, only price being counted. To relax that assumption, we can add into the utility function (3.1) a component representing the cost or aversion for congestion inducing a worse quality of experience and a smaller willingness to pay. This cost should increase with the total volume of traffic X sent on the network; we denote it by $c(\theta, X)$. That leads to

$$U(\theta, x) = V(\theta, x) - p - c(\theta, X). \tag{3.2}$$

If $x(\theta)$ is in general the volume used by a type-θ user, then

$$X = m \int x(\theta) f(\theta)\mathrm{d}\theta.$$

Recall that users are infinitesimal; hence each user will consider $c(\theta, X)$ as a constant when deciding his or her individual demand $x(\theta)$. Thus, for our flat-rate model, $x(\theta) = \bar{x}(\theta) 1\!\!1_{\{\tilde{V}(\theta)\geq p + c(\theta, X)\}}$. Hence

$$X = m \int \bar{x}(\theta) 1\!\!1_{\{\tilde{V}(\theta)\geq p + c(\theta, X)\}} f(\theta)\mathrm{d}\theta,$$

an equation in X that has a unique solution that we denote by $X(p)$, because the left-hand side is continuously increasing in X from 0 to ∞ while the right-hand side is continuously decreasing and non-negative. We get a total demand

$$D(p) = m \int 1\!\!1_{\{\tilde{V}(\theta)\geq p + c(\theta, X(p))\}} f(\theta)\mathrm{d}\theta$$

and a revenue $R(p) = pD(p)$ for which we can look for a maximum in p. In general, all this can be solved numerically. However, the next example provides a case for which an analytical solution can be derived.

Example 31 Coming back to Example 30, with $\bar{x}(\theta) = \bar{x}$, $V(\theta, x) = \theta\sqrt{x}$ and a parameter θ coming from a uniform distribution over $[0, 1]$, if we additionally assume that $c(\theta, X) = c(X) = X$, the equation in X becomes

$$X = m\bar{x} \int_0^1 \mathbb{1}_{\{\theta \geq (p+X)/\sqrt{\bar{x}}\}} \, d\theta,$$

i.e., $X = m\bar{x}(1 - (p + X)/\sqrt{\bar{x}})$ if $p \leq \sqrt{\bar{x}}$ and $X = 0$ otherwise (i.e., the price is too high to attract customers). We then get $X = m\bar{x}[1 - p/\sqrt{\bar{x}}]^+/(1 + m\sqrt{\bar{x}})$, and hence a demand $D(p) = X/\bar{x} = m[1 - p/\sqrt{\bar{x}}]^+/(1 + m\sqrt{\bar{x}})$; i.e., the same expression as without congestion effects, but multiplied by the constant $1/(1 + m\sqrt{\bar{x}})$. The revenue of the provider is then

$$R(p) = pD(p) = mp\left[1 - \frac{p}{\sqrt{\bar{x}}}\right]^+ \frac{1}{1 + m\sqrt{\bar{x}}},$$

and is again maximized for $p = \sqrt{\bar{x}}/2$, leading to a demand level of $m/(2(1 + m\sqrt{\bar{x}}))$ and a maximal revenue of $m\sqrt{\bar{x}}/(4(1 + m\sqrt{\bar{x}}))$.

3.4 Volume-based pricing

Usage-based pricing, and in particular its volume-based variant, is seen by Internet network access providers as an innovation and a means to promote network investment. If such a usage-based pricing is applied at the access, access providers will more easily be able to compute their access prices in order to pay their charges to the (peer) network providers for the wholesale traffic, which we recall is volume-based. It is also as a consequence seen as a way to "calm down" their intentions to be non-neutral with respect to traffic, something user associations and regulation bodies did fight against. While broadband access providers are leaning towards this type of charging scheme, it has commonly been used by wireless operators for data (and even voice). In this case, congestion can clearly be understood and felt now and then by users; hence there is a much wider acceptance of usage-based pricing.

Volume-based pricing consists in paying a price that increases with the amount of volume x you are consuming, by means of a function $p(x)$. It might be linear, $p(x) = \beta x$, but not necessarily: we might consider a concave function (such as $p(x) = \beta\sqrt{x}$) to still make it increasing but imposing a lower marginal cost for each additional unit of data, or the opposite, convex (such as $p(x) = \beta x^2$), to really try to prevent high-consumption behaviors.

Cap-based pricing is a special class of volume-based pricing. It consists in charging a fixed fee if you consume less than a given amount of data, with a penalty (an additional charge per unit of volume, or a decrease of quality) if you exceed the cap. With an overcharge β per unit of volume and a price p_0 under the cap x_0, the price paid for consuming x equals

$$p(x) = p_0 + \beta \max(0, x - x_0). \tag{3.3}$$

Another option is to use m cap values x_i $(i = 0, \ldots, m-1)$ with respective prices p_i $(i = 0, \ldots, m-1)$ and charge

$$p(x) = \sum_{i=0}^{m-1} p_i \, \mathbb{1}_{\{x_{i-1} < x \le x_i\}} + \beta \max(0, x - x_{m-1}),$$

with $x_{-1} = 0$; i.e., there is still an extra charge over the last cap. Such cap-based pricing schemes are specific volume-based ones making use of step functions.

Caps (as well as volume-based pricing in general) are claimed by providers to be introduced to manage (i.e., limit) congestion: they are expected to reduce demand/traffic with respect to flat-rate pricing. Even if it is true in general, opponents rightfully respond that this does not directly tackle congestion peaks, which usually happen during particular periods. When congestion actually occurs, nothing special is done. Cost-recovery motivation is also questioned in [241], because there is a limited marginal cost (in terms of management and maintenance) associated with a given unit of volume, if congestion is not considered. Bits are not perishable goods like power in electricity management for example.

To mathematically describe volume-based pricing, let us come back to the model introduced in the previous section, with a total mass m of users, the set of users being indexed by a parameter θ, and each type-θ user having a valuation function $V(\theta, x)$ when consuming a volume of data x. The utility function (3.1) now becomes

$$U(\theta, x) = V(\theta, x) - p(x) \tag{3.4}$$

when a volume-based charging $p(x)$ is applied. In this model again, no congestion cost is felt by users. A user of type θ determines the volume $x(\theta)$ maximizing his or her utility, i.e.,

$$x(\theta) = \arg\max_{x \ge 0} V(\theta, x) - p(x). \tag{3.5}$$

This allows one to compute the total volume of traffic

$$X = m \int x(\theta) f(\theta) \mathrm{d}\theta$$

and the revenue

$$R(p) = m \int p(x(\theta)) \mathrm{d}\theta.$$

With that expression, the provider can try to maximize revenue by determining the best function p or the best parameters in a parametric family. For example, with linear volume-based pricing $p(x) = \beta x$, the idea is to find the value of β maximizing $R(p)$. In a cap-based scheme (3.3), the idea is to determine the cap x_0, subscription fee p_0, and linear overcharge β, still with the same goal of maximizing $R(p)$.

We can remark here that the cap-based scheme is a generalization of both linear (corresponding to $x_0 = 0$ and $p_0 = 0$) and flat-rate (corresponding to $x_0 = \infty$) pricing.

Therefore, cap-based will always be at least as good as strict (linear) volume-based and flat-rate pricing.

When looking at linear volume-based pricing, we can also directly relate the revenue $R(p)$ to the total volume of traffic,

$$R(p) = m \int \beta x(\theta) \mathrm{d}\theta = \beta X.$$

When the valuation functions v and density f are known, the derivations of optimal strategies for the provider can be obtained, as described in the next examples.

Example 32 The case of a linear volume-based price $p(x) = \beta x$ is treated in [217] with the specific valuation function $V(\theta, x) = \theta\sqrt{x}$ and θ distributed uniformly over [0, 1], given that there is again a maximum volume $\bar{x}$ that can be consumed. The volume determination (3.5) becomes $x(\theta) = \arg\max_{0\leq x\leq\bar{x}} \theta\sqrt{x} - \beta x$; i.e., $x(\theta) = \min(\theta^2/(4\beta^2), \bar{x})$. As soon as $\theta \leq \theta' = 2\beta\sqrt{\bar{x}}$, a user θ will not consume the maximum amount of data. The total volume of traffic is therefore

$$X = m\left[\int_0^{\min(\theta',1)} \frac{\theta^2}{4\beta^2}\,\mathrm{d}\theta + \int_{\min(\theta',1)}^1 \bar{x}\,\mathrm{d}\theta\right].$$

It follows that

$$X = \begin{cases} m/(12\beta^2) & \text{if } \theta' \geq 1, \text{ i.e., } \beta \geq 1/(2\sqrt{\bar{x}}) \\ m\bar{x}\left(1 - \frac{4}{3}\beta\sqrt{\bar{x}}\right) & \text{otherwise} \end{cases}$$

and the revenue is

$$R(p) = \begin{cases} m/(12\beta) & \text{if } \beta \geq 1/(2\sqrt{\bar{x}}) \\ m\bar{x}\beta\left(1 - \frac{4}{3}\beta\sqrt{\bar{x}}\right) & \text{otherwise.} \end{cases}$$

Optimizing the revenue in terms of β, after the differentiation of $R(p)$ when it is differentiable, gives a solution $\beta^* = 3/(8\sqrt{\bar{x}})$ and a corresponding revenue $R(p) = (3/16)m\sqrt{\bar{x}}$. In comparison with the flat-rate revenue $m\sqrt{\bar{x}}/4$ in Example 30, volume-based pricing leads to a (slightly) smaller revenue.

We do not mean by the above example that linear pricing will always yield a smaller revenue for the provider; it just happens to be the case here for the considered functions and parameters. Our aim was merely to illustrate that such an analysis can be performed.

Example 33 We can also investigate the most appropriate cap-based pricing (3.3) exactly with the same data as in the previous example. We of course assume $\bar{x} > x_0$, since otherwise the pricing scheme would behave like a flat rate. The volume determination (3.5) gives

$$x(\theta) = \arg\max_{0\leq x\leq\bar{x}} \theta\sqrt{x} - p_0 - \beta\max(0, x - x_0).$$

For $x \leq x_0$ the objective function is increasing, so the maximum is attained at $x = x_0$. If $x \geq x_0$, the maximum is attained, as for linear volume-based charging, at

$\min(\theta^2/(4\beta^2), \bar{x})$ when this is above x_0, and at x_0 otherwise. But this is obtained under the condition that the utility is non-negative (where the subscription fee p_0 can intervene); otherwise a user of type θ has no interest in joining the network, i.e. $x(\theta) = 0$. This gives $x(\theta) = \min(\max(x_0, \theta^2/(4\beta^2)), \bar{x})$, provided that $U(\theta, \min(\max(x_0, \theta^2/(4\beta^2)), \bar{x})) \geq 0$, and $x(\theta) = 0$ otherwise.

The revenue $R(p)$ can be decomposed into three situations:

1. If $1/(4\beta^2) \leq x_0 \leq \bar{x}$, $x(\theta)$ will always be x_0 provided that $U(\theta, x_0) = \theta\sqrt{x_0} - p_0 \geq 0$, so we obtain

$$R(p) = m \int_0^1 p_0 \mathbb{1}_{\{\theta\sqrt{x_0} - p_0 \geq 0\}} \, \mathrm{d}\theta = m p_0 \max\left(0, 1 - \frac{p_0}{\sqrt{x_0}}\right).$$

2. If $x_0 \leq 1/(4\beta^2) \leq \bar{x}$, small-$\theta$ users consume x_0 and, above a threshold $2\beta\sqrt{x_0}$, they consume $\theta^2/(4\beta^2)$, leading to

$$R(p) = m \int_0^{2\beta\sqrt{x_0}} p_0 \mathbb{1}_{\{\theta\sqrt{x_0} - p_0 \geq 0\}} \, \mathrm{d}\theta + m \int_{2\beta\sqrt{x_0}}^1 \left(p_0 + \beta\left(\frac{\theta^2}{4\beta^2} - x_0\right)\right) \mathbb{1}_{\{\theta^2/(2\beta) \geq p_0 + \beta(\theta^2/(4\beta^2) - x_0)\}} \, \mathrm{d}\theta.$$

We remark that the first integral is $m p_0 \max(0, 2\beta\sqrt{x_0} - p_0/\sqrt{x_0})$. The condition in the indicator function included in the second integral gives an integration interval $\left[\bar{\theta}, 1\right]$, with $\bar{\theta} := \min\left(1, 2\beta \max(\sqrt{x_0}, \sqrt{\max(0, p_0/\beta - x_0)})\right)$. The second integral then equals $m((p_0 - \beta x_0)(1 - \bar{\theta}) + (1 - \bar{\theta}^3)/(12\beta))$.

3. If $x_0 \leq \bar{x} \leq 1/(4\beta^2)$, the upper bound $\bar{x}$ has to be used to give

$$R(p) = m \int_0^{2\beta\sqrt{x_0}} p_0 \mathbb{1}_{\{\theta\sqrt{x_0} - p_0 \geq 0\}} \, \mathrm{d}\theta + m \int_{2\beta\sqrt{x_0}}^{2\beta\sqrt{\bar{x}}} \left(p_0 + \beta\left(\frac{\theta^2}{4\beta^2} - x_0\right)\right) \mathbb{1}_{\left\{\theta\sqrt{\theta^2/(4\beta^2)} \geq p_0 + \beta(\theta^2/(4\beta^2) - x_0)\right\}} \mathrm{d}\theta + m \int_{2\beta\sqrt{\bar{x}}}^1 (p_0 + \beta(\bar{x} - x_0)) \mathbb{1}_{\{\theta\sqrt{\bar{x}} \geq p_0 + \beta(\bar{x} - x_0)\}} \, \mathrm{d}\theta.$$

The first integral is the same as in the previous case. The second one can be derived similarly also, and equals $m((p_0 - \beta x_0)(2\beta\sqrt{\bar{x}} - \tilde{\theta}) + ((2\beta\sqrt{\bar{x}})^3 - \tilde{\theta}^3)/(12\beta))$, with

$$\tilde{\theta} := 2\beta \min\left(\sqrt{\bar{x}}, \max(\sqrt{x_0}, \sqrt{\max(0, p_0/\beta - x_0)})\right).$$

Finally, the third integral equals $m(p_0 + \beta(\bar{x} - x_0))(1 - \hat{\theta})$, with

$$\hat{\theta} := \min\left(1, \max\left(2\beta\sqrt{\bar{x}}, \frac{p_0 + \beta(\bar{x} - x_0)}{\sqrt{\bar{x}}}\right)\right).$$

We plot the cap-based revenue when p_0 varies and for different values of x_0 in Figure 3.1, and when x_0 varies and for different values of p_0 in Figure 3.2. The curves are obtained when considering $\beta = 3/(8\sqrt{\bar{x}})$, the optimal parameter of the linear volume-based pricing, and taking $\bar{x} = 100$ and $m = 1$. We also plot each time the optimal flat-rate and

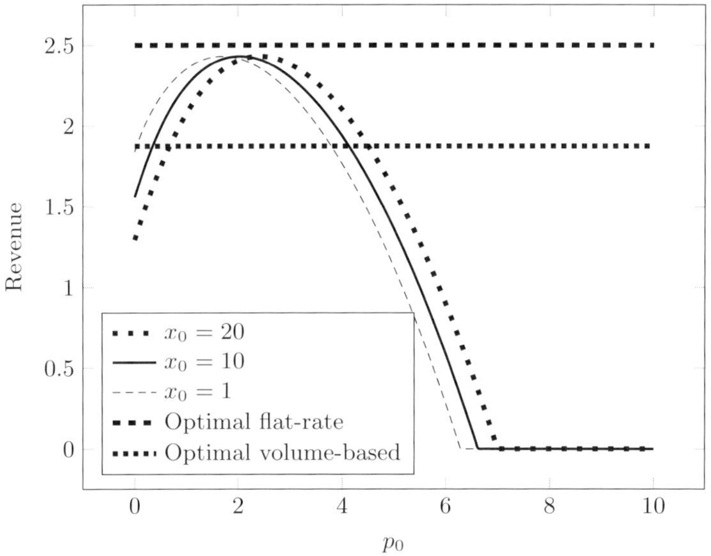

Figure 3.1 Provider revenue, with cap-based pricing in terms of p_0 for various values of x_0.

volume-based pricing for an easy comparison of revenues. Interestingly, flat-rate pricing always yields the largest revenue, *at least* with those forms of valuation functions and densities, whatever the parameters of the cap-based pricing (but with β being fixed). We have been able to find another set of parameters ($\beta = 5 \times 10^{-4}$, $x_0 = 10$, and $p_0 = 4.955$) yielding the same revenue as the optimal flat rate, but not a larger one. We have therefore illustrated here that flat-rate pricing can be the most valuable scheme even for providers; even though, for some other values, this could be different.

The case when there is congestion can be investigated as in the previous section, by adding a congestion cost $c(\theta, X)$ as in (3.2).

We remark also that, even if such an optimization is helpful for the provider to determine its prices and caps, this can hardly be dissociated from the competitive context (which will be addressed in general in Chapter 5). This can be observed in practice. In Canada, for example, Bell implemented a different cap depending on the province: in 2011, it was 25 GB in Ontario and 60 GB in Quebec, because the competitor Videotron proposed a 60-GB cap in the latter region [241]. Competition impacts the values, but can also impact the pricing scheme which is implemented.

However, with the difficulty for access providers of investing in infrastructures with flat rates (in part because there is a volume-based charge at the inter-domain level, creating some compatibility issues) and the reluctance of users to accept volume (cap)-based prices, there is some room to imagine the development of (sometimes individualized) pricing schemes that are based on the value for the generated service and/or the externalities provoked by a user's traffic. We review in the next section the main propositions in the research literature.

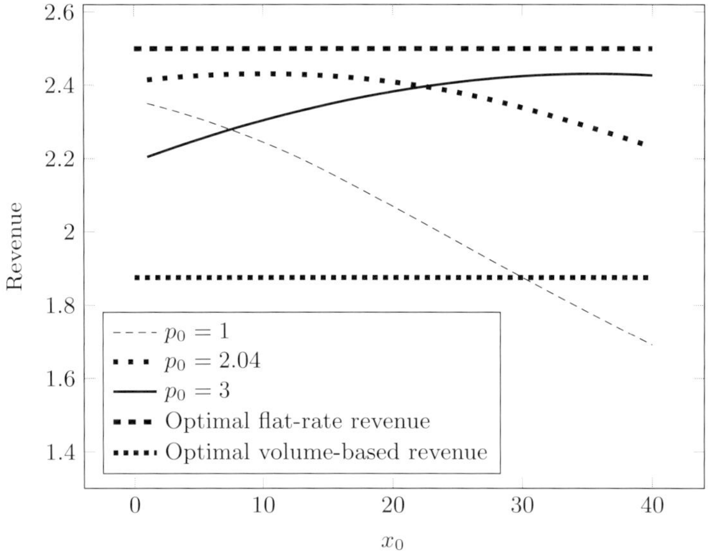

Figure 3.2 Provider revenue, with cap-based pricing in terms of x_0 for various values of p_0.

3.5 Congestion and value-based pricing

Historically, pricing schemes can be organized in two categories for telecommunication connections: schemes for connections *with* versus *without* resource reservation. We can remark that the absence of reservation in the Internet has been a reason for its success, because it needs only very limited and decentralized management. This is likely to continue. Even historical reservation-based applications, like voice and video, are more and more leaning towards no reservation. That may sound strange at first sight, because applications require different QoS, with, for example, video needing small delays and packet losses, voice requiring small delays but tolerating some losses, e-mail affording delay (within a given bound), and FTP being impacted instead by the average throughput. But if applications or users were treated (and priced) differently in a fair way, according to their value for service or QoS, the network could better satisfy users. This is the idea to be developed here: create incentives such that each user does not always choose the best QoS for his application and such that the final result is a fair utilization of the bandwidth. However, the pricing has to be simple, both for users' understanding (and therefore confidence) and from a management point of view to keep the advantages of engineering efficiency. We thus have a trade-off between engineering efficiency/simplicity and economic efficiency, since more complex schemes give the designer more degrees of freedom, allowing one to get closer to economic efficiency.

For surveys on pricing propositions, the reader can have a look at [77, 85, 94, 135, 254, 256, 280, 288].

We choose here to organize the propositions in the following categories.

- Those dealing with charging models for *guaranteed* services. Connections are established through connection-acceptance control (CAC) and reservation of resources.

Although it is unlikely to happen in general because it is too costly in terms of management, such schemes might be applied for some specific customers (the army, banks, hospitals) requiring security and no risk. Protocols such as MPLS [258] and RSVP [309] can be used here.

- The main trend is nonetheless about pricing without a strictly guaranteed QoS, but for which it is *statistically* more or less good and differentiated. Here again, there are several families of schemes.

  Decomposition of the traffic into several classes, which can be treated differently: according to priorities, on distinct subnetworks, etc. When priorities are applied, the price per class can be fixed (then it is called *posted*) or it may depend on the network condition (then it is called *non-posted*). Priority is of interest only in cases of congestion.

  Auctions have been proposed. Users can submit bids for priority at the packet level (such that only bids greater than some cutoff values are admitted), or bids for bandwidth over periods of time. The former is costly in terms of management and the latter reduces the management overhead and could be applied only for big customers (big content providers or content delivery networks). Bids for bandwidth over periods of time do not ensure that a winner will again get the same amount during the next period.

  There is also some activity around pricing that is based directly on the negative externality that your presence imposes on others. It is based on delay or losses in wired networks, or on interference in wireless ones.

  A last group of pricing schemes consists in charging for elastic traffic according to transfer rates. Users decide their payment and receive as transmission rate what the network allocates to them.

We do not mean to be exhaustive regarding the proposals, but believe that the main families are represented in that list.

3.5.1 Pricing and connection acceptance control

With full generality, strict guarantees of QoS can hardly be applied in the Internet, due to scalability and cost issues. However, resource reservation works in the traditional telephony network and is relevant for local networks. It is of particular interest for applications with stringent real-time requirements, such as real-time audio and video services. The idea is to provide resource reservation through CAC over a best-effort IP network [112, 113, 279], a method that was first imagined for ATM networks. One proposal is to combine resource allocation with pricing in order to get a statistical sharing of resources among users. Focusing on **effective bandwidth** [149] therefore seems a relevant way to proceed, but the outcomes have to be estimated.

To understand what the effective bandwidth is, let $X_j(0, t)$ be the (random) load (volume) created by a user j (among J users) during the time interval $(0, t)$. The X_j are assumed independent, and with stationary increments. The effective bandwidth allows one to superpose the J sources and to ensure that the probability that the sum of loads exceeds the capacity C of the link/network is controlled; i.e.,

$$\mathbb{P}\left[\sum_j X_j(0,t) > C\right] \le \exp(-\gamma) \tag{3.6}$$

for some γ. The effective bandwidth is

$$\alpha_j(s,t) = \frac{1}{st}\log \mathbb{E}\left[\exp(sX_j(0,t))\right]; \tag{3.7}$$

and there exist some s and K, depending on C and γ, such that (3.6) holds when the condition $\sum_j \alpha_j(s,t) \le K$ on the (deterministic) effective bandwidth is satisfied. The parameter t thus tunes the interval length, while s plays with the variations of X_j: when $s \to \infty$, $\alpha_j(s,t)$ tends to the peak rate (this corresponds to large values of γ), but when $s \to 0$, it tends to the mean rate. If this effective bandwidth is controlled, so is the probability of exceeding the capacity C. The peak rate and mean rate can be controlled thanks to a **token bucket** (or an equivalent algorithm called **leaky bucket**): to enter the system, a packet has to obtain a token. If one is available, the packet is served (and the token used); otherwise, it is lost. Tokens are stored in a queue with limited capacity h, and, every T time units, a token is generated. If the token queue is full, any new one is destroyed; otherwise it is stored. In that way, the mean rate is controlled by that of generation of tokens, while the peak rate is such that no more than h packets pass between two instances of generation of tokens. For the special case of an on–off source j sending at (peak) rate h when on, and with mean rate r, the effective bandwidth (3.7) becomes

$$\alpha_j(s,t,r,h) = \frac{1}{st}\log\left[1 + \frac{r}{h}(\exp(sth) - 1)\right]. \tag{3.8}$$

One simple way to charge users can be to estimate the effective bandwidth, and charge in terms of that value. Getting this estimation *a priori* is difficult, though.

Now, assume that the peak rate h is known, but not the mean rate R (therefore the rate is random). If the mean rate is $R = r$, the effective bandwidth can be computed from (3.8). Define the charge for a connection with peak rate h as $f(h,r,R)$ for a measured mean rate R while the declared one is r. A user will try to minimize $\mathbb{E}[f(h,r,R)]$. If we want to charge an expected price per unit of time that is linear in terms of the declared effective bandwidth r, it results in taking $f(h,r,R) = a(h,r) + b(h,r)R$ as the tangent to $\alpha_j(s,t,R,h)$ at the point $R = r$. Then a user j chooses R as the *a priori* expectation of the mean rate of the connection (see Figure 3.3). It is therefore incentive-compatible.

If (a_j, b_j) is the couple $(a(h,r), b(h,r))$ for user j, define $\sum_j a_j + b_j X_j(0,t)/t$ as the *effective load*, with $X_j(0,t)/t = R$. An arriving new connection will be accepted only if the effective load does not exceed a threshold.

The strict reservation of resources (even if unused) may be done using RSVP [309] in the Internet. Another option has been proposed in [245, 246], where the price per call is determined in terms of the network conditions when it is initiated. It is in this sense a *dynamic* pricing scheme. Here k different classes of calls are considered, with exponential duration μ_m for a call of class m. A call fee u_m is charged when the call is initiated (and does not depend on the exact duration). For class m, the arrival of

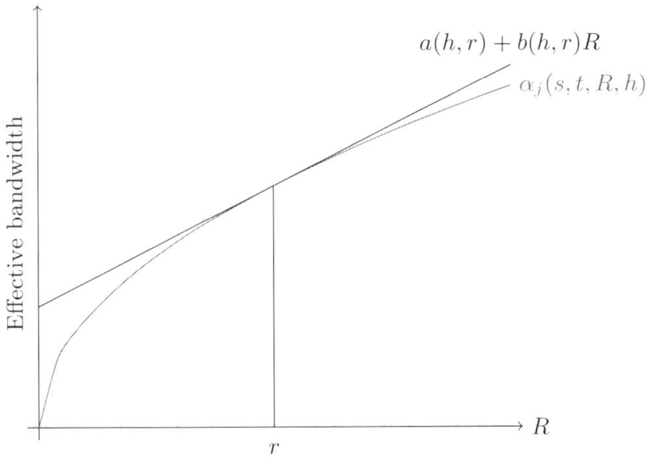

Figure 3.3 Pricing based on effective bandwidth.

calls follows a Poisson distribution with rate $\lambda_m(u_m)$, a continuous strictly decreasing function of u_m over an interval $[0, u_{m,\max}]$, and whose value is 0 above $u_{m,\max}$. Let $n_m(t)$ be the number of active class-m calls at time t, and let r_m be the number of bandwidth units required (reserved) for a class-m call. For a new call to be accepted, the total reserved resource should not exceed the total available resource R; i.e.,

$$\sum_{m=1}^{k} r_m n_m(t) \leq R.$$

Otherwise, new calls are rejected. The goal of a service provider is thus to determine the price vector $\boldsymbol{p} = (p_1, \ldots, p_k)$ as a function for the current state (vector) of reservations $\boldsymbol{n}(t) = (n_1(t), \ldots, n_m(t))$, in order to maximize the expected long-term revenue

$$\lim_{T\to\infty} \frac{1}{T}\mathbb{E}\left[\int_0^T \sum_{m=1}^{k} \lambda_m(p_m(t))p_m(t)\mathrm{d}t\right],$$

where the displayed dependence on t for $\boldsymbol{p}$ comes from the dependence on t of $\boldsymbol{n}$. Social welfare can similarly be the objective function. A dynamic programming method is used to obtain optimal and quasi-optimal prices [37]. Since $(\boldsymbol{n}(t))_{t\geq 0}$ is a (continuous-time) Markov chain whose transition rates are bounded by $\nu = \sum_{m=1}^{k} \lambda_m(0) + \mu_m \lceil R/r_m \rceil$, it can be uniformized to get a discrete-time one (considering the discrete times as the instants of changes of state). From this representation, we can end up with a **Bellman equation**. Bellman equations are driven by Bellman's principle of optimality, stating that "an optimal policy has the property that whatever the initial state and initial decision are, the remaining decisions must constitute an optimal policy with regard to the state resulting from the first decision" (see [37] for more details, definitions, and advantages of Bellman equations). As a consequence, and very sketchily, the optimal expected revenue per unit time J^* and $h(\boldsymbol{n})$ $\forall \boldsymbol{n}$, representing the relative reward in state $\boldsymbol{n}$, are such that their sum is equal to the *optimal* sum of (expected) revenue in the state plus the expected relative reward in the next state, the latter quantity being $\sum_{\boldsymbol{n}'} p_{\boldsymbol{n},\boldsymbol{n}'} h(\boldsymbol{n}')$ with $p_{\boldsymbol{n},\boldsymbol{n}'}$ the probability for transition from $\boldsymbol{n}$ to $\boldsymbol{n}'$. In our case we have the following Bellman equation, with J^* and $h(\cdot)$ as unknowns:

$$\begin{aligned} J^* + h(n) = \max_{p} \Bigg[& \sum_{m \notin C(\boldsymbol{n})} \lambda_m(p_m) p_m + \sum_{m \notin C(\boldsymbol{n})} \frac{\lambda_m(p_m)}{\nu} h(\boldsymbol{n} + \boldsymbol{e}_m) \\ & + \sum_{m=1}^{k} \frac{n_m \mu_m}{\nu} h(\boldsymbol{n} - \boldsymbol{e}_m) \\ & + \left(1 - \sum_{m \notin C(\boldsymbol{n})} \frac{\lambda_m(p_m)}{\nu} - \sum_{m=1}^{k} \frac{n_m \mu_m}{\nu} \right) h(\boldsymbol{n}) \Bigg], \end{aligned}$$

where $\boldsymbol{e}_m$ is the vector with coordinate m at 1 and 0 otherwise, and $C(\boldsymbol{n}) = \{m : \sum_{i=1}^{k} n_i r_i + r_m \leq R\}$, i.e., the set of classes for which an additional customer can be accepted. This system is known to have a unique J^*. The corresponding pricing vector $\boldsymbol{p}$ for each state $\boldsymbol{n}$ corresponds to the one maximizing the expected revenue. We remark that the revenue in state $\boldsymbol{n}$ is $\sum_{m \notin C(\boldsymbol{n})} \lambda_m(p_m) p_m$, and that $\lambda_m(p_m)/\nu \times (n_m \mu_m/\nu)$ is the probability of arrival (departure) of a class-m customer, while $1 - \sum_{m \notin C(\boldsymbol{n})} \lambda_m(p_m)/\nu - \sum_{m=1}^{k} n_m \mu_m/\nu$ is a "no-event" probability in the uniformization procedure because the rate ν is larger than the sum of rates in state $\boldsymbol{n}$. The Bellman equation can be solved, and the resulting optimal policy obtained, using classical dynamic programming algorithms [37]. This can be extended to general loss networks (non-exponential holding times) and to the case where the system has prior knowledge of connection times, at arrival [169]. A conclusion of the analysis, though, is that when demand statistics are slowly varying, time-of-day pricing will often suffice: the optimal pricing strategy can be closely approached by a static price (independently of the state). That means static pricing over periods of time.

3.5.2 Multiclass pricing

The Internet community has tried to design some architectures to cope with congestion. The main one is Diffserv, whose goal is to provide service differentiation to different users. In particular, Diffserv requires some scheduling procedures, and pushes the complexity to the edges of the network.

Paris Metro pricing

A first proposal, which is very simple to implement, and therefore very promising, is to partition a network into several logically separate networks (or classes), each having a fixed fraction of the capacity of the entire network [236]. All networks would behave independently like the current Internet, using TCP and UDP protocols. As a consequence, there would be no formal guarantee of QoS, but, charging different rates for different classes served in the same way, it is likely that self-regulation will lead to situations where the most expensive classes are less congested.

The name **Paris Metro pricing** (PMP) stems from the rules of the Paris Metro up to 1991, under which there were two classes of cars in trains, with exactly the same quality of seats. Since the ticket prices were different, the cars for the most expensive class were usually less congested (but there was no guarantee that this would happen), leading to a better perceived QoS.

This proposal has triggered numerous works on modeling and analysis in the literature [54, 115, 257]. To describe a mathematical model, we mainly follow and base our analysis on the unifying work [54]. The questions we aim at answering are as follows.

- Is it of interest for the provider to offer several classes of service instead of just one?
- If the answer is positive, how many classes should be defined, at which prices, and with which partitioning of the available capacity? It was initially suggested in [236] that only a few (three or four) subnetworks be implemented in order to minimize losses arising from not aggregating all the traffic.

We assume that there are k classes of services proposed by the provider, with capacity C_m allocated to class m. Of course, $C = \sum_{i=1}^{k} C_m$.

Users are assumed to be characterized by their type (or value) θ, representing their sensitivity to congestion, such that a type-θ customer using class m has a utility equal to

$$U_\theta(m) = V - p_m - \theta c(Q_m, C_m),$$

with p_m the access charge to be served at class m, V the valuation for accessing the service, and $c(Q_m, C_m)$ a function measuring the congestion perceived at class m when its capacity is C_m and Q_m is the mass of users choosing class m. We assume without loss of generality that there is a total mass 1 of users. Let $Q_0 = 1 - \sum_{m=1}^{k} Q_m$ be the mass of users choosing to opt out of the service proposed by the provider. We assume without loss of generality that $V \geq p_1 > p_2 > \ldots > p_k$. A user of type θ will therefore choose the service class maximizing his utility, i.e.,

$$m(\theta) = \arg\max_{1 \leq i \leq k} U_\theta(i),$$

or none if this leads to a negative utility. We thus have a two-level (Stackelberg) game, where the provider first decides the number of classes, prices, and distribution of capacity among classes, and then users distribute themselves over classes, selecting the most appropriate for them. Note that the provider can play by anticipating what the distribution of users among classes will be later on. Also, assuming that users are infinitesimal, the distribution of the θ values will be represented by a distribution $F(\theta)$ with density $f(\theta)$. The spreading of users among classes will be a Wardrop equilibrium (see the previous chapter): at equilibrium no user will have an interest in changing class. Obviously, at equilibrium, the $c(Q_m, C_m)$ will be increasing with m for classes with users; otherwise a class will be both more congested and more expensive, and hence of no interest for any user. Focusing on classes with users at equilibrium, a given user of type θ will prefer class m over class i if

$$V - p_m - \theta c(Q_m, C_m) \geq V - p_i - \theta c(Q_i, C_i),$$

i.e., if $p_m - p_i \leq \theta(c(Q_i, C_i) - c(Q_m, C_m))$. With monotonicity of $c(Q_m, C_m)$, class m will be preferred if

$$\theta \geq (p_m - p_i)/(c(Q_m, C_m) - c(Q_i, C_i)) \text{ when } i > m$$

and if

$$\theta \leq (p_m - p_i)/(c(Q_m, C_m) - c(Q_i, C_i)) \text{ when } i < m.$$

This defines thresholds $\theta_1 > \theta_2 > \ldots > \theta_k > \theta_{k+1} = 0$ such that, at equilibrium, for $1 \leq m \leq k$, $\forall \theta \in (\theta_{m+1}, \theta_m)$, class m is preferred, while none is chosen for $\theta > \theta_1$. We then have $Q_m = F(\theta_m) - F(\theta_{m+1})$.

The computation of the θ_ms is done using the fact that, at a threshold, the user is indifferent between the two classes (and indifferent between using the provider and not doing so for θ_1). Hence we need to solve a system of k equations with k unknowns.

Example 34 To illustrate how this user equilibrium can be solved, let $c(Q, C) = Q/C$, let F be a uniform distribution over [0, 1], and fix $k = 2$ as the number of offered classes. Then $Q_1 = \theta_1 - \theta_2$ and $Q_2 = \theta_2$.

At θ_2, the user is indifferent between choosing class 1 and choosing class 2; that is, $p_1 + \theta_2 c(Q_1, C_1) = p_2 + \theta_2 c(Q_2, C_2)$, giving

$$p_1 + \theta_2(\theta_1 - \theta_2)/C_1 = p_2 + \theta_2^2/C_2.$$

At θ_1, the user is indifferent between choosing class 1 and choosing no class; thus $V = p_1 + \theta_1 c(Q_1, C_1)$, giving

$$V = p_1 + \theta_1(\theta_1 - \theta_2)/C_1.$$

From the last equation, $\theta_2 = \theta_1 - (V - p_1)C_1/\theta_1$. This leads to an equation of degree 4 in θ_1, which can be solved numerically. For instance, assume $V = 2$, $C_1 = C_2 = 1/2$, $p_1 = 1.5$, and $p_2 = 0.1$. This gives $\theta_1 = 0.67082$ and $\theta_2 = 0.29814$.

On top of that user-distribution game, the provider has to decide the number of classes and the corresponding prices and capacities. The goal could be to maximize the revenue

$$\sum_{m=1}^{k} p_m Q_m,$$

or, alternatively, to maximize the social welfare

$$\sum_{m=1}^{k} \int_{\theta_{m+1}}^{\theta_m} (V - \theta c(Q_m, C_m)) f(\theta) \mathrm{d}\theta.$$

This optimization step can be performed numerically, by computing first the user equilibrium for each configuration of the provider's decisions. It is interesting to note that the relevance of PMP depends on the actual form of the perceived congestion cost as highlighted in [54]: if it is a *partition-preferred congestion function* such that $c(Q, C) \geq c(\alpha Q, \alpha C)$ for all $0 < \alpha < 1$, such as the load function $c(Q, C) = Q/C$ we used in the previous example, then the differentiated pricing will produce an increased revenue or an increased social welfare if the parameters are properly selected. On the other hand, if it is a *multiplexing-preferred congestion function* such that $c(Q, C) < c(\alpha Q, \alpha C)$ for all $0 < \alpha < 1$, then separating the network into subnetworks will *not* be of interest. A typical example is when the function c is taken as the response time of a queue, for instance in an M/M/1 queue with $c(Q, C) = 1/(C - Q)$ for $Q < C$ and ∞ otherwise, with Q and C representing here respectively the arrival rate of customers and the service rate of the queue. In that case, the best solution is to keep a single class.

Posted priority pricing, loss-based

Let us now turn our attention to *priority* pricing. A first idea is to apply *posted* prices, such that the price of each priority class is established in advance. In [42], each customer is assigned a quota for high-priority packets (according to his or her contract) and, if that quota is exceeded, must pay a penalty the next month. In [59], a priority flag is assigned to each packet according to the type of service, but also a reject flag for services that can bear some losses.

We are going now to describe two different priority pricing models depending on whether we look at systems with delays or with losses.

Consider first a system with losses like in [201, 202, 203]. Time is here discretized, divided into slots. The provider can serve C units of data during a slot. There are n users in contention for the service, and k classes of service with a strict priority rule; i.e., within a slot the provider transmits some priority-m traffic only if all traffic with priority $j > m$ has already been served (still in that slot). We remark that this ordering of priority classes is the opposite of the commonly used one, with class 1 corresponding here to the lowest priority. The common ordering will be used in the other models.

A user $i \in \{1, \ldots, n\}$ decides the number of traffic units to send to each traffic priority class: $d_i(m)$ for class m. The vector $\boldsymbol{d}_i = (d_i(1), \ldots, d_i(k))$ is the total allocation choice of user i, and $\boldsymbol{d} = (d(1), \ldots, d(k))$ is the aggregated allocation vector, with

$$d(m) = \sum_{i=1}^{n} d_i(m)$$

the aggregated traffic sent to class m.

The probability of being transmitted for a unit of traffic of class m is thus, according to the strict priority rule, as follows.

- $P_{\mathrm{tr}}(m, \boldsymbol{d}) = 1$ if $\sum_{j=m}^{k} d(j) \leq C$, meaning that all traffic units for high priorities up to m (included) can be served by the provider.
- $P_{\mathrm{tr}}(m, \boldsymbol{d}) = (C - \sum_{j=m+1}^{k} d(j))/d(m)$ if $\sum_{j=m+1}^{k} d(j) < C < \sum_{j=m}^{k} d(j)$ (with the convention $\sum_{j=k+1}^{k} = 0$), meaning that class m is the one at which demand exceeds (remaining) capacity. In that case, the served units (in class m) are chosen uniformly within the class. This is illustrated in Figure 3.4 for a single class.
- $P_{\mathrm{tr}}(m, \boldsymbol{d}) = 0$ if $\sum_{j=m+1}^{k} d(j) \geq C$, meaning that the provider's capacity has already been exceeded by demand from higher priority classes.

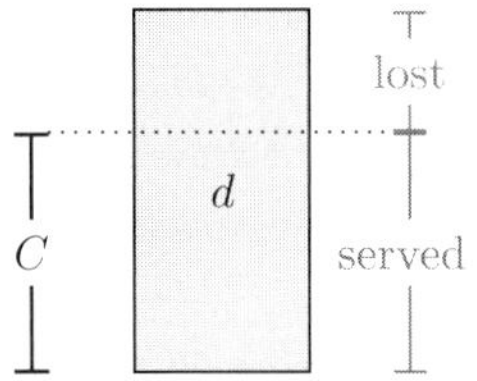

Figure 3.4 The loss model during a slot for a single class: only C traffic units can be served among the d submitted. Thus $d - C$ are lost, with the lost units being chosen uniformly among the d.

We denote by m_0 the class number such that $P_{\mathrm{tr}}(m, \boldsymbol{d}) > 0\ \forall m \geq m_0$ and $P_{\mathrm{tr}}(m, \boldsymbol{d}) = 0\ \forall m < m_0$, i.e., the class at which congestion occurs.

The charge per unit of traffic *sent to the provider* is p_m for class m, with, of course, a higher price for a higher priority; i.e., $p_m > p_j$ if $m > j$. In other words, a user is charged not for what is transmitted by the provider but for any submitted traffic unit. If a unit is lost, it will have to be retransmitted and will be charged again, hence providing a type of congestion pricing. The perceived cost for a transmitted unit is thus $p_m / P_{\mathrm{tr}}(m, \boldsymbol{d})$, $1/P_{\mathrm{tr}}(m, \boldsymbol{d})$ being the average number of retransmissions before a successful one (with the number of trials following a geometric distribution with probability of success $P_{\mathrm{tr}}(m, \boldsymbol{d})$).

User i's satisfaction with the service is denoted by $U_i(x_i)$ and assumed to depend on the throughput x_i during a slot, which is given by

$$x_i = \sum_{m=1}^{k} P_{\mathrm{tr}}(m, \boldsymbol{d}) d_i(m).$$

The utility function U_i is assumed to be increasing, bounded, strictly concave, and twice differentiable. Thus user i will choose the amount sent to each class in order to maximize his or her net utility

$$U_i(x_i) - \sum_{m=1}^{k} d_i(m) p_m,$$

where $\sum_{m=1}^{k} d_i(m) p_m$ is the total amount charged to user i. For user i, the function $D_i(p) = \arg\max_{x \geq 0} U_i(x) - xp$ is defined, representing the traffic volume user i would submit when the price per unit traffic is p and all traffic is transmitted with probability 1, i.e., when the *perceived* price per traffic unit is p. From the assumptions on U_i, $D_i(p) = (U_i')^{-1}(p)$ if $p \leq U_i'(0)$ and 0 otherwise (see Subsection 2.1.2). Let $D(p) = \sum_{i=1}^{n} D_i(p)$ be the aggregated demand function at price p. D is therefore continuous, and strictly decreasing on its support.

We further assume that $D(p_1) > C$, otherwise only the first class would be used, and that $D(p_m) > 0\ \forall 1 \leq m \leq k$, otherwise class m would not be used and could be removed from the analysis. Consequently, one can define the clearing price p^* as the price such that demand equals capacity; i.e., $p^* := D^{-1}(C)$.

The outcome for a user is influenced by all other users' decisions, because the throughput user i will get depends on the total demand vector $\boldsymbol{d}$.

For given prices, we can look for a user equilibrium, i.e., a demand vector such that no user has an interest in changing strategies (meaning demands submitted to each class). It is assumed in [203] as in our previous models above that there are many small users, so that changes in the allocation of an individual user have little effect on the transmission probabilities. The user equilibrium concept is then again the Wardrop equilibrium. It can then be proved that a user equilibrium $\boldsymbol{d}^*$ is such that $m_0 = \max\{m : D(p_m) > C\} = \max\{m : p_m < p^*\}$, and we have the following.

(i) if $D(p_{m_0+1}) < C$ then the equilibrium is unique, and only class m_0 is used. The total amount of traffic sent exceeds the capacity, so the served traffic equals C and the perceived cost equals p^*. Hence the success probability $P_{\text{tr}}(m_0, \boldsymbol{d}^*) = p_{m_0}/p^*$ on class m_0.
(ii) If $D(p_{m_0+1}) = C$ then there exist infinitely many equilibria where some traffic $x \in [0, C]$ is sent to class $m_0 + 1$ (with success probability 1), and traffic $(C - x)p_{m_0+1}/p_{m_0}$ is sent to class m_0 (with success probability p_{m_0}/p_{m_0+1}. However the perceived cost on both classes is unique and equals $p^* = p_{m_0+1}$.
(iii) In all cases the provider's resource is fully used, and

$$x_i^* = D_i(p^*), \tag{3.9}$$

meaning that at equilibrium the throughput of a user is equal to the demand of that user at price p^*.

The above result can be established by proving first that at most two classes are used at equilibrium. Assuming that at least two classes are used, let us denote the smallest (largest) class used by m_1 (m_2). We then have the following.

- m_1 is congested (otherwise the perceived cost would be $p_{m_1} < p_{m_2}$, and some traffic on class m_2 should switch to m_1), while m_2 is not (otherwise the success probability on class m_1 is 0, leading to an infinite cost: some traffic on class m_1 should switch to another class).
- $m_2 = m_1 + 1$: if this were not the case, there would be some class $\tilde{m}$, $m_1 < \tilde{m} < m_2$, but that class would be either congested (leading to a null success probability on class m_1, so the corresponding traffic should then switch to another class) or not congested (but then the perceived cost would be $p_{\tilde{m}} < p_{m_2}$, and some class-m_2 traffic should switch to $\tilde{m}$), both cases leading to a contradiction.

Hence no more than two classes can be used. Now, if exactly two classes m_1 and $m_1 + 1$ are used, then their perceived cost equals p_{m_1+1}, and the success probability on class m_1 is p_{m_1}/p_{m_1+1}. Class m_1 being congested, the overall served traffic $\sum_i x_i$ equals the capacity C, and also equals the total demand at the perceived price p_{m_1+1}; hence $D(p_{m_1+1}) = C$, which implies that $p_{m_1+1} = p^*$ and $m_1 = m_0$. We remark that any traffic distribution described for that situation (case (ii)) is an equilibrium; as a particular case, when $x = 0$ in case (ii) only class $m_0 + 1$ is used.

On the other hand, if $D(p_{m_0+1}) < C$ then only one class, m_1, is used. But $m_1 \leq m_0$, otherwise $D(p_{m_1}) \leq D(p_{m_0+1}) < C$; i.e., class m_1 is not congested, which means that some traffic on class m_1 should switch to a lower-priority class such as class m_0 (there remains some unused capacity; hence that traffic would be served with probability 1 at a smaller cost). Secondly, $m_1 \geq m_0$, otherwise $D(p_{m_1}) > C$, meaning that class p_{m_1} is congested and the amount of traffic served equals C, and thus the perceived cost is $p^* = D^{-1}(C) > p_{m_0}$; some traffic would be better off joining the higher-priority class m_0. This ends the proof.

An interesting consequence is that in all cases the provider's revenue is $R = p^* D(p^*) = p^* C$, and is independent of the price vector $(p_m)_m$. Indeed, the perceived cost on the class(es) used (m_0 and possibly $m_0 + 1$) equals p^*, which gives

$$\begin{aligned} R &= d(m_0)p_{m_0} + d(m_0+1)p_{m_0+1} = d(m_0)P_{\text{tr}}(m_0, \boldsymbol{d}^*)p^* + d(m_0+1)p_{m_0+1} \\ &= d(m_0)\frac{C - d(m_0+1)}{d(m_0)}p^* + d(m_0+1)p^* \\ &= Cp^* = p^* D(p^*), \end{aligned}$$

where in the second line we used the fact that $d(m_0 + 1)$ is non-zero only if $p_{m_0+1} = p^*$. We remark that, even though pricing does not change the provider's revenue, it changes the losses experienced by users (that is, their QoS).

From the above analysis, it seems useless to offer more than two classes, since at most two will be used. But it is not known *a priori* which ones will be used, and proposing more will help the system to adapt to changes in demand characteristics.

Posted priority pricing, delay-based

We consider now that excess traffic is not lost, but stored in a queue in order to be served when the server is available. Congestion is then endured through delay instead of losses being suffered. We consider here a network proposing two classes of traffic (to simplify the analysis) with a strict (preemptive) priority of service, with class 1 served before class 2, and two types of flows, coming from two types of applications. The model is mainly the one proposed in [197] and also analyzed in [133].

As is often done for modeling analysis, the network is here abstracted as a single bottleneck queue, the node at which congestion is happening (the most). We will later on, for the sake of simplicity, assume this queue to be an M/M/1 queue with service rate μ.

The two types of flows will be called *voice* and *data*, with parameters respectively indexed by v and d, to highlight the fact that one type (voice) is more sensitive to delay than the other (data). Their arrival rates of traffic are λ_{v} and λ_{d} per user. The utility of a user depends on the *average delay* D he or she experiences and on the per-packet price p:

$$U_{\text{d}}(D) = D^{-\alpha_{\text{d}}} - p \quad \text{and} \quad U_{\text{v}}(D) = D^{-\alpha_{\text{v}}} - p,$$

where $\alpha_{\text{d}} < \alpha_{\text{v}}$ because voice users have a strong preference for small delays, while data users are less sensitive to high delays than voice users.

The network provider has to decide the per-packet prices p_1 and p_2 associated with each class. We thus again have a two-stage Stackelberg game with

(i) first the network provider deciding the prices p_1 and p_2; and then
(ii) the users entering the system, or not, depending on whether they find the service valuable, which is determined by the prices decided in (i).

The game is also analyzed by backward induction, with the network provider anticipating what the users' choice will be when making its decision.

Again, we assume that users are *infinitesimal*, in the sense that a single user's decision has no impact on the outcome of others, so that the equilibrium notion will again follow the Wardrop principle. We can consider two types of class allocations for users of each type:

- the case of *dedicated* classes, where voice users are forced to go to class 1 (the highest-priority one) and data users forced to go to class 2 to stick to their relative preferences; and
- the case of *open* classes, where we let users choose by themselves.

According to the Wardrop principle, a user will enter the system (in a class he is allowed to enter) as soon as his utility is positive, or leave the system if it is negative. Let n_{v} and n_{d} be the numbers (or, more accurately, the mass) of voice and data users in the system, decomposed into $n_{\mathrm{v}} = n_{\mathrm{v},1} + n_{\mathrm{v},2}$ and $n_{\mathrm{d}} = n_{\mathrm{d},1} + n_{\mathrm{d},2}$, i.e., into the numbers in each class. At Wardrop equilibrium, considering a type d or v and a class its members can access, either

- there is no user because of a non-positive utility if they enter, or
- there are some, but no more are likely to come (meaning that the utility of doing so is non-positive) and none are likely to leave (meaning that the utility of doing so is non-negative; on their leaving some new users would enter).

Formally, $\forall j \in \{\mathrm{v}, \mathrm{d}\}$ and $\forall i \in \{1, 2\}$:

- either $n_{j,i} > 0$ and $U_j(D_i) = 0$
- or $n_{j,i} = 0$ and $U_j(D_i) \leq 0$,

with D_1 and D_2 the average delays of classes 1 and 2. If we consider an M/M/1 queue with strict priority queuing, the average delays are [197]

$$D_1 = \frac{1}{\mu - (n_{\mathrm{v},1}\lambda_{\mathrm{v}} + n_{\mathrm{d},1}\lambda_{\mathrm{d}})}$$

and

$$D_2 = \frac{\mu}{(\mu - (n_{\mathrm{v},1}\lambda_{\mathrm{v}} + n_{\mathrm{d},1}\lambda_{\mathrm{d}}))(\mu - n_{\mathrm{v}}\lambda_{\mathrm{v}} - n_{\mathrm{d}}\lambda_{\mathrm{d}})},$$

where the packet arrival rate for class i ($i \in \{1, 2\}$) is $n_{\mathrm{v},i}\lambda_{\mathrm{v}} + n_{\mathrm{d},i}\lambda_{\mathrm{d}}$.

We first consider the case of dedicated classes with voice allocated to class 1 and data to class 2, enforcing $n_{\mathrm{d},1} = n_{\mathrm{v},2} = 0$. Since voice users have a strict (preemptive) priority over data users, they do not see them in the queue, and make their choice independently of the number n_{d} of data users. Therefore, from a dynamic point of view, the mass $n_{\mathrm{v}} = n_{\mathrm{v},1}$ of high-priority users will "increase" until the utility becomes zero, i.e., $(D_1)^{-\alpha_{\mathrm{v}}} = (1/(\mu - n_{\mathrm{v}}\lambda_{\mathrm{v}}))^{-\alpha_{\mathrm{v}}} = p_1$. This gives the equilibrium value

$$n_{\mathrm{v}}^* = \frac{\mu - p_1^{1/\alpha_{\mathrm{v}}}}{\lambda_{\mathrm{v}}} \tag{3.10}$$

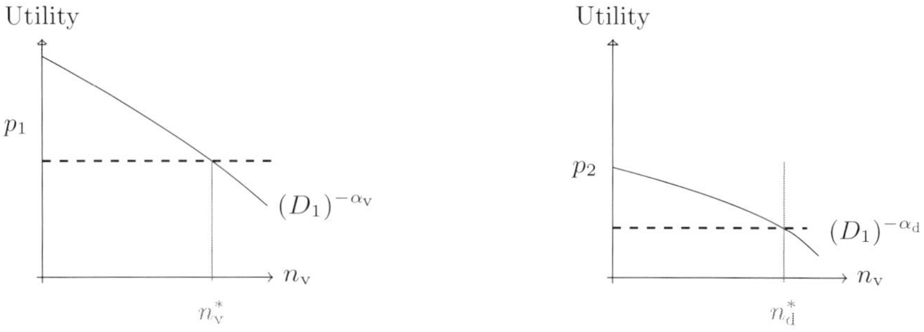

Figure 3.5 The case of dedicated classes. Wardrop equilibrium determination.

if $p_1 < \mu^{\alpha_v}$ and 0 otherwise (because then $U_v < 0$ already when no customer is in the queue; i.e., the service is too expensive). This is illustrated by the left-hand part of Figure 3.5. Then, this value being fixed, the equilibrium mass $n_d^* = n_{d,2}^*$ can be computed similarly, as the solution of

$$(D_2)^{-\alpha_d} = \left(\frac{\mu}{(\mu - \lambda_v n_v^*)(\mu - \lambda_v n_v^* - \lambda_d n_d)}\right)^{-\alpha_d} = p_2,$$

as illustrated by the right-hand part of Figure 3.5, provided that the price is low enough for there to be a non-negative utility when no data customers are present. This gives

$$n_d^* = \begin{cases} \lambda_d^{-1}((p_1)^{1/\alpha_v} - \mu(p_2)^{1/\alpha_d}/(p_1)^{1/\alpha_v}) & \text{if } p_2 < (p_1)^{2\alpha_d/\alpha_v} \text{ and } p_1 \le \mu^{\alpha_v} \\ \lambda_d^{-1}(\mu - (p_2)^{1/\alpha_d}) & \text{if } p_2 < \mu^{\alpha_d} \text{ and } p_1 > \mu^{\alpha_v} \\ 0 & \text{otherwise.} \end{cases} \tag{3.11}$$

What about the case of open classes, such that the users decide to which class they send their flow (voice and data can then share a class)? Consider first the high-priority class 1. If there are $(n_{v,1}, n_{d,1})$ customers in that class, their respective utilities are

$$U_v(n_{v,1}, n_{d,1}) = D_1^{-\alpha_v} - p_1 \quad \text{and} \quad U_d(n_{v,1}, n_{d,1}) = D_1^{-\alpha_d} - p_1,$$

with $D_1 = 1/(\mu - (n_{v,1}\lambda_v + n_{d,1}\lambda_d))$. We remark that the curve $U_v(n_{v,1}, n_{d,1}) = 0$ (in $(n_{v,1}, n_{d,1})$) is always above the curve $U_d(n_{v,1}, n_{d,1}) = 0$ if $p_1 > 1$, whereas the opposite situation pertains if $p_1 < 1$. Those two situations are illustrated in Figure 3.6. The reason is that the first curve, by straightforward computations, is found to be equivalent to $\mu - (n_{v,1}\lambda_v + n_{d,1}\lambda_d) = p_1^{1/\alpha_v}$ while the second corresponds to $\mu - (n_{v,1}\lambda_v + n_{d,1}\lambda_d) = p_1^{1/\alpha_d}$. From the assumption that $\alpha_v > \alpha_d$, $p_1^{1/\alpha_d} > p_1^{1/\alpha_v}$ if and only if $p_1 > 1$; hence the conclusion. From this, we can describe the Wardrop equilibrium, and how to reach it (we will also understand the uniqueness, since starting from any point drives to that point). A formal proof is provided in [197], but we prefer to provide here the intuition thanks to the dynamics of the population evolution. This is illustrated in Figure 3.6 for both cases, $p_1 < 1$ and $p_1 > 1$, but we are going to describe it only when $p_1 > 1$ (the left-hand side of Figure 3.6), since the other case is obtained similarly. Start from a point, say (0, 0), but

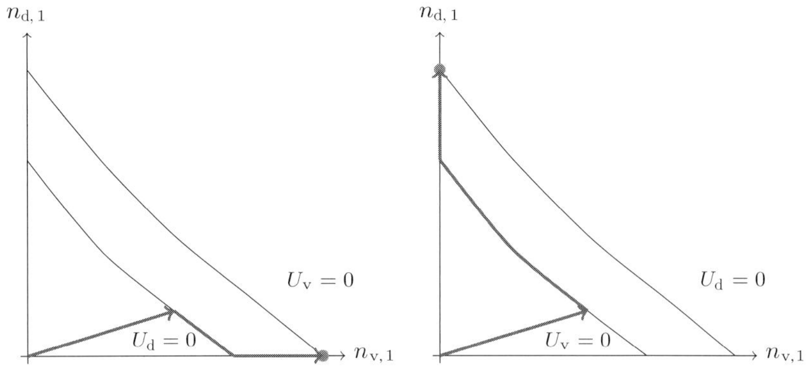

Figure 3.6 The case of open classes. Wardrop equilibrium determination. Left, $p_1 > 1$; right, $p_1 < 1$. Utilities are positive below the curves, and negative above.

any other will lead to the same result. Utilities are positive, so new customers apply for service. Customers (of any type) will enter until the curve $U_d(n_{v,1}, n_{d,1}) = 0$ is reached. But then, we still have $U_v(n_{v,1}, n_{d,1}) > 0$, meaning that voice users still enter. This drives to a negative utility for data users, so that some leave at the same time. We thus "slide" on the curve $U_d(n_{v,1}, n_{d,1}) = 0$ as shown on the left-hand side of Figure 3.6, up to $n_{d,1} = 0$. Then new voice customers can still enter up to the moment when $U_v(n^*_{v,1}, 0) = 0$. Only voice customers use class 1, $n^*_{v,1} = (\mu - p_1^{1/\alpha_v})/\lambda_v$ as in the dedicated-classes case of (3.10) (assuming that $p_1 < \mu^{\alpha_v}$, otherwise there is no customer at all). If $p_1 < 1$, class 1 is used only by data users, with $n^*_{d,1} = (\mu - p_1^{1/\alpha_d})/\lambda_d$.

For the low-priority class, with the values $n^*_{v,1}$ and $n^*_{d,1}$ being fixed, a similar analysis can be performed, leading to the same conclusions depending on $p_2 < 1$ or $p_2 > 1$, and we end up with a number of customers as in (3.11), but with index v (d) for class 2 if $p_2 > 1$ ($p_2 < 1$) and the same thing for the index used for class 1, depending on the value of p_1.

We thus have an equilibrium with four situations for $(n^*_{v,1}, n^*_{v,2}, n^*_{d,1}, n^*_{d,2})$:

- $p_1, p_2 > 1$, only voice users in the system;
- $p_1, p_2 > 1$, only data users in the system;
- $p_1 > 1$, $p_2 < 1$, voice users in class 1, data users in class 2; and
- $p_1 < 1$, $p_2 > 1$, data users in class 1, voice users in class 2. (We remark, though, that it may seem a bit strange to charge more for the low-priority class.)

The revenue of the provider at equilibrium is

$$R = p_1(\lambda_v n^*_{1,v} + \lambda_d n^*_{1,d}) + p_2(\lambda_v n^*_{2,v} + \lambda_d n^*_{2,d}),$$

and the prices p_1 and p_2 can be determined to provide a maximum. The optimal values are explicitly given in [197] (we do not provide them here, or the proofs, which are quite technical and long).

Non-posted priority pricing

The second sub-class of priority pricing is *non-posted* priority pricing, where the price of the packet class depends on the traffic level. In [125], an adaptive priority pricing depending on the context (similar to the principles in [150] and described in Subsection 3.5.5) is used.

Expected capacity

Finally, let us mention the *expected-capacity* theory developed by Clark in [57], where packets are flagged *in* or *out*, and are served without priority except in the case of congestion, when *out* packets receive a congestion pushback.

3.5.3 Auctions

There have been many proposals to apply auctions as a pricing mechanism for users in the Internet. First, bidding for priority was proposed in [210, 211] in a proposition called **smart market**. All users submit a bid attached to each packet they send, and only bids greater than some cutoff value are admitted. The bid expresses the user's willingness to pay. If the network is uncongested, the price is zero, whereas, if there is congestion, the charge is the willingness to pay of the lowest-priority packet admitted. This charging scheme is a variation of the Vickrey auction (see Subsection 2.4.8 for a description of VCG auctions). A drawback of this type of pricing scheme is that it is costly in terms of management, and is therefore admitted to be difficult to apply to the Internet.

In [165, 271, 272, 273], auctions for packets are replaced by auctions for bandwidth during intervals of time to reduce the management overhead. This type of auction can be applied in small networks, or when users are big companies, but can still hardly be applied to end-users. They are worth a mention, though.

To illustrate those schemes, we focus on the scheme developed in [184, 185] because it presents several advantages with respect to the one in [271], and is based on VCG auctions. Assume that a network resource with capacity Q has to be shared among and sold to users. (This scheme is also extended in [186] to the allocation on a tree network, corresponding to the topology of an access network – the so-called last mile, where congestion is likely to occur.) A (new) user i, when entering the game, submits m_i two-dimensional bids of the form $s_i^{(j)} = (q_i^{(j)}, p_i^{(j)})$, where $\forall i \in \{1, \ldots, m_i\}$, $q_i^{(j)}$ is the requested quantity of resource, and $p_i^{(j)}$ is the corresponding proposed unit price. This allows user i to provide the unit prices he or she would be willing to pay for several levels of bandwidth, in the bid $\boldsymbol{s}_i = ((p_i^{(1)}, q_i^{(1)}), (p_i^{(2)}, q_i^{(2)}), \ldots, (p_i^{(m_i)}, q_i^{(m_i)}))$. Without loss of generality, we assume that bids are ordered such that $p_i^{(1)} \leq p_i^{(2)} \leq \ldots \leq p_i^{(m_i)}$. Define $\mathcal{I}$ as the set of users and $\boldsymbol{s} = (\boldsymbol{s}_i)_i$ as the bid profile. Given the multiple bids of all available players, the resource owner has to allocate to every i an amount $a_i(s)$ of bandwidth, and to compute a total charge $c_i(s)$.

Each user i is modeled through a quasi-linear utility function

$$U_i(s) = V_i(a_i(s)) - c_i(s),$$

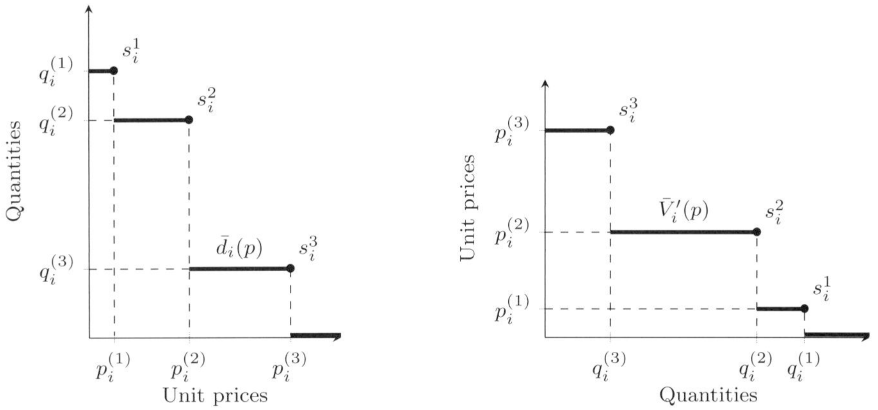

Figure 3.7 Pseudo-marginal valuation and pseudo-demand functions associated with the multi-bid $\mathbf{s}_i$.

where V_i is player i's valuation function, which is assumed to be non-decreasing and concave. A careful choice of the allocation and charging rules will allow one to drive the system to a desired output.

From player i's multi-bid $\mathbf{s}_i$, the auctioneer can build the so-called **pseudo-marginal valuation function** $\bar{\theta}_i'$ and **pseudo-demand function** $\bar{d}_i$ given by

$$\bar{\theta}_i'(q) = \max_{1 \le m \le m_i} \{p_i^{(m)} : q_i^{(m)} \ge q\} \quad \text{if } q_i^{(1)} \ge q, \qquad 0 \text{ otherwise;}$$

$$\bar{d}_i(p) = \max_{1 \le m \le m_i} \{q_i^{(m)} : p_i^{(m)} \ge p\} \quad \text{if } p_i^{(m_i)} < p, \qquad 0 \text{ otherwise.}$$

The pseudo-marginal valuation function $\bar{\theta}_i'$ is the empirical (stair-step) approximation of the marginal valuation function derived from the declared unit prices for quantities of resource. The pseudo-demand function is the generalized inverse function of this pseudo-marginal valuation function, describing the empirical (stair-step) approximation of requested quantities in terms of the unit price. Those functions are illustrated in Figure 3.7. They are good approximations of the true marginal valuation and demand functions, provided that user i has bidded *truthfully*, i.e., such that $q_i^{(m)} = V'(p_i^{(m)})$ for all m. The *aggregated pseudo-demand function* associated with the multiple bid profile s,

$$\bar{d}(q) = \sum_{i \in \mathcal{I}} \bar{d}_i(p),$$

is then an approximation of the total demand, derived from the declared valuations (see Figure 3.8).

To compute the allocations, we first compute the *pseudo-market clearing price* $\bar{u}$, representing the highest unit price p such that the aggregated pseudo-demand is strictly above the available capacity Q of the link:

$$\bar{u} \triangleq \begin{cases} \max\{p : \bar{d}(p) > Q\} & \text{if } \bar{d}(0) > Q \\ 0 & \text{if } \bar{d}(0) \le Q. \end{cases} \tag{3.12}$$

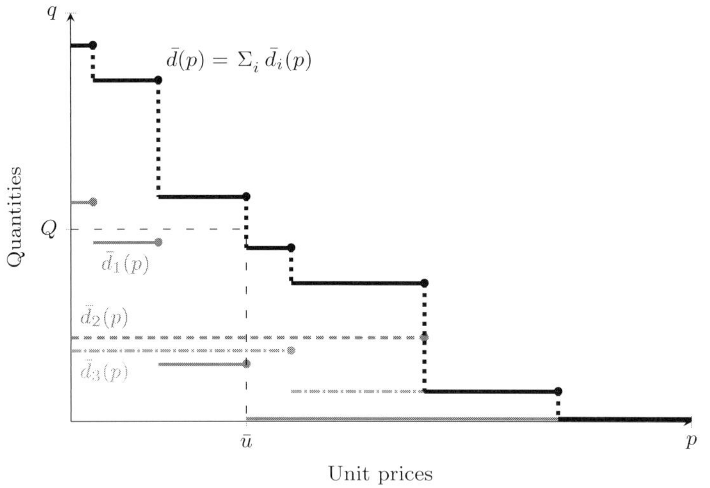

Figure 3.8 Pseudo-marginal valuation and pseudo-demand functions associated with the multi-bid $\boldsymbol{s}_i$.

The pseudo-market clearing price is illustrated in Figure 3.8.

The multi-bid allocation is then chosen as

$$a_i(s) = \begin{cases} \bar{d}_i(\bar{u}^+) + \dfrac{\bar{d}_i(\bar{u}) - \bar{d}_i(\bar{u}^+)}{\bar{d}(\bar{u}) - \bar{d}(\bar{u}^+)}\left[Q - \bar{d}(\bar{u}^+)\right] & \text{if } \bar{d}(0) > Q \\ \bar{d}_i(0) & \text{if } \bar{d}(0) \leq Q. \end{cases} \tag{3.13}$$

In (3.13), $\bar{u}^+ = \lim_{Q_1 \to Q, Q_1 > Q} \max\{p : \bar{d}(p) > Q_1\}$ is the right limit of the market clearing price at Q. It is introduced in order to ensure that $\bar{d}(\bar{u}^+)$ does not strictly exceed the capacity Q (because otherwise $\bar{u} > \bar{u}^+$, which would be a contradiction) and thus that no capacity is over-allocated. But, because of that, it may happen that $\bar{d}(\bar{u}^+) < Q$; i.e., not all the capacity is allocated (due to the non-continuity of $\bar{d}$, see Figure 3.8). The expression (3.13) allows one to avoid that.

- If $\bar{u} = \bar{u}^+$, the complicated second part vanishes and $a_i(s) = \bar{d}_i(\bar{u})$, the declared pseudo-demand at the pseudo-market clearing price.
- If, on the other hand, $\bar{u} > \bar{u}^+$ (because of a discontinuity of $\bar{d}$ in $\bar{u}$), the supply exceeds the pseudo-demand, and the surplus $Q - \bar{d}(\bar{u}^+)$ is proportionally shared among users who introduced a bid at price $\bar{u}$, with weights $\bar{d}_i(\bar{u}) - \bar{d}_i(\bar{u}^+)$.

The allocation being defined, we now need to design a charging scheme. The pricing principle follows the second-price principle of VCG auctions described in Subsection 2.4.8: each user pays for the declared "social opportunity cost" he or she imposes on others:

$$c_i(s) = \sum_{j \in \mathcal{I} \cup \{0\}, j \neq i} \int_{a_j(s)}^{a_j(\boldsymbol{s}_{-i})} \bar{\theta}'_j,$$

i.e., the loss of (pseudo-)valuation the presence of i imposes on other players. The use of the pseudo-valuation comes from the fact that it is the function/value that is actually known by the resource seller.

Thanks to the VCG principle, it is possible to prove that the following properties are satisfied [184].

- *Incentive compatibility*. A player cannot do much better than revealing her or his true valuation, i.e., than submitting a truthful multi-bid (i.e., $q_i^{(m)} = V'(p_i^{(m)})$ for all m).
- *Individual rationality*. If player i submits a truthful multi-bid ($\boldsymbol{s}_i \in \mathcal{S}_i^{\mathrm{T}}$), he or she will end up with a non-negative utility.
- *Efficiency*. The allocation $a(s)$ is optimal up to a certain value, in terms of maximizing social welfare: any other allocation $\tilde{a}(s)$ will not improve $\sum_{i \in \mathcal{I}} V_i(a_i(s))$ by more than a bounded quantity.

How close to truthful and to efficient we are depends on the accuracy of the pseudo-marginal valuation function as an approximation of the marginal valuation function.

This pricing scheme presents several advantages, especially with respect to the auction scheme in [271].

- A player sends bids only once, when entering the game, and does not need to know the bid profile of competitors. This limits considerably the signaling requirements.
- There is no convergence phase needed, with users exchanging information and resubmitting bids to reach an equilibrium: if the network conditions change, new allocations and charges are automatically computed (thus there is no associated loss of efficiency because of a transient phase). This is analyzed in [185].
- The number m_i of bids that user i submits comes from a trade-off between economic efficiency (indeed, when users bid truthfully, the larger m_i, the better the approximations of marginal and demand functions by their "pseudo"-counterparts, and the smaller the distance to optimality) and computational complexity (the smaller m_i is, the simpler the computations).

Finally, let us remark that, when individual users have a limited impact on the pseudo-market clearing price, a simplified version of the scheme can consist in charging each buyer the same unit price $\bar{u}$, i.e., applying a uniform auction as developed in Section 2.4.

3.5.4 Interference-based pricing for wireless networks

The above pricing schemes do look at performance measures and charge per packet of the unit of throughput. But in wireless networks, such as those of the third generation (3G), performance is based on the power level at which you transmit information, using code-division multiple-access (CDMA) technology [291]. This power creates a negative externality on others, through interference. It has been proposed by many authors that one should base the charging scheme on this power. This approach provides a framework towards efficient resource control in CDMA networks. There is an extensive literature on that topic [11, 179, 204, 267, 268, 277, 278]. It often separates the case of **uplink** communications, i.e., communications from the mobile to the base station, from **downlink communications**, those going from the base station to the mobiles. The reason stems from the slightly different interference model.

We remark that, in all cases, the QoS measures are based on the **signal-to-interference-and-noise ratio** (SINR). In the uplink, the SINR for user i's flow is

$$\mathrm{SINR}_i = \gamma_i = \frac{w}{r}\frac{h_i\psi_i}{\sum_{j\neq i} h_j\psi_j + \sigma^2},$$

with

- w the spread-spectrum bandwidth on which the signal is transmitted,
- r the rate of transmission,
- ψ_i the power transmission of mobile i, which can be directly controlled by the mobile itself (the higher the power, the better the SINR, provided that the other mobiles do not increase their own power),
- h_i the path loss (or path attenuation), corresponding to the signal attenuation during its propagation through space ($h_i\psi_i$ is then the power received at the base station from mobile i), and
- σ^2 the background noise.

In this expression, the numerator is roughly $h_i\psi_i$, the power received at the base station and generated by i, divided by the noise (background plus signal power received from all other mobiles). In the downlink, on the other hand, the SINR is

$$\mathrm{SINR}_i = \gamma_i = \frac{w}{r}\frac{h_i\psi_i}{(1-\nu)h_i\sum_{j\neq i}\psi_j + \sigma^2}$$

where ν is the orthogonality factor of i, with $1-\nu$ measuring the orthogonality loss between signals transmitted simultaneously on a CDMA downlink.

From the SINR, we can derive almost all major QoS metrics, and therefore the level of satisfaction of users. The valuation of user i can be written as $V_i(\psi_i, \gamma_i)$, in terms of the powers, but the powers of competitors are abstracted in the SINR. The utility function is, still taking quasi-linear form,

$$U_i = V_i(\psi_i, \gamma_i) - \alpha_i\psi_i,$$

where the second term expresses the cost charged to user i as a linear function of his or her power ψ_i (i.e., with a unit price α_i).

Different valuation functions can be found in the literature. For example in [267, 268], the metric considered is derived from the number of bits successfully transmitted per joule,

$$V_i(\psi_i, \gamma_i) = \frac{r}{\psi_i}(1 - 2\mathrm{BER}(\gamma_i))^{\ell} = \frac{r}{\psi_i}(1 - e^{-\gamma_i/2})^{\ell}, \tag{3.14}$$

where $\mathrm{BER}(\gamma_i)$ is the bit error rate and ℓ is the packet length. Another example can be found in [11], where the valuation function can be interpreted as being proportional to the Shannon capacity,

$$V_i(\gamma_i) = \beta_i \ln(1 + \gamma_i), \tag{3.15}$$

with β_i a coefficient specific to each user. In other words, the valuation is simply linear in the throughput the user can get, and the charge is again linear in the power, $\alpha_i \psi_i$.

In both cases, a Nash equilibrium can be characterized. On top of that game, the provider can decide the pricing variables α_i for each user i. It could be the same α for all users, or different αs for different classes, or even an α_i per customer, if the valuations are properly evaluated. The goal can be to maximize the revenue, or potentially social welfare.

3.5.5 The Kelly mechanism

The work initiated by Frank Kelly and his co-authors [114, 148, 150, 152], but often called the *Kelly mechanism*, can be summarized as a simple and scalable mechanism for bandwidth allocation and pricing in networks. Note that the mechanism focuses on allocation and pricing; no routing decision is made (routing is assumed fixed here).

The model is the following. Consider a set of $\mathcal{J}$ resources (links) with a transmission capacity C_j for resource j. A route $\mathcal{R}$ is a non-empty subset of $\mathcal{J}$ and $\bar{\mathcal{R}}$ is the set of possible routes. Let $A_{j\mathcal{R}} = 1$ if $j \in \mathcal{R}$ and 0 otherwise, and define $\boldsymbol{A}$ as $\boldsymbol{A} = (A_{j\mathcal{R}})_{j\in\mathcal{J},\mathcal{R}\in\bar{\mathcal{R}}}$. If each route is associated with a user $\mathcal{R}$, let $V_{\mathcal{R}}(x_{\mathcal{R}})$ be the valuation function of that user when his or her flow rate is $x_{\mathcal{R}}$. As "usual," the function $V_{\mathcal{R}}$ is assumed to be an increasing, strictly concave, and continuously differentiable function. Let $\boldsymbol{V} = (V_{\mathcal{R}}(\cdot), \mathcal{R} \in \bar{\mathcal{R}})$ and $\boldsymbol{C} = (C_j, j \in \mathcal{J})$ be respectively the vectors of valuations and capacities. From the system point of view, the goal is to determine the vector $\boldsymbol{x} = (x_{\mathcal{R}})_{\mathcal{R}\in\bar{\mathcal{R}}}$ of allocations maximizing the sum of valuations of users (hence maximizing social welfare),

$$\mathrm{SW} = \sum_{\mathcal{R}\in\bar{\mathcal{R}}} V_{\mathcal{R}}(x_{\mathcal{R}}). \tag{3.16}$$

The Kelly mechanism proposes a decentralized way to solve (3.16). Each user $\mathcal{R}$ submits a bid $w_{\mathcal{R}} \geq 0$, representing the amount he or she is willing to pay per unit time, in order to maximize his or her utility

$$V_{\mathcal{R}}(x_{\mathcal{R}}) - w_{\mathcal{R}}. \tag{3.17}$$

The rate $x_{\mathcal{R}}$ is then computed as $x_{\mathcal{R}} = w_{\mathcal{R}}/\lambda_{\mathcal{R}}$, where $\lambda_{\mathcal{R}}$ is the charge per unit flow and unit time for user $\mathcal{R}$. The network then determines the allocation and payment by solving the problem (called the network problem)

$$\max_{x} \sum_{\mathcal{R}\in\bar{\mathcal{R}}} \omega_{\mathcal{R}} \log x_{\mathcal{R}} \tag{3.18}$$

subject to $\boldsymbol{A}\boldsymbol{x} \leq \boldsymbol{C}$ and $\boldsymbol{x} \geq 0$, posing $x_{\mathcal{R}} = 0$ if $w_{\mathcal{R}} = 0$. This formulation is very convenient because it allows one to compute optimal flow rates very easily. Indeed, it is shown in [148, 150, 151] that there always exist vectors $\boldsymbol{\lambda}$, $\boldsymbol{\omega}$, and $\boldsymbol{x}$ satisfying $\omega_{\mathcal{R}} = x_{\mathcal{R}}/\lambda_{\mathcal{R}}$ $\forall\mathcal{R} \in \bar{\mathcal{R}}$ such that $\omega_{\mathcal{R}}$ maximizes (3.17), $\boldsymbol{x}$ maximizes (3.18), and, furthermore, $\boldsymbol{x}$ is the unique solution maximizing (3.16). A point of interest in (3.18) is that the network does not need to know the valuation function of users, just their bids. In this network problem, the vector $\boldsymbol{\mu} = (\mu_j)_{j\in\mathcal{J}}$ of Lagrange multipliers of the capacity constraints and the solution $\boldsymbol{x}$ satisfy

$$x_{\mathcal{R}} = 0 \ \forall \mathcal{R} \in \bar{\mathcal{R}} \quad \text{such that} \quad w_{\mathcal{R}} = 0,$$
$$\lambda_{\mathcal{R}} = w_{\mathcal{R}}/x_{\mathcal{R}} = \sum_{j\in\mathcal{R}} \mu_j \ \forall \mathcal{R} \in \bar{\mathcal{R}} \quad \text{such that} \quad w_{\mathcal{R}} > 0,$$
$$\mu_j \geq 0 \text{ with equality if } \sum_{\mathcal{R}:j\in\mathcal{R}} x_{\mathcal{R}} < C_j.$$

In other words, the charge per unit flow $\lambda_{\mathcal{R}}$ for user $\mathcal{R}$ is the sum of the Lagrange multipliers (shadow prices) of all links forming his or her route; and a charge at a node is imposed only if all the capacity is used. In (3.17), users are price-taking buyers, in the sense that the users do not take into account the influence of their bid on the final charge – they do not play strategically. The problems are solved in a decentralized way, with the intermediate nodes sending signals (the shadow prices) to the users, to which the users adapt. Consider the system of differential equations

$$\frac{\mathrm{d}}{\mathrm{d}t} x_{\mathcal{R}}(t) = \kappa_{\mathcal{R}} \left(\omega_{\mathcal{R}}(t) - x_{\mathcal{R}}(t) \sum_{j\in\mathcal{R}} \mu_j(t) \right), \tag{3.19}$$

where

$$\mu_j(t) = p_j\left(\sum_{s:j\in s} x_s(t)\right)$$

is the shadow price per unit flow through j, $p_j(y)$ is the price charged by resource j when the load through it is y, and $\kappa_{\mathcal{R}}$ is the linear rate of increase (or decrease) of $x_{\mathcal{R}}$ in response to the difference between the bid $\omega_{\mathcal{R}}(t)$ and the charge $x_{\mathcal{R}}(t)\sum_{j\in\mathcal{R}} \mu_j(t)$.

The motivation behind these equations is the following. If resource j generates a continuous stream of feedback signal at rate $yp_j(y)$ when the total flow through resource j is y, that resource j sends a proportion $x_{\mathcal{R}}/y$ of these feedback signals to a user $\mathcal{R}$ with a flow of rate $x_{\mathcal{R}}$ through resource j, and that user $\mathcal{R}$ views each feedback signal as a congestion indication requiring some reduction of flow $x_{\mathcal{R}}$, then it is a flow-control algorithm. It can be shown using Lyapunov functions that the system of differential equations has a unique value x such that $x_{\mathcal{R}} = \omega_{\mathcal{R}}/(\sum_{j\in\mathcal{R}} \mu_j)$ arbitrarily closely approximates the maximization of problem (3.18).

In the case of a single resource (i.e., $\mathcal{J}$ has a single element), the mechanism simply allocates the full capacity C. The principle is to allocate a rate x_r to user r proportional to his bid w_r, with

$$x_r = \frac{w_r}{\sum_{i\in R} w_i} C.$$

The price per unit of resource is then $\lambda_r = x_r/w_r = C/(\sum_{i\in R} w_i)$ and actually does not depend on r.

This mechanism can be used as a congestion-pricing mechanism to achieve proportional fairness when considering $V_{\mathcal{R}}(x_{\mathcal{R}})$ of the form $m_{\mathcal{R}} \log x_{\mathcal{R}}$ for all $\mathcal{R}$. It is easy to check (see [150]) that the resulting vector of rates $\boldsymbol{x}$ is *proportionally fair*; that is, if $\boldsymbol{x} \geq 0$ and $\boldsymbol{A}\boldsymbol{x} \leq \boldsymbol{C}$, and for any other feasible vector $\boldsymbol{x}^*$ of (3.16), the aggregate proportional change is zero or negative:

$$\sum_{\mathcal{R}\in\bar{\mathcal{R}}} \omega_{\mathcal{R}} \frac{x^*_{\mathcal{R}} - x_{\mathcal{R}}}{x_{\mathcal{R}}} \leq 0.$$

Proportional fairness provides a trade-off between maximizing the total network throughput and allowing all users at least a minimal level of service, as introduced in Subsection 2.1.3.

This proposal has inspired a lot of activity. Among other things, it had been argued that the mechanism is dealing with **price-taking players**: users are short-sighted, they maximize their utility without anticipating the consequence of their action on the prices of the links. The case of strategic buyers, or **price-setting players** (carrying out this anticipation) is considered in [143], and the loss of resulting efficiency is analyzed. The mechanism is then combined with VCG auctions in [305] to provide an efficient allocation for strategic buyers at Nash equilibrium.

3.6 Economics of bundling

Bundling is often defined as selling different products together, as a single unit. The interest is that a bundle may appeal much more to customers than when items are sold individually. Bundles are usually made cheaper than buying items separately when a separate sale is offered. The typical bundle offered in telecommunications is the triple-play offer, including Internet access, television, and (fixed) telephony. We now see bundles including in addition wireless telephony.

When offering n products, the provider selling services can offer to sell as a group any subset of goods. A bundle is thus a subset of the n products, with $2^n - 1$ choices (not including the empty set). For $n = 2$ products, labeled 1 and 2, the providers can propose "menus" with prices for $\{1, 2\}$ (prices separately for the two products), $\{(1, 2)\}$ (offering only the bundle), $\{1, 2, (1, 2)\}$ (selling them separately but offering also the bundle), $\{1, (1, 2)\}$ (2 offered only in the bundle) or $\{2, (1, 2)\}$ (1 offered only in the bundle). The set of possibilities for the menu choice increases dramatically with the number of goods.

To illustrate the interest of bundling, consider for instance two customers, say A and B, having the opportunity to buy Internet access (indexed by I) and television access (indexed by T). Assume that providing those services to the customers can be done at no marginal cost for a monopolist ISP (a reasonable assumption), and that the valuations for Internet and television access are respectively $v_{\mathrm{A,I}} = 7$ and $v_{\mathrm{A,T}} = 3$ for A, and $v_{\mathrm{B,I}} = 4$ and $v_{\mathrm{B,T}} = 6$ for B. We moreover assume that the valuations for getting both services are the sums of the two valuations. What are the optimal prices p_{I} and p_{T} that the provider should charge for the services? If $p_{\mathrm{I}} \leq 4$, both customers will subscribe to the Internet, while if $4 < p_{\mathrm{I}} \leq 7$ only A will subscribe, and if $p_{\mathrm{I}} > 7$ neither will. The revenue curve the provider will get for the Internet is therefore $R_{\mathrm{I}}(p_{\mathrm{I}}) = 2p_{\mathrm{I}}\mathbb{1}_{\{p_{\mathrm{I}}\leq 4\}} + p_{\mathrm{I}}\mathbb{1}_{\{4<p_{\mathrm{I}}\leq 7\}}$,

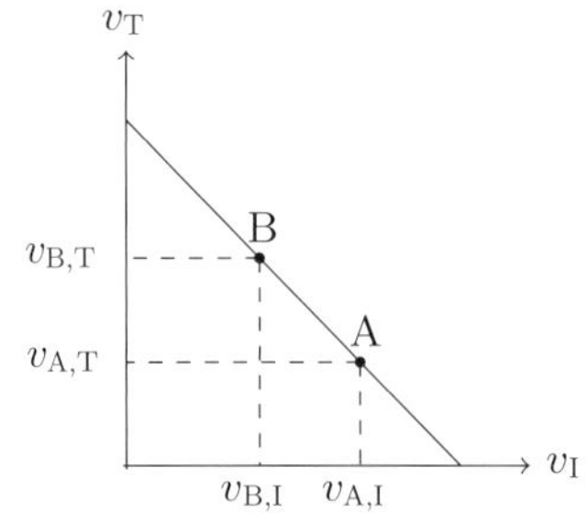

Figure 3.9 Valuations of users A and B, and bundle pricing.

which is maximized for $p_I^* = 4$. By proceeding similarly for television, the revenue generated from that service is $R_T(p_T) = 2p_T \mathbb{1}_{\{p_T \leq 3\}} + p_T \mathbb{1}_{\{3<p_I \leq 6\}}$ maximized for $p_T^* = 3$ or $p_T^* = 6$. The total optimal revenue is then $R_I(p_I^*) + R_T(p_T^*) = 8 + 6 = 14$. If, on the other hand, the ISP proposes a bundle price of 10 for the two services, both customers will subscribe and the revenue generated will be 20, much higher than if no bundle is proposed. Bundling here extracts all potential surplus from the two customers. This is characterized in Figure 3.9, where the two customers are represented by their valuations $(v_{j,I}, v_{j,T})$, for $j \in \{A, B\}$, and the drawn curve is the maximum bundle price to attract both customers on both services. We can also check that, if one were charging separately for television and Internet, attracting both customers would then imply setting prices of at most $v_{B,I}$ for the Internet and $v_{A,T}$ for television.

This leads to the complicated question of whether bundles should be proposed or not. In general, there are several practical interests and concerns brought into play by bundling.

- Bundling allows one to increase the market for low-value products when they are attached to a high-value one.
- On the other hand, the seller has to take care not to "cannibalize" with bundles, i.e., to actually reduce profits because including too many items in the bundle may reduce the sales of those products: the income gained by selling bundles may be less than that which can be obtained when some items are sold separately or in other sets of bundles.

There are conditions under which bundling is optimal. Explicit sufficient conditions on distributions are provided in [209] for when only two products are considered, with random reservation values for the two goods for customers. As a particular case, it is proved in [209] that bundling is always optimal when valuations for the goods are independently distributed. Determining whether bundling is optimal is already a difficult problem for two products, and the existing analyzing tools are shown in [207] to be difficult to generalize even for three goods, because of the numerous sub-bundles that have to be looked at, with their potential influence on other proposed bundles. Actually, the determination of optimal bundles and prices is known to be an NP-hard problem. However, it is proved in [207] that the "menu" of bundles (the set of bundles that are proposed), if it is to be optimal in terms of revenue generation, should make every two goods *connected* by a *bundle path*, in the sense that there exists a sequence of bundles with the first good in the first bundle and the other good in the last and such that there is a common good in any successive pair of bundles.

Another type of works looks at the case of bundling a very large number of goods [31, 227]. This is typically what happens for online content, cable television, Netflix services, an e-library, or copyrighted music. Bundling a large set of products, from a mathematical perspective, smoothes out the uneven preferences that people have for different items in the bundle: the random part is made less important (by reducing the variance), in a type of "law of large numbers" as is experienced in statistics. It may be profitable to just ask for subscription fees for the service and let the users access all the programs/items rather than selling them per use. In this sense, the flat-rate pricing of the Internet can also be seen as a form of bundling for the access to all the available web pages, for example. The key issue is the marginal cost of providing the (electronic) items being almost zero.

4 Economics at the content and application level

4.1 A bit of history

We briefly introduced in the first two chapters the birth, development, and history of the Internet network, but mainly as an access and connectivity "tool" (corresponding also to the goal of the telephone network). But connectivity and voice are not the only applications, and the Internet is not the only network of interest. The cable (television) network is another example. Now, the Internet is providing a lot of different types of interactions and access to whatever type of data, tool, or software one requires for leisure, at work, and in daily life; the possibilities seem limited only by our imagination.

The Internet was initiated by the Defense Department in the USA in order to share research data. For security reasons, i.e., to make it less vulnerable to attacks, the network was designed to be decentralized, a key characteristic that actually, albeit unintentionally, became a reason for its economic development and success.

While the infrastructure was developing, there was an increasing demand for automatic information sharing between scientists working in different universities and institutes all over the world. The **World Wide Web** (WWW) was released on August 6, 1991, by CERN (the European Organization for Nuclear Research), and designed to meet that demand. We remark that the terms World Wide Web and Internet are often used indiscriminately, but the Internet is usually more about the communications infrastructure, while the World Wide Web is about the mass of documents that can be found on the network, connected by hyperlinks. To better navigate on the World Wide Web, the first web browser, Mosaic (which later became Netscape), was released in 1993.

It is important to remark that this part of the history, i.e., the initially designed military and academic networks, is not insignificant because it meant that no business model was at the heart of the design of the network, hence the many economic issues that have to be solved now, and that are not always in line with the initial spirit of the network. The first noticeable commercial online service, CompuServe, appeared in the USA in 1979 (long before the official opening to commercial entities in 1992) and had a dominating position throughout the 1980s. It allowed the exchange of e-mails between subscribers, and made it possible to chat, to access news, forums, etc. At that time, the services were, of course, text-based only.

But the list of applications available through the Internet was not limited to browsing to access data on web pages, chat, and e-mail: new applications appeared, such as voice communications over IP (first version in 1996), television (first attempt in 1994), video on demand (YouTube was not the first one, but was created in 2005), file sharing (peer-to-peer networks; Napster was released in 1999), teleconferencing, online gaming, social networking, home networking, remote monitoring (for security and medicine), e-shopping, etc. Applications are flourishing and their number is experiencing a tremendous increase. One wants to know what the next **killer application** will be, that is, the next application that will be a huge success and potentially yield a lot of money to its creator.

All those applications have to provide revenues to their managing companies. This chapter aims at investigating how such revenue is produced. Moreover, new types of services, and therefore of actors, have appeared and have to be discussed too. Let us first mention **grid computing**, that is, the sharing and pooling of (potentially geographically distant and heterogeneous) computer resources through the Internet, in order to provide more efficiency to the "customer." Grid computing became popular during the second half of the 1990s. The terminology **cloud computing** appeared about ten years later; this was based on basically the same principle: all kinds of computing can be delivered totally over the Internet. Cloud computing has become very popular thanks to the ease of sharing data (photographs, files of any kind) and storing them (in the cloud). Companies are also increasingly leaning towards renting what they need in terms of facilities instead of buying them for occasional use. Among the most well-known cloud computing companies, we can cite as examples Amazon Web Services (offering infrastructures and application services for any type of project), Rackspace (a cloud platform for building websites and applications or offering a private cloud), Microsoft Azure and Google Cloud Platform (to develop applications), and storage and sharing facilities such as Microsoft Skydrive, Google Drive, Dropbox, etc. A related set of actors is made of **content delivery networks**, which deploy servers in data centers, in order to offer increased performance and increased availability, and serve a large amount of content.

The sharing of files among end-users has also been extremely popular thanks to peer-to-peer (P2P) systems. With peer-to-peer networking, each user can be either a client or server for others; there is no need for a central server. Its extreme success was also probably in part due to the ease of access to copyrighted material (music, movies) for free. The file-sharing system Napster, which was released in 1999, has been very successful and is usually considered a killer application for P2P. Napster uses a client–server structure for some tasks such as searching file location(s); it was actually shut down in 2001 (because it was easy to track down that some files were actually copyrighted). Since then, other networks such as Gnutella have been using a P2P structure for almost all tasks.

We will also focus on a specific type of service provider, namely **search engines**, which play an important role by allowing end-users to access applications or some content without necessarily knowing of their existence beforehand, by just composing a set of relevant keywords on the engine.

Our goal in this chapter is to investigate the economic models of such application service providers, considering how they can be designed or improved. This is of particular importance, since, quoting [161], "although it would seem that digital content was one of the most obvious products for sale, content providers continue to struggle to find business models which will bring in adequate revenue." The value and supply chain for content delivery is still evolving in telecommunications. Many content data or applications are free on the Internet (because that is at least up to now the perception from users of how the Internet should behave), so the displaying of advertisements thanks to the inclusion of hyperlinks or banners on the content is a key tool that deserves a lot of attention in terms of revenue generation (and is the subject of Section 4.2). Those ads can be smartly chosen in terms not only of the money the advertisers wish to spend for them to be displayed, but also in terms of the relevance for users. In many cases, though, especially for smartphones, a careful choice has to be made between launching a free application with advertisements and a paid application (this will be treated in Section 4.3). The specific economic cases of cloud computing, P2P networks, and content delivery networks will be treated in the last sections.

To close this introduction, we think that it is insightful to illustrate the intertwining of economic roles by briefly describing the economic development and expansion of a key and dominant company in telecommunication networks: Google. Google was created in 1998, and it initially proposed a search engine, more precisely software located on a web site on which any user can compose keywords, from which a ranked list of documents available on the World Wide Web is returned. The list is obtained from a link-analysis algorithm called **PageRank**, which assigns a weight to documents, to measure their importance and relevance with respect to the keywords. This algorithm was perceived as very quick and efficient, hence the success of Google, which is now almost a monopoly, with more than 60% of the searches worldwide in 2010. But how can Google make money from this free tool? In addition to the links proposed by the PageRank algorithm, there is actually a list of (at most eleven) *advertisement* links, clearly identified as such, at the top and on the right of the page, for which Google is paid as soon as one of the links is clicked by the user. Such **sponsored links** are illustrated in Figure 4.1. This ad business will be discussed in Section 4.2. Selling ads allows one to better understand users' behaviors and expectations, from which the engine can be refined and yet more revenue extracted from ads. But Google has expanded its business to much more than just a search engine activity. It also proposes tools such as Gmail (e-mail services), Google Calendar, YouTube (where advertisements are also displayed), Google News, Google Earth and Google Maps, the operating system Android, etc. Most of them are free, and some of them are even free of ads. There might be a reason why Google is providing all this for free:[1] for Android, for instance, it makes the operating system industry more competitive (imposing cost reductions on competitors), and actually increases its revenues through content itself because of increasing demand due to decreasing use costs. This is what Android does with respect to Apple, but also what Google Docs does with respect to Microsoft, and why we now have Google Voice. It also allows better control over the whole supply chain. Thanks to those price reductions due to competition on all means of access, Google can earn more on content via searches and YouTube, in markets where it is dominant.

[1] See http://elidourado.com/blog/theory-of-google/ for more explanation.

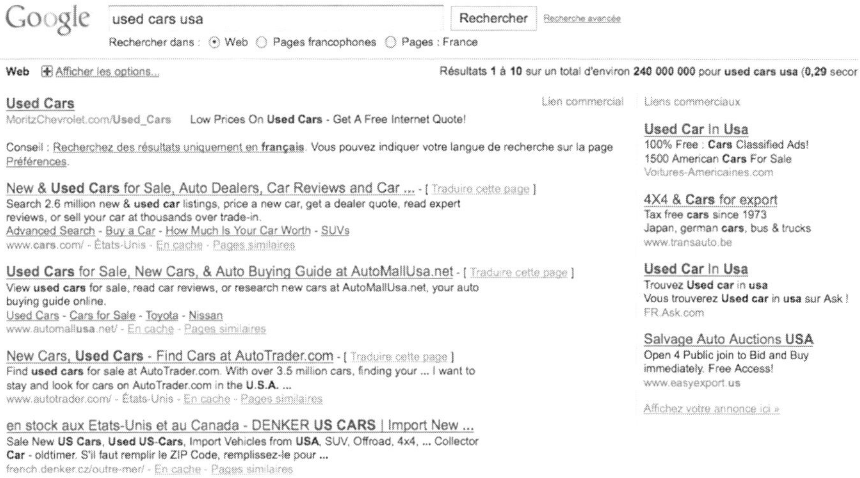

Figure 4.1 What is typically displayed as a result of a Google search; sponsored links are at the top and on the right of the page.

Android is an interesting example from another perspective: Google also generates revenues from Google Play (the Google application store for Android) to compete with Apple's App Store, and does not have to pay now for being the default search engine in the browser of mobile phones; it is then able to get more revenues from ads. This of course has to be pondered with the costs of entering this mobile market. Because of this new business, Google needs to share the operating system/store revenues with network carriers and device vendors, a reason why it also looks for an increasing influence on those devices' and carriers' businesses. The navigator Google Chrome is in exactly the same way a manner to "naturally" integrate Google searches into the users' navigation and as a consequence increase revenue from advertisements. Finally, with Google needing computing power for its PageRank and searching algorithm, it owns more than 900,000 servers (this was estimated in 2011). They are used also for cloud computing and storage capacities, which can be leased.

4.2 Advertising

On the Internet, most (but not all) applications and content data are available for free. This is an important part of the flavor of full connectivity and freedom of the network. Exceptions are, for example, video on demand, wireless applications, restricted access for extended content at news web sites, some applications, etc. Owing to this free access, service providers usually get revenue thanks to advertising. Ads can be seen on the screen at web pages, in videos, or on search engine outputs (see again Figure 4.1). There are different ways to display ads: a banner or a video can be incorporated within a text, or superimposed on a video, or a new page can open in front of the current one (a so-called **pop-up**). Vertical banners are also increasingly replacing horizontal ones

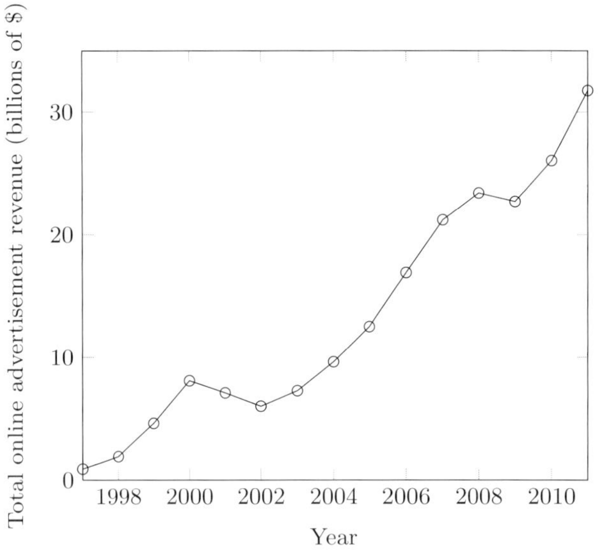

Figure 4.2 The evolution of advertisement revenue in the USA.

on web sites, because they are permanently seen even when readers are scrolling down the page. **Interstitial ads** are similarly becoming increasingly popular. The principle is that the ad appears as a full page before the actual content for some (limited) time. The viewer then focuses on the advertisement while waiting for the content. It is interesting to note that there used to be and still are network access providers offering free access in exchange for ad banners on the screen of your computer. The number of free Web access providers has dramatically declined recently though, and they usually offer a lower QoS than "traditional" ones.[2]

Advertising has been an increasing business on the Internet. Focusing on the USA, it is known to have yielded $0.907 billion in 1997, $26.04 billion in 2010, and $31.74 billion in 2011 (an increase of 22% in only one year).[3] The evolution of the revenue is illustrated in Figure 4.2, where the effects of the dot-com crash of 2001 and the subprime crisis are visible. Major growth is still expected for the near future. Mobile advertising increased by 149% in 2011, and is experiencing the largest boom due to smartphones. Search ads account for about half of the revenue. It is interesting to note that Google did not believe at its creation that advertising would be its major source of revenue (but believed in server resources). Ad slots used to be sold by a human team, and nothing was automated. Advertisers were charged according to the number of views. Now, advertising slots are in most cases allocated through automated auctions, which we discuss in the next subsection. In addition, online advertising has become more contextually relevant and makes use of the user's known preferences to a larger extent than does television or newspaper advertising. Those preferences are known from your

[2] See http://www.freedomlist.com/ for a list.

[3] For updated numbers, see the reports of the Internet Advertising Bureau at http://www.iab.net/.

history and social behavior if you are identified on a social network (e.g., Facebook) or at a merchant's web site (e.g., Amazon), but, more easily, advertisers use **cookies**, i.e., files recording on your own device your visits to particular websites, from which information advertisements can be better targeted, in the seller's and maybe the user's interest. Geo-localization can also be of interest here. As an example, Facebook ads are placed on the right-hand side of every Facebook page. They appear below the section titled "People you may know" – they use the history of your social activity and the things you "liked" to choose the most relevant ads.

To make clicked ads more efficient, the notion of a **landing page** has been introduced: the principle is to display the web page of the advertiser differently when its ad is clicked, depending on the composed keyword and/or the user's history.

4.2.1 Auctioning for advertising slots: basic principles

We explain in this subsection the basic principles and steps driving an auction for advertisements at a search engine or on a web page. Assume that a search engine, a (free) application, or a web site has slots to fill for advertisements.

- First, potential advertisers have to select keywords (for search engines) or web sites (based on semantics, in order to target the most appropriate content and users) on which they wish to place their ads. Thus the advertisement can be more efficient in terms of sales and/or brand image improvement. This type of strategy has always been applied in the ad industry, with, for instance, spots on TV associated with programs that are more popular within a target segment of the population or shown at a time of day at which the target audience is likely to be watching, but it can be automated and applied in telecommunications at a much finer granularity thanks to cookies, social networking, keywords, and again semantic analysis.
- Advertisers must then declare how much they are willing to pay to have their ads displayed, through a bid. Their declaration is not necessarily *truthful* in the sense that the advertisers can be strategic and declare (play) something that will lead to more revenues, and depends on their overall budget. This will be discussed in the next subsection. When being truthful is always the best strategy, the so-called incentive-compatibility property described in Subsection 2.4.1 is satisfied.
- The slots are allocated from the declarations (or bids) of advertisers following an auction mechanism. There might be several criteria to determine the winner, who is not necessarily the one with the highest bid, as we will see. The allocation of slots is based on a *ranking* procedure. Different slots may have different impacts on the user: coming back to Figure 4.1, a user interested in a search will tend to look at the sponsored link at the top of the page, or the first ones on the right. The further down the list your ad/link is, the less likely it is to be looked at and clicked. Bidding sufficiently high to be better positioned is thus of interest, but not being positioned too highly at the risk of spending too much on an advertisement. Another question is this: who actually runs the auction? The web site or application owner could do it in principle, but would need particular skills to be efficient and a large database of

potential advertisers following specifically this web site or application. Alternatively, an auction authority/tool can take care of running the auction (see Subsection 4.2.6). The owner need just provide his address to the tool, which will use its centralized database of advertisers, and will choose the ones which are most related to the content.

- A fee charged to the advertiser (that will go to the slot owner) is defined. This will also be described more in the next subsections. The charge can be imposed depending on different types of actions, and this will also be discussed later on. Advertisers may be charged

  each time their ad is displayed (**pay-per-impression**),
  each time the ad is clicked (**pay-per-click**), or
  each time the click results in a real sale (**pay-per-transaction**).

  A general notion of **pay-per-action** is starting to be introduced to cover all those concepts. As a last issue, how much should one charge then? Here too, there are several possibilities that we will discuss and that have already been described in Subsection 2.4.3.

  (1) The first-price principle, where advertisers pay their bid if they have won a slot.
  (2) The so-called **generalized second price** (GSP), where they pay

    the bid of the advertiser just below them in the ranking if the ranking strategy is based on bids; or
    the price such that the revenue generated corresponds to the (declared) revenue of the advertiser below them if the ranking is based on the revenue that can be generated from the allocation.

  (3) Another option is to make use of VCG auctions: each advertiser pays the opportunity cost that his presence imposes on all other advertisers.

In the next subsection, we describe more precisely how the auctions between advertisers are run, and what properties they satisfy. We will then discuss the choice of pay-per-click or pay-per-view, or pay-per-action in general, before presenting the commercial tools available to help advertisers.

4.2.2 Auctions between advertisers

Assume in this part that some advertisers are competing for advertising slots. We will mainly discuss slots available at a search engine where several are usually available, but slots at any other web page can be similarly looked at, the main issue being then whether or not to compete for a particular slot depending on the content and/or the user, but not directly on a keyword (this will be the topic of a later subsection). The questions we aim at investigating are as follows.

- Who should be allocated which slot?
- How much should they be charged for that?

The question is how to design a mechanism with relevant properties (see Subsection 2.4.1) in terms of revenue for the engine or site displaying the ads, satisfaction for the users and advertisers, etc. A survey on sponsored search auctions can be found in [180].

We consider a set $\mathcal{N} = \{1, \ldots, n\}$ of n advertisers competing for a set $\mathcal{K} = \{1, \ldots, k\}$ of k advertisement slots on a web page (which can be banners) or on the results page corresponding to a keyword for a search engine. We will assume that $n > k$ so that there is real competition for the slots.

Each advertiser $i \in \mathcal{N}$ has a valuation v_i associated with the "considered action" (impression, click, sale). This valuation represents the maximum price i is willing to pay (we will see later on that this price can be per view, per click, per transaction, etc.). The selection of which advertisers are allocated which slots is made through an auction. Thus each advertiser i submits a bid b_i that is not necessarily his valuation v_i. The vector $\boldsymbol{b} = (b_1, \ldots, b_n)$ is the bid profile.

The k slots do not have the same probability of being "looked at" by the end-users; therefore, they do not have the same "importance" for advertisers. The same thing happens for the advertisers, which are not likely to be seen with the same importance. A general term to denote these importances is the **quality score**. The quality score is a measure of how relevant the ad will be to a person seeing it on a given web page or search results page. This depends not only on the ad itself, but also on the slot, and even possibly on the selection of other advertisers displayed, because your ad may be disregarded if there are other very relevant ads. Looking back to the example displayed in Figure 4.1, the slots at the top of the page are more likely to have an impact than those at the bottom because they attract the eyes first, with perceived importance usually decreasing on going down the ranking. This quality score is used to define the charge for the advertisers. A standard measure for the quality score is the **click-through rate** (CTR), which is usually considered in the context of sponsored search auctions, for example, but also on many web pages. The CTR is the probability that a given ad will be clicked when displayed (it is therefore required that there be a hyperlink on the ad). Even though being clicked is not necessarily the goal of advertisers (see Subsection 4.2.4 for more discussion on this), we will focus in this part on the CTR as the quality score. This will be done without loss of generality, since it otherwise suffices to replace the CTR by any other quality measure. To simplify the analysis, we will consider that the CTR is **separable**: the probability that an ad of advertiser $i \in \mathcal{N}$, occupying slot $s \in \mathcal{K}$, receives a click is the product $w_{is} = q_i \theta_s$, separating the effects of the slot and of the advertiser. Here

- θ_s represents the probability that a user considers the ad on slot s (so as to click it if the ad seems sufficiently relevant; slots are, without loss of generality, ordered such that $\theta_1 \geq \ldots \geq \theta_n$); and
- q_i represents the attractiveness of bidder i (sometimes called the bidder's CTR).

As pointed out above, the CTR usually also depends on who has been allocated the other slots, but we neglect this effect here. How CTRs are determined will be discussed a bit in Subsection 4.2.5.

With all those data, the auctioneer decides which advertisers will be allocated the different slots. The principle is to rank the advertisers in decreasing order of the quality score times the bid, $w_{is}b_i$. More exactly, the advertiser maximizing $w_{i1}b_i$ is first selected, then the one maximizing $w_{i2}b_i$, and so on. With the separability condition $w_{is} = q_i\theta_s$, this amounts to ranking the advertisers according to q_ib_i and just allocating the slots in that order (which is not true in general without this separability assumption). Ranking using this quality score is typically what is currently implemented, whereas initially Yahoo! was rather ranking according to the bid only, which just means that $w_{is} = 1$ was considered $\forall i, s$ in the expressions. This particular case will also be investigated. Let (s) denote the index of the advertiser selected for slot s. We remark that, if the charge is p_s at the sth slot, then the revenue generated by the auctioneer with a pay-per-click scheme is $\sum_{s=1}^{k} w_{(s)s}p_s$.

The allocation having been determined, the last question is that of when and how much to charge advertisers. The pay-per-click scheme, where advertisers pay each time their ad is clicked, is most commonly implemented, but other possibilities exist (see again Subsection 4.2.4). We will consider that scheme here, but the results can easily be changed to another scheme by just adjusting the values (dividing by the CTR for a pay-per-view scheme, for example).

The amount charged to advertisers could be the bid (the first-price principle) or follow VCG auctions, but what is most commonly implemented is the *generalized second-price* (GSP) auction scheme. This scheme consists in making you pay not what you have declared, but a price equivalent to the minimum bid to maintain your position in the ranking; that means comparing with the bidder ranked below you (hence the name "second price" corresponding to when there is only one slot). To make it easier to grasp, let us be explicit about what happens when one is ranking by bids only, or by CTR times bid, i.e., by revenue, since the revenue associated with slot s is $w_{(s)s}p_s$.

- If we rank by bid, the winner of any slot $s \leq k$ is charged $b_{(s+1)}$, the $(s+1)$th largest bid, because bidding less would mean losing the sth slot.
- If we rank by revenue, under the separability assumption, the price p_s charged for the sth slot is such that bidding less than p_s would make you lose the sth slot to the $(s+1)$th bidder, i.e., the minimum price is such that $q_{(s)}\theta_s p_s \geq q_{(s+1)}\theta_s b_{(s+1)}$. This gives

$$p_s = b_{(s+1)}\frac{q_{(s+1)}}{q_{(s)}}.$$

  The intuition behind this ranking is that some advertisers may have a very low CTR q and hence generate a low revenue even if their bids are higher than those of competitors. The scheme avoids displaying them in this case.

Example 35 To illustrate the schemes, consider a simple example with $k = 3$ slots, where $\theta_1 = 1/2$, $\theta_2 = 1/4$, and $\theta_3 = 1/5$, and $n = 5$ advertisers with the bids and CTRs given in Table 4.1. If we rank per bid, the three slots are obviously allocated to the first three advertisers. According to the GSP principle, advertiser 1 will pay the bid of advertiser 2, i.e., $p_1 = b_2 = 9$ (while bidding $b_1 = 10$), advertiser 2 will pay the bid of advertiser 3 (that is, $p_2 = b_3 = 6$, while bidding 9), and advertiser 3 will play the bid of the fourth

Table 4.1 An example with five bidders

Advertiser i	Bid b_i	CTR q_i	Product $b_i q_i$
1	10	0.05	0.5
2	9	0.1	0.9
3	6	0.12	0.72
4	5	0.15	0.75
5	4	0.2	0.8

ranked bidder, advertiser 4, i.e., $b_4 = 5$ (while bidding 6). The expected revenue per search generated by the scheme is then

$$\sum_{s=1}^{3} \theta_s q_{(s)} p_s = \sum_{s=1}^{3} \theta_s q_{(s)} b_{(s+1)} = \frac{1}{2} \times 0.45 + \frac{1}{4} \times 0.6 + \frac{1}{5} \times 0.6 = 0.52.$$

Now, let us see how it works if we rank according to revenue, i.e., according to $q_i b_i$. Those values are given in the fourth column of Table 4.1, hence advertiser 2 is allocated the first slot, advertiser 5 the second, and advertiser 4 the third. The ranking is thus very different from what happens on allocating according to bids only. The price per click for the sth slot is $p_s = b_{(s+1)} q_{(s+1)} / q_{(s)}$, giving $p_1 = 8$, $p_2 = 3.75$, and $p_3 = 4.8$. The prices *per click* are therefore not monotonic, but the total expected charges (price times CTR) are. The revenue generated by the scheme is

$$\sum_{s=1}^{3} \theta_s q_{(s)} p_s = \sum_{s=1}^{3} \theta_s q_{(s+1)} b_{(s+1)} = \frac{1}{2} \times 0.8 + \frac{1}{4} \times 0.75 + \frac{1}{5} \times 0.72 = 0.7315.$$

We can also remark that, even if it is true here, ranking per revenue and applying the GSP scheme, quite counter-intuitively, does not always lead to a higher revenue than ranking per bid. This was already highlighted in [192]. To simply illustrate this, change the value of CTR q_2 to 1 in Table 4.1. This does not change the ranking and price per click when ranking per bid, but the revenue is now $\frac{1}{2} \times 0.45 + \frac{1}{4} \times 6 + \frac{1}{5} \times 0.6 = 1.87$. Similarly, the ranking is not changed for the revenue-based ranking, but the revenue is no longer $\frac{1}{2} \times 0.8 + \frac{1}{4} \times 0.75 + \frac{1}{5} \times 0.72 = 0.7315$. The CTR of the winner does not impact the charge here.

Notice that, if just one slot is offered, GSP is then equivalent to a VCG auction. Indeed, the allocation with VCG selects (i.e., ranks) advertisers in order to maximize

- the declared valuation (bid) of the winner for the bid-based ranking, hence selecting the largest bidder like GSP; for the price to be paid, it is equal to the negative externality (loss of declared value) that the winner imposes on others, hence the second largest bid (only the second highest bidder's allocation is changed due to the winner's presence, because he or she would have won otherwise); and
- the (declared) generated revenue for the revenue-based ranking, hence selecting the advertiser maximizing $q_i b_i$ like GSP, and the *total charge* imposed on the winner is

the loss of declared revenue of other players, i.e., the value $\theta_1 q_{(2)} b_{(2)}$, which, translated into a price per click for the winner, gives $\theta_1 q_{(2)} b_{(2)}/(\theta_1 q_{(1)}) = b_{(2)} q_{(2)}/q_{(1)}$.

In both cases, the charges are the same as for GSP. But, with more than one slot, this is no longer true, as illustrated below.

Example 36 Coming back to Example 35 with three slots, and focusing on revenue-based ranking only (because this is the most commonly applied type of ranking), the allocations with VCG are the same as for GSP but the charges are

- for the winner of the first slot, advertiser 2, the loss of (declared) revenue due to his presence that is imposed on all others, i.e.,

$$b_3 q_3(\theta_3 - 0) + b_4 q_4(\theta_2 - \theta_3) + b_5 q_5(\theta_1 - \theta_2)$$
$$= \frac{0.72}{5} + 0.75\left(\frac{1}{4} - \frac{1}{5}\right) + 0.8\left(\frac{1}{2} - \frac{1}{4}\right) = 0.3815,$$

  because nothing changes (there is no allocation) for advertiser 1, advertiser 3 would have got the third slot had advertiser 2 not been present, and advertisers 4 and 5 would have been one slot up, and hence experience some loss also;
- similarly for the winner of the second slot, advertiser 5,

$$b_3 q_3(\theta_3 - 0) + b_4 q_4(\theta_2 - \theta_3) = 0.72 \times \frac{1}{5} + 0.75\left(\frac{1}{4} - \frac{1}{5}\right) = 0.1815;$$

- and for the winner of the third slot, advertiser 4,

$$b_3 q_3(\theta_3 - 0) = 0.72 \times \frac{1}{5} = 0.144.$$

Those charges for the winners of the first two slots are different from those which occur using GSP in Example 35 (which have to be multiplied by the CTR to give the equivalent total charges). Summing those values, the total expected revenue per search using VCG would be 0.707, which is less than the 0.7315 when using GSP.

The questions are then as follows. (i) What is it in the best interest of advertisers to do when GSP is used? Should they be truthful? If not, what should they bid? This is indeed a non-cooperative game for slots between advertisers. (ii) What is then the solution of the bidding game; is there a Nash equilibrium? Is it unique? (iii) Shouldn't VCG be implemented instead of GSP?

To answer the first question, GSP is, unfortunately, not incentive-compatible, precisely because VCG prices are the unique truthful prices corresponding to the given allocations (see Chapter 2). To give a simple example, consider three advertisers competing for two slots, such that with truthful bids advertiser 1 is allocated the first slot and advertiser 2 the second slot. Thus $q_1 v_1 \geq q_2 v_2 \geq q_3 v_3$. Let us check whether advertiser 1 has an interest in changing his or her bid to a value $b_1 < v_1$. With a truthful bid, advertiser 1 gets an average benefit from each click of $v_1 - p_1 = v_1 - v_2 q_2/q_1$, and hence a total benefit $\theta_1 q_1(v_1 - v_2 q_2/q_1)$. If the advertiser decreases his or her bid b_1 in order to be allocated the second slot instead (i.e., so that his or her bid becomes just below

$q_2v_2/q_1 = p_1$), the total benefit becomes $\theta_2 q_1(v_1 - p'_2)$, with $p'_2 = q_3v_3/q_1$ the price that advertiser 1 would then have to pay per click. There is then an interest in decreasing the bid if $\theta_1 q_1(v_1 - v_2q_2/q_1) < \theta_2 q_1(v_1 - q_3v_3/q_1)$, i.e., if

$$\theta_1(q_1v_1 - q_2v_2) < \theta_2(q_2v_1 - q_3v_3).$$

Example 37 With values $q_1v_1 = 4$, $q_2v_2 = 3$, $q_3v_3 = 1$, and $\theta_1 = 1/4 > \theta_2 = 1/5$, the above condition is satisfied.

Basically, the idea is that, if the CTRs θ_s for two consecutive slots are similar, while the difference between consecutive valuations is large (in the opposite direction), then there is an interest in lowering the bid.

Bidding is thus more complicated with GSP for advertisers, because being truthful is not optimal, and a "better" strategy should be determined. There can be many Nash equilibria, especially given that bidders can play repeatedly (see Subsection 2.3.7). The game is not static. Players can learn information about others and play strategically given that knowledge. Actually, a Bayesian–Nash equilibrium of GSP is not known in general, and GSP can be proved not to be efficient in the sense that it does not maximize the total value extracted from the slots. But, under our assumption that CTRs are separable, there exists a **locally envy-free equilibrium** such that GSP allocations and payments correspond to the VCG ones. A locally envy-free equilibrium is a bid profile in which every bidder allocated position s has no incentive to switch to position $s-1$ or $s+1$ through a bid change. He or she may be better off switching to another position, but not to the neighboring positions. The reason for looking at those equilibria is that an advertiser would try to move his or her bid from his or her valuation in any direction, and, if switching to a neighbor slot is not attractive, might not change at all if not playing "fully" strategically. For more about this, please see [91, 163, 289].

Why not implement VCG instead of GSP, given the nice properties satisfied by the former (existence and uniqueness of a dominant-strategy equilibrium, efficiency, incentive compatibility) and not by the latter? Actually, it can be proved that, for a given set of bids $(b_i)_i$, the expected revenue the seller of slots can get is at least as high when using GSP as in the dominant-strategy equilibrium of the VCG auction [91]. The proof is quite quick, proceeding by induction, as follows. For the last assigned slot (the kth), it is as if there were a single slot and we have already seen that the charges imposed by GSP and VCG are the same. Now, consider the sth slot $(1 \leq s \leq k-1)$. Let $p_j^{(\mathrm{GSP})}$ and $p_j^{(\mathrm{VCG})}$ be respectively the charges per click of GSP and VCG for each slot j. Our induction assumption is $p_{s+1}^{(\mathrm{GSP})} \geq p_{s+1}^{(\mathrm{VCG})}$. Then the difference in charging between slots s and $s+1$ with VCG is the difference of opportunity costs between the two positions:

$$\begin{aligned}
\theta_s q_{(s)} p_s^{(\mathrm{VCG})} - \theta_{s+1} q_{(s+1)} p_{s+1}^{(\mathrm{VCG})} &= b_{(s+1)} q_{(s+1)} (\theta_s - \theta_{s+1}) \\
&\leq b_{(s+1)} q_{(s+1)} \theta_s - b_{(s+2)} q_{(s+2)} \theta_{s+1} \\
&= \theta_s q_{(s)} p_s^{(\mathrm{GSP})} - \theta_{s+1} q_{(s+1)} p_{s+1}^{(\mathrm{GSP})}.
\end{aligned}$$

So $\theta_s q_{(s)} p_s^{(\mathrm{GSP})} \geq \theta_s q_{(s)} p_s^{(\mathrm{VCG})} + \theta_{s+1} q_{(s+1)} (p_{s+1}^{(\mathrm{GSP})} - p_{s+1}^{(\mathrm{VCG})}) \geq \theta_s q_{(s)} p_s^{(\mathrm{VCG})}$, whence the result. More remarkably, it can be proved that the revenue at a GSP-based equilibrium is larger than the one at the equilibrium when using VCG [91]. This is, from the auction designer's point of view, a strong argument in favor of GSP.

As for other arguments for using GSP over VCG, it is generally admitted that VCG is hard to explain to typical advertisers; GSP is simpler to grasp. Also, bidding truthfully remains "natural" under the GSP rules (even if it is not a dominant strategy).

4.2.3 Extensions of the basic auction model

The model described above presents a simplified version of the auction problem, whose solution is already not so easy to find. But there are many other constraints that our simplified model does not take into account. We are going to review a few of them in this subsection.

First, on which keyword/page should an advertiser bid, and what are the corresponding submitted and declared valuations? Tools such as Google AdSense (see Subsection 4.2.6) help to partly address this question. Another major issue is that each advertiser has usually a limited budget for a given period of time. It is important and interesting to design bidding strategies that take this constraint into account. However, there is a limited literature on this, because of the complexity of the problem, especially when the budget is for multiple keywords/pages. We can refer the reader to [51, 126, 223] for details, but the main mathematical framework is the **knapsack problem**, a combinatorial optimization problem in which advertisers have to decide where to place bids and with which values, with the constraint that the total budget is not exceeded. The goal is to maximize the revenue; we are not going to develop this point in this book but rather refer to the above references.

Another noticeable point is that most works assume that an Internet user will make a search only once and will never come again, or that a web page will be visited only once by a user. But it is reasonable to assume that a user may come back to a page to look again at its content, and even recompose a search to look for new or additional information, or because he or she does not remember the previous results. In that case, it may be relevant to allocate the slots *randomly*, so that a returning user will see new ads. Of course, the probability of their being displayed would depend on the CTR and on the bids of the advertisers, so a careful choice of those probabilities may produce a higher expected revenue than the current deterministic strategy. This is what is investigated in [189] and summarized here. In order to highlight the potential benefits from a random allocation, consider just one commercial slot, with two advertisers, named, say, 1 and 2, competing for that slot. Each advertiser, $j = 1, 2$, submits a bid b_j for being clicked, has a probability $\pi_j = \pi_j(b_1, b_2)$ of an ad being displayed, and pays a price $p_j = p_j(b_1, b_2)$ if the ad is clicked. The most interesting part of the model concerns the users' behavior: we assume that there are two types of users, namely type-A users, who are interested in purchasing the good sold by advertiser 1, but can potentially click on the ad of advertiser 2 without purchasing it eventually, and, symmetrically, type-B users who are potential

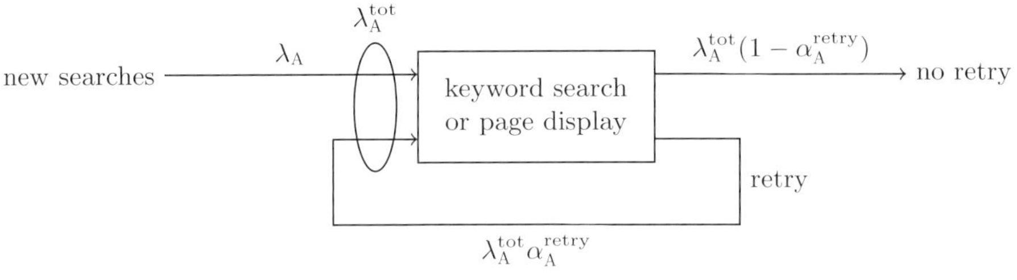

Figure 4.3 Type-A users' behavior.

buyers for advertiser 2 only. Let λ_i for $i \in \{A, B\}$ be the average number of *first* requests per time unit of type-i users. Focusing on type-A users (type-B notations are symmetric), the probability of clicking on the ad of advertiser $j \in \{1, 2\}$ is denoted by $c_{A,j}$, and $h_{A,1} > 0$ is the probability that the user purchases the good after clicking on the ad of advertiser 1 if it is displayed (recall that $h_{A,2} = 0$ to simplify the model). Therefore the probability that a type-A user buys the item sold by advertiser 1 is $\pi_1 c_{A,1} h_{A,1}$, and the probability is 0 for advertiser 2's product. Denote by R_A the probability of a not-buying type-A user coming back later (with the same probabilities of clicking, buying, etc.). The overall probability of retrying is $\alpha_A^{retry} := R_A(1 - \pi_1 c_{A,1} h_{A,1})$. All the parameters are described in Figure 4.3. From the figure, we can easily see that the *total* rate of type-A users' requests λ_A^{tot}, including new requests and all retries, is

$$\lambda_A^{tot} = \lambda_A + \alpha_A^{retry}\lambda_A^{tot},$$

giving

$$\lambda_A^{tot} = \frac{\lambda_A}{1-\alpha_A^{retry}} = \frac{\lambda_A}{1 - R_A(1 - \pi_1 c_{A,1} h_{A,1})}. \tag{4.1}$$

Using $\pi_1 + \pi_2 = 1$, the auctioneer's mean revenue per time unit is thus

$$\begin{aligned} U &= p_1 \underbrace{\pi_1(\lambda_A^{tot} c_{A,1} + \lambda_B^{tot} c_{B,1})}_{\text{number of clicks on ad 1}} + p_2 \underbrace{\pi_2(\lambda_B^{tot} c_{B,2} + \lambda_A^{tot} c_{A,2})}_{\text{number of clicks on ad 2}} \\ &= \lambda_A \frac{\pi_1(p_1 c_{A,1} - p_2 c_{A,2}) + p_2 c_{A,2}}{1 - R_A(1 - \pi_1 c_{A,1} h_{A,1})} + \lambda_B \frac{\pi_1(p_1 c_{B,1} - p_2 c_{B,2}) + p_2 c_{B,2}}{1 - R_B(1 - c_{B,2} h_{B,2} + \pi_1 c_{B,2} h_{B,2})}. \end{aligned} \tag{4.2}$$

A random policy will be better than a deterministic one if there is a value of π_1 such that $0 < \pi_1 < 1$ yielding a larger U than when $\pi_1 \in \{0, 1\}$ (corresponding to the two choices, always selecting advertiser 1 versus always selecting advertiser 2). The value π_1^* maximizing U is characterized in [189], and conditions for which that $\pi_1^* \in (0, 1)$ are obtained. A simple derivation of U with respect to π_1 actually allows one to get the result, noting that U is a continuous, differentiable, and strictly concave function of π_1 on $[0, 1]$. The following example illustrates the application of such a randomization and compares it with a deterministic choice.

Example 38 Consider the simple case where user types are symmetric, and assume that users never click on an ad that does not correspond to their type, i.e., $c_{A,2} = c_{B,1} = 0$. Use the probability values $c_{A,1} = c_{B,2} = 1/2$, $h_{A,1} = h_{B,2} = 1/2$, and $R := R_A = R_B = 0.8$. Take rates $\lambda_A = 1$ and $\lambda_B = 0.8$ and prices per click of $p_1 = 1$ and $p_2 = 0.8$. The revenue (4.2) is maximized at $\pi_1^* = 2/3$, giving 1.4. If we compare it with the optimal revenue when only one ad (that of provider 1, here) is displayed, which equals 1.25, a gain of 12% is observed. This gain is even increased for a larger value of R.

We can remark here that changing the allocation procedure means that advertisers may play strategically and not necessarily truthfully, and that we have not discussed the pricing rule either. In [189], the advertisers' willingness to pay (their revenue) is expressed in terms of the probability of their ad being displayed:

$$V_1(\pi_1) = \lambda_A^{\text{tot}} \pi_1 c_{A,1} h_{A,1} v_1$$

for advertiser 1, for example, where v_1 is the income that the advertiser makes on each sale (recall that type-B users never buy goods from advertiser 1). We present a VCG auction (in order to satisfy truthfulness) that is interpreted as an auction for an infinitely divisible good with total quantity 1 – the total sum of probabilities to be displayed. A conclusion, when comparing the revenues and social welfare at equilibrium of the bidding game among advertisers with those obtained using a deterministic GSP rule, is that the search engine revenue is (again) lower with the VCG randomization scheme, but neither the social welfare nor the advertisers' revenue is lower. Thus deterministic GSP is better than randomized VCG from the seller's point of view. But our purpose was not to design a scheme, just to illustrate that randomization might be of interest and deserves attention.

Another often raised issue concerns the separability assumption of the CTR, as the product of the probability of the ads being clicked and a quality factor corresponding to the slot itself [6, 83, 153]. It is considered more relevant to assume instead that the quality of a well-placed ad produces an externality on the other ads. A Markovian model can thus be used to represent the users' behavior: following [6, 83], we can assume that there are two parameters characterizing what a user does when reading the ad of i:

- r_i, representing the probability that he or she will click on the ad; and
- t_i, representing the probability that he or she will look at the ad below no matter whether the ad i has been clicked or not.

This allows one to model that the ads are read one after the other from the top one to the last one until reading is stopped, depending on the ads seen previously. For a given ranking, the probability of an ad being looked at for the jth slot is given by $\prod_{s=1}^{j-1} t_{(s)}$, and the CTR is therefore $\prod_{s=1}^{j-1} t_{(s)} r_j$. In [6], it is shown that ranking the bidders according to

$r_i b_i/(1 - t_i)$ is optimal if we are trying to maximize the total expected (clicked) *declared* valuations

$$\sum_{j=1}^{k} \prod_{s=1}^{j-1} t_{(s)} r_{(j)} b_{(j)}.$$

The proof follows from swapping arguments between consecutive slots. If the VCG pricing rule is used for the advertisers, bidders are better off being truthful ($b_i = v_i$ is a dominant strategy for each bidder i; i.e., incentive-compatibility is satisfied), and the scheme is thus optimal in terms of total expected valuations. We can remark then that the rule no longer corresponds to the one which ranks by CTRs times bids. Using the same ranking but with GSP pricing instead of VCG, truthful bidding is not the best solution for advertisers; [83] shows the existence of a Nash equilibrium (nothing is said about uniqueness, though) in the bidding game, and bounds the total expected valuation with respect to an optimal (undetermined) ranking.

Similarly, as another complication of the model, the conversion rate (that is, the probability that clicking on a link does result in a sale for the advertiser) has often been empirically shown to depend on the allocated slot. This fact has been acknowledged in [40, 110], and the model has been redefined accordingly, with the probability of conversion h_{is} for advertiser i if the ad is in slot s determined from another separability equation, $h_{is} = h_i h^s$, with h_i and h^s being respectively advertiser- and slot-dependent parameters. The GSP pricing rule can be analyzed under this new assumption.

Actually, a lot of possible variations or extensions can be or have been designed. To learn more about them, we refer the interested reader to [180].

4.2.4 Pay-per-click or pay-per-view?

The natural question for an advertiser is the following: should I opt for a pay-per-click or pay-per-impression advertisement campaign? AdWords will usually implement pay-per-click, but the option is open on many other businesses, such as banners on web sites, Facebook and other social networks, etc. The two types of bids may even compete within a single system; the ranking can then be performed with a CTR of 1 for pay-per-impression bids in the above model. It was reported in 2008 by the Interactive Advertising Bureau [140] that 39% of ads were priced by pay-per-view, 57% by pay-per-click (or an equivalent), and the rest by a mixture of those. Google (with AdSense, see Subsection 4.2.6) also offers owners of web sites a choice between pay-per-click and pay-per-impression. It depends on the contract (at the web-site owner and advertiser levels). Pay-per-impression, or rather cost-per-impression from the advertiser's point of view, is often referred to as **cost-per-mile** (CPM), where *mile* is to be understood in the sense of *thousand*: the advertiser pays per thousand impressions. But again, when the choice is allowed, how should one choose between the two? The issue is still a matter of debate.

- CTRs for banner ads are estimated to be on average only 0.3% [161]. The rate is similar for sponsored searches and is considered to be even lower on Facebook. As a consequence, pay-per-click is often considered a more effective cost model because advertisers pay only for relevant traffic from interested readers. At the beginning of online advertising, ads were bought by impression only. But advertisers were in general skeptical about the relevance of impressions for which they had to pay. Thus Google introduced pay-per-click in 2002. It is interesting to note that Yahoo! was initially implementing a pay-per-impression scheme, but followed Google in pay-per-click for its sponsored searches.

- Many companies are not interested in direct sales from clicks on the ads, but rather in **brand awareness**. We can, for instance, think of Coca-Cola or McDonald's: advertisers of this type just want their name to be displayed so that knowledge of their brand can be increased; they have a very limited financial gain from their web site being visited and no direct sale from clicks. For this reason, because a click is not always representative of the advertiser's goal, the concept of pay-per-action (which can be an impression or a sale, but not only that) is the newest feature and is (still) in a beta version of the Google AdWords program.

Thus the answer depends on the real goal of advertisers (ads leading to direct sales or improved brand awareness): in principle, direct-sales-prone advertisers should prefer pay-per-click and brand-awareness-prone advertisers pay-per-impression. However, there may also be a strategic computation of the gains: if you are not interested in clicks, like Coca-Cola for example, because brand awareness is your goal, pay-per-impression should be the relevant (or fair) scheme, but actually it is possible that pay-per-click will bring more impressions if the bid is properly chosen, and more revenue/awareness in the end. Similarly, it is not obvious that, whatever the CTR, an advertiser interested in direct sales is better off using the pay-per-click method. There is the belief that exposure counts over the long run, such that an ad that is displayed often thanks to pay-per-impression can become attractive and grab attention, generating more clicks. Many believe that, for an advertising campaign, paying per impression is generally less expensive. In this sense the question of pay-per-click versus pay-per-impression is not necessarily incentive-compatible: the optimal choice does not reveal the objective of the campaign. An online advertising campaign can be related to a TV advertising one: paying per view is indeed close to TV advertising, where the price per commercial depends on the audience; advertisers can target segments of the population corresponding to the program aired in the same way as Internet advertisers target content or keywords. It can be shown (see, for example, [82]) using a simple model that risk-neutral advertisers may have different preferences, between pay-per-click and pay-per-view, from publishers and consumers.

The decision for a web-site owner to go with pay-per-view or pay-per-click can be analyzed like in [101, 198], by which our next model is inspired. Without going into the details of the auction/pricing process, assume that the price per view is p_{v}, and that the price per click is p_{c}. Indeed, web publishers are price takers as a consequence of the auctions between advertisers. A web publisher has to decide the proportion $0 \leq \alpha \leq 1$ of banners which will be priced by pay-per-view ($1 - \alpha$ will use pay-per-click) and the total amount a of advertising to be displayed on the page (or, similarly, the number of slots for sponsored links on a search engine). To simplify the analysis, α and a are abusively assumed to be real values. Let $\lambda(a)$ and $c(a)$ be respectively the average number of visits per unit of time to the web site and the average number of clicks per unit of time (note that we may have more than one click from a visit to a page if several ads are found to be of interest). Those functions are assumed to depend on the amount of advertising and $\lambda(\cdot)$ to decrease with this amount: the fewer ads, the better according to the readers. Note that in [101, 198] $c(\cdot)$ is also assumed to be decreasing, which does not seem to be a very representative assumption in general. Of course, there is a maximum amount of advertising that can be included, $\bar{a}$. The revenue per time unit of the web publisher is given by

$$R(a, \alpha) = \lambda(a)\alpha a p_{\mathrm{v}} + (1 - \alpha) p_{\mathrm{c}} c(a).$$

Indeed, for the revenue generated from pay-per-view: the number of visits (page displays) is $\lambda(a)$; αa ads are pay-per-view and pay p_{v}. For the second part, there are $c(a)$ clicks overall, but only $1 - \alpha$ of the clicks are for pay-per-click ads (assuming the CTR to be the same for the two types of ads), and the price is p_{c} each time. To get the optimal value, we differentiate with respect to a and α:

$$\frac{\partial R}{\partial \alpha} = \lambda(a) a p_{\mathrm{v}} - p_{\mathrm{c}} c(a), \tag{4.3}$$

$$\frac{\partial R}{\partial a} = \left(\lambda(a) + \frac{\mathrm{d}\lambda(a)}{\mathrm{d}a} a\right) \alpha p_{\mathrm{v}} + (1 - \alpha) p_{\mathrm{c}} \frac{\mathrm{d}c(a)}{\mathrm{d}a}. \tag{4.4}$$

From the first equation, at the maximum (a^*, α^*), either $\alpha^* = 0$ if $\lambda(a^*)a^* p_{\mathrm{v}} - p_{\mathrm{c}} c(a^*) < 0$, or $\alpha^* = 1$ if $\lambda(a^*)a^* p_{\mathrm{v}} - p_{\mathrm{c}} c(a^*) > 0$, or α^* can be in $[0, 1]$ if $\lambda(a^*)a^* p_{\mathrm{v}} - p_{\mathrm{c}} c(a^*) = 0$. This equation can be rewritten as $a^* p_{\mathrm{v}} = p_{\mathrm{c}} c(a^*)/\lambda(a^*)$; i.e., the repartition is such that the revenue per visit from the two types of ads is equivalent. We thus investigate the three possibilities for α^* in the second equation:

- $\alpha^* = 0$ gives $\partial R/\partial a = p_{\mathrm{c}}\, \mathrm{d}c(a)/\mathrm{d}a < 0$. That would mean no advertisement; hence this cannot be a solution.
- $\alpha^* = 1$ gives $\partial R/\partial a = (\lambda(a) + a\, \mathrm{d}\lambda(a)/\mathrm{d}a) p_{\mathrm{v}} = \lambda(a) p_{\mathrm{v}}(1 + e_\lambda(a))$, with $e_\lambda(a) = \mathrm{d}\lambda(a)/\mathrm{d}a \times a/\lambda(a)$ the elasticity of the number of visits when the advertising amount is a (see Subsection 2.1.2 for the definition of elasticity). An elasticity always smaller than -1 would give a negative derivative, in which case again $\alpha^* = 1$ cannot be a solution. An elasticity always larger than -1 leads to $a^* = \bar{a}$. Otherwise, a^* would be the point such that $e_\lambda(a^*) = -1$.
- $\alpha^* \in (0, 1)$ gives $\partial R/\partial a = \lambda(a)\alpha^* p_{\mathrm{v}} + (a\alpha^* p_{\mathrm{v}}\, \mathrm{d}\lambda(a)/\mathrm{d}a + (1 - \alpha^*) p_{\mathrm{c}}\, \mathrm{d}c(a)/\mathrm{d}a)$. Here again, we need to look at the sign of this function to check whether a solution is possible. It seems intractable to solve the problem in its full generality, but it can be simplified for specific (practical) functions, or treated numerically.

The solution thus depends on the form of the functions λ and c.

Example 39 Consider $\lambda(a) = (\bar{a} - a)$ for $a \leq \bar{a}$ and 0 if $a \geq \bar{a}$, and $c(a) = (\bar{a} - a)/2$ also for $a \leq \bar{a}$ and 0 if $a \geq \bar{a}$. In this case, the average number of clicks per visit is $c(a)/\lambda(a) = 1/2$, i.e., the users click on an ad on average half of the time, independently of the number of ads displayed.

From our above analysis, we will get an allocation with both pay-per-view and pay-per-click ads (i.e., $\alpha^* \in (0, 1)$) if $a^* p_{\mathrm{v}} = p_{\mathrm{c}} c(a^*)/\lambda(a^*)$, i.e., $a^* = p_{\mathrm{c}}/(2p_{\mathrm{v}})$, assuming that $p_{\mathrm{c}}/(2p_{\mathrm{v}}) \leq \bar{a}$.

Equation (4.4) gives (for $a \leq \bar{a}$) $\partial R/\partial a = \bar{a}\alpha p_{\mathrm{v}} - (1-\alpha)p_{\mathrm{c}}/2 - 2\alpha p_{\mathrm{v}} a$. On replacing a by $a^* = p_{\mathrm{c}}/(2p_{\mathrm{v}})$ in the optimality condition $\partial R(a)/\partial a = 0$, we get $p_{\mathrm{c}}/(2p_{\mathrm{v}}) = \bar{a}/2 - (1-\alpha)p_{\mathrm{c}}/(4\alpha p_{\mathrm{v}})$, giving

$$\alpha^* = \frac{1}{\bar{a}(2p_{\mathrm{v}}/p_{\mathrm{c}}) - 1}.$$

This actually gives $\alpha^* < 1$ if $\bar{a} p_{\mathrm{v}}/p_{\mathrm{c}} > 1$, i.e., if the pay-per-view price is high enough with respect to the pay-per-click price. We omit the border cases here.

The case when the owner of the web site is the price setter (hence the price is not derived from an auction as above) and there is imperfect competition, i.e., such that there is a distinct market for pay-per-click and pay-per-view ads, is analyzed in [102].

Some interesting practical comparisons of pay-per-click and pay-per-view advertisement campaigns (from the advertiser's point of view) can be found on the web, for example on Facebook or LinkedIn social networks, involving spending the same amount of money with both schemes and comparing the numbers of impressions and clicks.[4]

In order to deal simultaneously with the two types of advertisers (looking for direct sales or striving to increase brand awareness), the two types pay-per-view and pay-per-click can be mixed to give **impression-plus-click** pricing [117]. Consider a single advertisement slot. The principle is to define a contract made of two components p_{v} and p_{c}, where p_{v} is the price that the advertiser will pay if the ad is displayed, and p_{c} is the additional price to be paid if the ad is clicked. An advertiser i is characterized by his or her utility

$$u_i(p_{\mathrm{v}}, p_{\mathrm{c}}) = (m_i + q_i \ell_i) - (p_{\mathrm{v}} + q_i p_{\mathrm{c}}),$$

where q_i is the estimated perceived CTR of advertiser i on the slot, m_i the value for the ad being displayed, and ℓ_i his or her additional value for the ad being clicked. The ad publisher has, on the other hand, a utility

$$u^{(\mathrm{p})}(p_{\mathrm{v}}, p_{\mathrm{c}}) = p_{\mathrm{v}} + q_i^{(\mathrm{p})} p_{\mathrm{c}},$$

with $q_i^{(\mathrm{p})}$ the CTR of advertiser i estimated by the publisher ($q_i^{(\mathrm{p})}$ is possibly different from q_i). It can be remarked that an advertiser with the same CTR q_i and the utility

[4] See, for example, http://www.profitworks.ca/blog/243-cost-per-click-vs-cost-per-impression-google-adwords-facebook-linkedin.

parameters $(m', \ell') = (m_i - \kappa, \ell_i + \kappa/q_i)\ \forall\kappa \in [-q_i\ell_i, m_i]$ will have the same utility and can be considered equivalent. In the auction scheme, each advertiser i is asked to submit two extremal contracts (specifying his or her utility function to the web publisher) $(p^{\max}_{v,i}, p'_{c,i})$ and $(p'_{v,i}, p^{\max}_{c,i})$ such that $p^{\max}_{v,i}$ and $p^{\max}_{c,i}$ are respectively the maximum prices per impression and per click which he is willing to pay. The web publisher then computes for each advertiser i the maximum revenue he or she (seemingly) can get from the extremal contracts: $R_i = \max(p^{\max}_{v,i} + q^{(p)}_i p'_{c,i}, p'_{v,i} + q^{(p)}_i p^{\max}_{c,i})$. The advertiser with the largest R_i is allocated the slot. In terms of payment, the contract producing the second highest revenue $R_{(2)}$ is considered, where we index by (i) the advertiser producing the ith largest revenue. There are then two cases.

- If $R_{(2)}$ comes from $(p^{\max}_{v}, p'_{c})$, it may happen that this second contract has a higher $p^{\max}_{v,(2)}$ than that of the winner $p^{\max}_{v,(1)}$. The final contract is determined by decreasing $p^{\max}_{c}$, but in such a way as to keep the same revenue. The final contract is thus $(p^{\max}_{v,(1)}, (R_{(2)} - p^{\max}_{v,(1)})/q^{(p)}_{(1)})$.
- Otherwise, $R_{(2)}$ comes from $(p'_{v}, p^{\max}_{c})$. Similarly, we ensure that we will end up with valid bounds by making a final contract $((R_{(2)} - p^{\max}_{c,(1)})q^{(p)}_{(1)}, p^{\max}_{c,(1)})$.

Under mild conditions, this mechanism is dominant-strategy incentive-compatible for advertisers [117].

4.2.5 Learning

There is a need to correctly estimate quantities such as CTRs and valuations in order to avoid misunderstandings and inefficient allocations. For example, as illustrated by the last paragraph of the previous subsection, CTRs might be differently estimated by the advertiser and the auction designer. Actually, if the advertiser's estimated CTR is higher than the publisher's, then the advertiser would expect to pay more than the publisher would expect to receive, resulting in a non-optimal allocation (in terms of the auctioneer's revenue or advertisers' cumulated utility). It is also shown in [178] that, if the CTRs are not properly estimated, advertisers may have an incentive to increase their bids by some value (called the *value of learning*). Therefore, not only efficiency but also incentive properties can be improved if a correct evaluation is performed. It is also important to achieve a trustful agreement between the advertiser and the publisher: for example, several lawsuits have been filed by would-be advertisers who claim that they have been victims of an arbitrary process implemented by a quasi-monopoly, because they suspect that the publisher has been abusively overcharging by artificially increasing the CTR.

Because auctions are held repeatedly, the auctioneer/slot seller and the advertisers have an interest in learning the values and behaviors of other actors to better adjust their next decisions. For more on this issue, the reader can look at [180, Section 5] and the references therein. Notice that learning the valuations can also lead to improved revenues for the auctioneer implementing GSP, which in general is not incentive-compatible, such that bidders might under-bid with respect to their valuation; estimating the valuations can allow one to adjust prices.

Let us briefly look at an important metric, the CTR: it can be estimated by calculating the ratio between the number of successes n and the total number of impressions N,

$$\mathrm{CTR} \approx \frac{n}{N}. \tag{4.5}$$

We can in this way estimate the CTR q_i of advertiser i and θ_s of slot s by looking at their average values (independently of the slot and advertiser, respectively). The CTR $w_{i,s}$ of i at slot s can also be specifically targeted. Instead of (4.5), an exponential averaging can be used to put more weight on the latest impressions.

But we need a sufficient number of impressions to accurately estimate the CTR. To give an idea, assuming all impressions to be independent, the number n of clicks follows a binomial distribution with parameters (CTR, N). A confidence interval that is based on the central limit theorem to estimate the CTR with a 95% confidence level (the probability of being within the interval) is of the form

$$\left(\frac{n}{N} - 1.96\sqrt{\mathrm{CTR}(1-\mathrm{CTR})}/\sqrt{N},\ \frac{n}{N} + 1.96\sqrt{\mathrm{CTR}(1-\mathrm{CTR})}/\sqrt{N}\right),$$

and hence of half-width $1.96\sqrt{\mathrm{CTR}(1-\mathrm{CTR})}/\sqrt{N}$. The relative half-width (or error) is obtained by dividing this by the CTR, giving $1.96\sqrt{(1-\mathrm{CTR})}/\sqrt{\mathrm{CTR}\times N}$. If, for instance, the CTR is 3%, and we desire a relative error of at most 5%, this means that the number of impressions needs to be at least sufficiently large that

$$1.96\sqrt{(1-0.03)}/\sqrt{0.03 \times N} \leq 0.05,$$

i.e., $N \geq 4{,}969$, a large number. It will be reduced if we limit the precision or if the CTR is larger. Then what if an advertiser is relatively new and his or her CTR has not been estimated properly? The auctioneer must then find a balance between exploration, to better estimate the CTR (by displaying the associated ad), and exploitation, to display the "best" ads.

4.2.6 Existing tools/companies

There are many companies and tools helping advertisers to place their ads on search results or on banners on web pages. The most well known are probably Google AdWords (and Google AdSense to place ads on web sites other than Google ones), Yahoo! Search Marketing and Microsoft AdCenter. One of the main advantages of such big companies for advertisers is that they give access to a huge market (i.e., a huge number of searches or web sites), thus allowing one to better target ad placement in terms of content and viewers. This makes it easier to reach your targeted audience.

If you are a web site, you need only, instead of running complicated auctions yourself, apply to one of the above companies, which will, once your site has been approved, send you the (simple) necessary HTML code to paste into your page. You will have the freedom to customize the appearance of the ad. Even if your content includes the possibility to offer searches, Google, for example, proposes "AdSense for Search," providing a Google search box that can be included on the site and present advertisements too, which will earn you money if clicked.

From the advertiser's side, the companies attempt to assist as much as possible those wishing to spend online ad money. Google AdWords is the largest tool because most searches are performed using Google's search engine. It is possible for advertisers to select the ad content (text or image), to target specific regions or languages, to target some time of day, and to perform many other segmentation procedures. There is usually an amount an advertiser has to spend to register for the tool (this is as low as $5 on Google AdWords, for example) and a minimum price to be paid (a reserve price) if there is a click. When designing your advertising campaign, you need to assign your budget to keywords, but there are tools (for example, Google Analytics for Google AdWords) helping you to determine the best-performing keywords. Some advertisers do not know much about online advertising and do not want to spend time on this; companies (the same ones as above, but see also, for example, 7Search[5]) offer to review their advertising campaign for them and select words and phrases that will produce the best return on investment: they just require from advertisers their preferred maximum bid amount. Advertisers can usually get (real-time) reports on their ads, with descriptions of clicks, of costs, etc. Yahoo! is proposing two types of subscription plans to advertisers: FastTrack, which, for $199 in 2009, assists you with recommendations on keywords and customizes your ads; and SelfServe, which is free, but lets you manage the whole process. Thus the tools/companies differ in the way they assist advertisers, but also in their payment models, which can consist in charges after a certain amount of clicks (Google) or a deposit in advance that will be then used (Yahoo!).

4.3 Paid applications versus free applications with advertisements

In the previous section, we were discussing how *free* applications/web sites could earn money thanks to advertisements. Now, for any type of application, there is a choice to be made between being paid for usage, or the application being free but then including advertisements to earn money. The former option is implemented by numerous applications, but a typical example is online newspapers. Most Internet users were reluctant to pay a fee to consult a news web site (behind what is called a **paywall**), but this seems to have changed thanks to tablets, bringing more audience and new revenue streams, even if the industry is experiencing economic difficulties.

Another important example is smartphone applications. Here too, users are in general reluctant to download a paid application: free apps accounted for 89% of total downloads in 2012, or more precisely 40.1 billion compared with 5 billion paid downloads,[6] and most downloaded paid applications cost less than $3. This trend of going for free is expected to expand. Many application stores exist for the various operating systems: Apple's online store (App Store) for iPhone, iPod touch, and iPad applications, Android

[5] http://7search.com/.

[6] Gartner research data.

Market for Android OS, OVI Store by Nokia for Symbian, BlackBerry App World, Windows Store by Microsoft, etc. The question to be investigated is this: should an application designer go for a free application with advertisements, or a paid one? The model we are going to describe is a very simplified one, which is intended to give some hints on such an analysis. It pertains to the stream of revenue-sharing contracts between app stores and application designers. This issue is of importance because the average revenue per app has declined while the number of apps has increased (mainly because of fierce competition), especially in the mobile market. Advertising, which is managed by the app store, is then an easy option. In most cases, the store takes 30% of the revenue generated by the application (the only exception is 20% for BlackBerry App World).

Let us first look at the case when the application is a paid one, with a unit price p. Let α be the proportion of revenue the store gets from the application, c_{ap} the application's development cost, and n the number of applications sold. The utilities of the store and the application designer are respectively

$$R_{\text{st}}(p) = \alpha n p, \tag{4.6}$$

$$R_{\text{ap}}(p) = (1-\alpha) n p - c_{\text{ap}}. \tag{4.7}$$

Let $F(p)$ be the expected proportion (or complementary distribution) of interested users willing to download the application at a price p (decreasing, with $F(0) = 1$ and $F(\infty) = 0$), and let Λ be the total number, or mass, of potentially interested users. In other words, if X represents the willingness to pay of a randomly picked user, $F(p) = \mathbb{P}[X \geq p]$. Then $n = \Lambda F(p)$. We assume that there is a mass at 0 because some users will download/use the application only if it is free. In (4.6) and (4.7), the optimal price is the same for the application designer and the store (if α is fixed), with just the constraint of needing to have a non-negative revenue in (4.7). Hence we end up with the maximization of $F(p)p$, provided that the solution p^* satisfies $R_{\text{ap}}(p^*) \geq 0$; i.e., $p^* F(p^*) \geq c_{\text{ap}}/(\Lambda(1-\alpha))$. This type of maximization has been discussed in Section 3.3.

We investigate now the case of free applications with ads. The general revenue formulations of (4.6) and (4.7) are still valid, but with some redefinitions of variables. We consider now a proportion α' of advertisement revenue instead of α being taken by the store. Instead of p, p' represents here the advertisement revenue from the application over its whole life (to use the same time unit). In addition, the number of users n can be considered as the total number $\Lambda F(0) = \Lambda$ multiplied by a factor $\mu \in [0, 1]$ representing the proportion of users willing to see ads, because some are strictly opposed to ads. We thus get

$$R'_{\text{st}} = \alpha' \mu \Lambda p', \tag{4.8}$$

$$R'_{\text{ap}} = (1-\alpha') \mu \Lambda p' - c_{\text{ap}}. \tag{4.9}$$

What is the best solution for the store and the application designer? From the store's viewpoint, advertisement is preferred if $R'_{\text{st}} \geq R_{\text{st}}(p^*)$, i.e., if $\alpha' \mu / \alpha \geq F(p^*) p^* / p'$. From the application designer's viewpoint, advertisement is preferred if $R'_{\text{ap}} \geq R_{\text{ap}}(p^*)$, i.e., if $(1-\alpha')\mu/(1-\alpha) \geq F(p^*) p^* / p'$. There are therefore situations under which advertisement is preferred for both, or paid application for both, or any other combination, depending on the parameters. Assume, for instance, that $\alpha = 0.3$, $\mu = 0.8$, and $\alpha' = 0.4$, and that $F(p^*)p^* = 1$ (without loss of generality; this is just a unit redefining factor). Thus the app designer will prefer advertisement if $p' \geq 1.45833$, whereas the store will prefer it if and only if $p' \geq 0.9375$. This gives here the price intervals $[0, 0.9375]$ over which both prefer a paid application, $[0.9375, 1.45833]$ over which only the application developer prefers a paid application, and $[1.45833, \infty)$ on which advertisement is better for both. The choice depends on who is expected to take the decision. Of course, the parameter values have to be estimated beforehand in order to get valuable results.

As a last remark regarding an issue that has not been treated in our very simplified model, the rank of an application in app stores' lists corresponding to keywords or topics is a key issue (users do not scroll down much) for its success and benefits. Ranking is usually based on a combination of the number of sales and price of the application, hence being directly related to the benefits for the store.

Another notion of applications, called **freemium**, is a mixture of free and premium (or paid). One offers a basic service for free, and then tries to incentivize users to apply additional features, for a fee. The most well-known example of such applications is Skype, the proprietary voice-over-IP service and software application. It allows users to communicate by voice using a microphone and a webcam. Instant messaging is also available for free, as is videoconferencing. Users are allowed to reach regular phone lines, or to do group video, but at a cost in both cases. Similarly, Hulu is an online video company from which users can watch television for free, but watching older or other programs involves a payment. In the smartphone applications industry, especially when talking about smartphone games, freemium applications are frequent. The most common practice is to offer an application with advertisements for free, and the ad-free app can be purchased for a fee. Add-ons or a full version of the application/game are available at a cost in many other cases once the user has been "attracted."

4.4 Economics of clouds/grids

Cloud computing is generally defined as the use of hardware and software computing resources that are delivered as a service over the Internet.[7] The cloud can offer access to processing power, storage capacity, sharing capacity, etc.

Most discussions today on cloud computing are about technical issues: protocols to access resources, performance, security, etc., though economic issues and relations between actors will have an important impact on the cloud's evolution. From the point of view of economics, cloud computing is very attractive to consumers and companies: it allows users to consume, and therefore pay for, only what they need and when they

[7] See http://en.wikipedia.org/wiki/Cloud_computing.

need it. The main interest of cloud computing lies in the economies of scale arising from mutualizing resources (see [130] for more details), namely the following.

- Supply-side savings due to large-scale data centers lowering the cost per server. For example, we can get electrical power savings, since power is a non-negligible part of the costs. Indeed, data centers are located in areas where electricity is cheaper. In addition, power use is optimized in a data center (unused computing devices being turned off, while locally managed computers are often always wastefully turned on). Moreover, economies of scale help to lower the labor (fewer employees) and acquisition (through bundles) costs.
- Demand is aggregated and thus has a reduced variability. The required resources can then be better planned.

Outsourcing is claimed to save maintenance costs as well as costs of purchases for limited use. Fears about security and access to data by unauthorized people are probably hindering the expansion of cloud computing; but this issue will not be treated in this book, since it falls outside its primary interest.

The benefits of cloud computing at the governmental level have been investigated in [8] for the USA, taking into account the savings in terms of infrastructure, but also the migration costs. The benefit-to-cost ratios (discounted net benefits divided by discounted investment costs) range from 5.7 to 15.4 depending on the type of department or agency (and its requirement in terms of public, hybrid, or private cloud computing because of confidentiality issues), thus cloud computing is often assumed to be worth the cost.

What are the differences among public, private, and hybrid clouds? A public cloud is made of resources (CPU or storage capacities mainly) available to the general public over the Internet. It can be used for non-sensitive data. A private cloud is a cloud operated for a single organization. A hybrid cloud is a combination of public and private clouds, allowing some fault tolerance without the need for Internet connectivity thanks to some local access. A typical example of hybrid cloud use is for online shops, for which the sensitive information such as credit card information is kept internally (on private clouds), while the account management service is managed on the public cloud (using encryption).

Like for access networks (see Chapter 3), cloud usage can be on a pay-per-use basis or payment can be made through subscription. We illustrate below the main trends in cloud pricing. The value-chain creation and relations in Internet technology platforms will be discussed a bit in Chapter 6, and are described in [127]. Comparing pricing mechanisms between competing cloud infrastructure vendors is not easy because each vendor proposes different packages of CPU, storage, RAM, etc. This lack of easy comparison is probably an expected outcome: there is no standardized measure of resource power. For example, Amazon (one of the main sellers of cloud computing) uses a metric called "EC2 compute units" or ECUs, as a measure of virtual CPU power. An ECU is based on an equivalent of a 2007 Intel Xeon or AMD Opteron CPU running at 1–1.2 GHz. Amazon has kept this type of metric even though those CPUs are no

Table 4.2 Rackspace cloud servers pricing schemes under Windows as of February 2013 (RAM is the main differentiator)

RAM	Price/hour	Price/month	Additional features
1 GB	$0.08	$58.40	Space 40 GB, public network 30 Mbps
2 GB	$0.16	$116.80	Space 80 GB, public network 60 Mbps
4 GB	$0.32	$233.60	Space 160 GB, public network 100 Mbps
8 GB	$0.58	$423.40	Space 320 GB, public network 150 Mbps
30 GB	$1.56	$1,138.80	Space 1.2 TB, public network 300 Mbps

Table 4.3 Amazon cloud servers on-demand pricing schemes under Windows as of February 2013

Denomination	Price/hour	Features
Default (small)	$0.115	1.7 GB RAM, 1 ECU, storage 160 GB, moderate I/O performance
Medium	$0.230	3.75 GB RAM, 2 ECUs, storage 410 GB, moderate I/O performance
Large	$0.460	7.5 GB RAM, 4 ECUs, storage 850 GB, high I/O performance
Extra-large	$0.920	15 GB RAM, 8 ECUs, storage 1,690 GB, high I/O performance

longer being run. Rackspace, on the other hand, uses RAM memory size as its main differentiator.

Table 4.2 describes a range of prices as presented on the web site of Rackspace (this cloud vendor being chosen arbitrarily) under Windows as stated in February 2013, to give an idea of the type of pricing mechanism applied. A similar range of prices exists under Linux rather than Windows (but in that case the prices are less than half those in Table 4.2), using Windows plus SQL Web or Windows plus SQL Standard. In Table 4.2, the prices are first linear before becoming concave. Note that this initial linear trend is not the case for Rackspace under Linux, Windows plus SQL Web, or Windows plus SQL Standard, for which a strictly concave curve is instead seen.

Amazon uses a pricing scheme with similar properties, the motto being "pay only for what you use." There is no minimum fee. New customers can even start using the cloud for free, receiving each month and during one year, 750 hours of ECU running Linux/Unix Micro instance usage and of ECU running Microsoft Windows Server Micro instance usage, plus again 750 hours of Elastic Load Balancing plus 15 GB data processing, etc. This initial free access is applied in order to attract customers. There exist several types of usages that are charged by Amazon.

- The cloud can be used (and charged) on-demand, by paying per hour of use with no long-term commitments or need to plan usage. Windows standard on-demand instances are charged as described in Table 4.3. Here too a linear-like pricing is used. Many other plans exist, called high-memory instances, high-CPU instances, cluster

Table 4.4 Amazon cloud servers reserved pricing schemes under Windows as of February 2013

Denomination	Upfront cost	Rate/hour
Default (small)	$69	$0.059
Medium	$138	$0.118
Large	$276	$0.235
Extra-large	$552	$0.47

global processing unit (GPU) instances, etc. Linux is charged for exactly the same list of instances, at about 56% the rate presented for Windows.

- A customer also has the possibility to use a scheme that is based on so-called reserved instances, such that you make a one-time upfront payment for each instance you want to reserve and receive a discount on the per-hour rate. Table 4.4 describes a one-year subscription, but a three-year subscription is also available, still for Windows. Reserved instances are available for exactly the same instances as the on-demand ones. For large consumptions, the gains with reserved instances can be substantial with respect to on-demand ones. A minimum consumption has to be ensured, though (or the risk of consuming less than expected must be pondered).

As noted in [23], the rate for "pay only for what you use" may seem expensive, even more expensive sometimes than buying servers. Using clouds has a major advantage though: it allows you to buy only what you need, when you need it, instead of buying resources to match peak utilizations, which will be barely used most of the time, and to avoid the risk of over-provisioning a network (a "peace of mind" effect).

Another interesting business model is the one from Dropbox, a cloud storage (and file-sharing) company. In this case, every individual is entitled to open for free an account with 2 GB of storage. Each time you invite a friend to open a new account, an additional 500 MB will be added. The principle is to attract consumers to the service. Then, if one wishes to get more space, one has to pay. For usual end-users, the prices (as of February 2013) are $99 for 100 GB, $199 for 200 GB, and $499 for 500 GB, all per year. Per month, it is a bit more expensive, namely $9.99 for 100 GB, $19.99 for 200 GB, and $49.99 for 500 GB. Thus a *linear* pricing is adopted. For professionals, another type of pricing is used (including services such as assistance, and the possibility to restore removed files), with a price defined on the basis of the number of associated licenses. The initial price includes five licenses and 1 TB for $795 per year. Any additional license brings 200 GB, at a price of $125 per license per year.

All this illustrates the range of pricing models that have been developed. A proper calibration of prices with respect to demand and usage needs to be performed. There is an extensive literature on this area. Just to illustrate this, on the question about leasing storage clouds or buying resources, we refer to [294], where a model including depreciation rates evaluates the best of the two options in terms of many parameters. In [276], the basic fixed prices for cloud usage (approximately linear in resources) are replaced by an adaptive pricing (in the spirit of the pricing models for network access developed in Chapter 3) with a reservation price made of two components: a fixed charge per unit and a congestion-based charge.

4.5 Economics of peer-to-peer systems

Peer-to-peer (P2P) systems have been important actors in the past decade (even though P2P traffic has been overtaken by video streaming, mainly YouTube, in terms of Internet volume traffic during the past few years).

What are P2P systems? A *peer* is a computer or any device that can be connected to others through the Internet. In P2P systems files are sent from one peer to the other without the need to store the file in a centralized server; hence each computer can be a server or a client. To be connected to a P2P system, a user just needs a file-sharing application. Several exist, and they can actually be grouped into two main categories. In both cases, the files are shared in a decentralized way, but the *catalog* of available files can be centralized or file discovery can be fully decentralized. A centralized list makes it very easy to reach files, but makes the list owner legally responsible for the content.

The first notable P2P application was Napster in 1999, which had millions of peers. Napster was shut down in July 2001 because it worked as mentioned above with a centralized list, illegally referring to copyrighted material. Many other networks have been developed, using a fully decentralized implementation, such as Gnutella, G2, the eDonkey network, etc. An intermediate architecture defines nodes with stable connectivity and more resources as **super peers**, such that peers connect to super peers and super peers forward queries only between themselves to avoid traffic flooding. This is what happens with Gnutella 2.0 and Kazaa. One can wonder, though, what the incentives for peers to act as super peers are. Altruistic behavior, small operational costs, and broadband connectivity are the main reasons.

P2P networks do not directly themselves form an economic actor, in the sense that they are made of a community sharing files or content in general for free, mainly because free access is part of the spirit of this type of community. But P2P networks have to be studied for several reasons: (i) they play a major role in the Internet, creating a non-negligible amount of traffic and attracting end-users; (ii) they have an impact on many content providers because copyrighted material is easily exchanged by P2P and therefore not sold by content owners; and (iii) P2P requires cooperation to work in practice, which is not happening particularly well with self-interested users – we need to see how it is addressed, sometimes including *virtual* money or any other set of possible rules to be imposed. We therefore briefly look at those aspects and refer the reader to the many existing references in the literature for more developments (see, for example, [19], which includes a nice state-of-the-art summary on the issue).

A user (or a peer) has an interest in participating in a P2P system if many others are there too: files/content can then be downloaded. However, **free riding** can be observed. Free riding consists in users participating in the system only to download files, and then leaving the system because there is no self-interest in staying just to share with others; there is even resource consumption in terms of bandwidth, storage, and energy, without any direct personal advantage. But if all peers act in that way, the system just does not work. It was, for example, highlighted in [4] that, in 2000, 70% of Gnutella users actually did not share any file, and 90% did not answer queries. Designing incentives for users to cooperate is thus a key issue, requiring an analysis through non-cooperative game theory of interest to ensure that the outcome of the game is as "good" as expected.

Among the main incentive schemes that have been designed, we can note a few options.

- The history of contributions can be tracked and used to limit the downloading capacity if a user's sharing is not considered sufficient. A complete history of transactions can be recorded by a central server, or a **reputation** score can be used, summarizing the user's previous behavior and quantifying how trustworthy she or he is in terms of cooperation, reliability, QoS offered, etc. Reputation has become popular in the Internet, not only for P2P networks, but also on ads web sites such as eBay or any equivalent store to give confidence before engaging in a transaction. Somewhat similarly, Kazaa reduces the free-rider problem by slowing down the downloads of users who are not sharing enough and getting a lot from the community. This is based on a participation level, computed as the upload in MB divided by download in MB times 100. Having a high level gives you first priority in downloading lines with other users, and shows that you are not a leech.
- Tokens (a kind of virtual money) can be awarded to users sharing their files, and "spent" when files are downloaded. If the number of tokens available is zero, the user cannot download any more. To allow the users to start with the system, a number of tokens is initially provided on subscribing to the P2P network. MojoNation and Karma use such schemes.
- **BitTorrent** has been introduced as an efficient P2P application for which there was no requirement to track the long-term behavior of nodes, but it is "enforced" that peers share their upload bandwidth when downloading files.

There exists an extensive literature on incentive modeling in P2P networks; see among others [47, 100, 176, 310]. In all models, peers are strategic players, trying to maximize their own utility by participating in the P2P system. To give an illustration, we look at the model presented in [47]. Here n peers are considered, and the utility function of peer i is

$$U_i = -c_i D_i + p(D_i/D_0) \sum_{j=1, j\neq i}^{n} B_{ij} D_j,$$

where the symbols have the following meanings.

- D_i is the *contribution* of i, considered in a general way, also being what peer i can play with (his or her strategic decision variable). It could be disk space, volume of data, etc. For an absolute measure of contribution D_0, representing a standard (and expected) value set by the system designer, $d_i = D_i/D_0$ is the dimensionless contribution of i.
- c_i is the cost for each unit of resource contributed by i, and B_{ij} measures how much a unit of contribution from j benefits i.

- $p(d_i)$ is a function representing how much a peer is rewarded by other peers as an increasing function of his or her contribution. For example, as an implementation, peer j can accept a request from peer i with probability $p(d_i)$, hence the expected reward we get in the expression of U_i. A choice $p(d) = d/(1 + d^\alpha)$ has the desirable properties of being an increasing function and that $p(0) = 0$ and $p(d) \to 1$ as $d \to \infty$.

A dimensionless utility function is then

$$u_i = U_i/(D_0 c_i) = -d_i + p(d_i) \sum_{j=1, j\neq i}^{n} b_{ij} d_j,$$

with $b_{ij} = B_{ij}/c_i$.

The function $p(d_i)$ incentivizes peers to contribute. Indeed, it is easy to see otherwise that, if the contribution of peers is not taken into account, considering $p(d_i)$ constant, then $d_i = D_i/D_0 = 0\ \forall i$ is the Nash equilibrium since u_i is maximized for $d_i = 0$ whatever the decisions of other peers.

Consider now, to simplify the analysis, the case of homogeneous peers, $b_{ij} = b\,\forall i \neq j$, and assume that $p(d) = d/(1 + d)$ (i.e., $\alpha = 1$ in the above setting). To determine the best response of peer i, we differentiate u_i with respect to d_i and get

$$\frac{\partial u_i}{\partial d_i} = -1 + bp'(d_i) \sum_{j=1, j\neq i}^{n} d_j = -1 + b\frac{\sum_{j=1, j\neq i}^{n} d_j}{(1 + d_i)^2}.$$

In the case of two peers, the best response of peer 1 with respect to d_2 (the other case is symmetric) is such that $(1 + d_1)^2 = bd_2$; i.e., $d_1 = \sqrt{bd_2} - 1$, if this value is positive, and $d_1 = 0$ otherwise. Solving the system $d_1 = \sqrt{bd_2} - 1$, $d_2 = \sqrt{bd_1} - 1$ gives the solution

$$d_1 = d_2 = d^* = (b/2 - 1) \pm \sqrt{(b/2 - 1)^2 - 1}.$$

Thus a strictly positive solution exists with the considered expression of $p(\cdot)$ only if $b \geq 4$, and actually two Nash equilibria exist if $b > 4$. For the case of n players, the derivative with respect to each contribution gives the equation characterizing the symmetric equilibrium $d^* = \sqrt{b(n-1)d^*} - 1$, whose solution is $d_i = d^* = b(n-1)/2 - 1 \pm \sqrt{b(n-1)/2 - 1^2 - 1}$.

The heterogeneous case is a bit more complicated to analyze (see [47] for more details), but is again the solution of the fixed-point equation

$$d_i^* = \sqrt{\sum_{j=1, j\neq i}^{n} b_{ij} d_j^*} - 1\ \ \forall i \tag{4.10}$$

and can be learned with peers playing successively by playing their best responses (4.10) in terms of the contributions of other peers (playing 0 if the above value is negative), until a fixed-point solution is reached.

What are the business models associated with P2P networking? How do companies make money? P2P networking was not designed initially to make money, and free access is clearly the main reason for its success. For example, no revenue-making procedure was initially associated with Napster. Gnutella is an open-source file-sharing protocol that can be downloaded for free, mainly to share music files. Kazaa, on the other hand, has always been intended to be a commercial enterprise. Its revenue is based on advertising, similarly to what we have developed in previous sections. As a last example, BitTorrent is currently looking for a way to turn file sharing into a source of revenue, something often referred to as "the pirate's dilemma" (turning pirates into legal and paying actors). To quote Matt Mason, BitTorrent's executive director of marketing,[8]

> With more than 100 million users worldwide using their ecosystem of software, devices and content, BitTorrent is focused on pushing advanced content delivery technologies forward in a way that makes sense for content creators and consumers alike. After seeing what's coming next from BitTorrent, I'm super excited about being a part of the effort to take the most efficient and powerful method of sharing digital information on Earth to new heights. [. . .] There's not a business model for content in the digital world. There's a new business model for every piece of content you release. [. . .] BitTorrent drove 124 million legal music downloads in the first six months of 2012, accounting for nearly one-third of the total 405 million music downloads on BitTorrent.

BitTorrent is therefore trying to design new ways to "consume" P2P networks, such as promoting movie-ticket sales, iTunes downloads, or advertising. New business models have to be imagined for those communities.

P2P networks are also often argued to impact the economy of other actors: the (free) P2P distribution of copyrighted files is usually said to affect the music and film industries. This is currently a subject of debate. On the one hand, there are fewer CD and DVD sales (a correlation with the arrival of Napster is often mentioned), meaning smaller revenues for producers and creators. As another potential consequence, it may also affect physical music stores and the associated jobs, even if the trend towards e-commerce might be another major reason for the business losses. The Motion Picture Association of America (MPAA) claims that American (movie) studios lost \$2.3 billion because of Internet piracy in 2005.[9] Similarly, the Recording Industry Association of America (RIAA) claimed that the number of CDs shipped in the USA fell from 940 million to 800 million between 2000 and 2002. On the other hand, opponents of this idea argue that it is quite difficult to determine the exact causes of the business losses, because of many related trends such as increasing online (legal) purchases, drops in the prices of CDs, etc. It is argued that the loss numbers are often exaggerated: for instance, the MPAA claim is based on the assumption that pirates would have otherwise bought the movie, which is probably not always the case (though just basing the potential sales on a level of 10% of illegal downloads, the (cumulated) loss was estimated at 240 billion euros by 2015 in Europe according to [221]). Moreover, it was also observed that the loss claimed by the RIAA was not experienced at all in the first years of rapid growth of P2P networking. It is even often claimed that file sharing can boost DVD or CD sales[10] because, as an indirect effect, users can learn more about music or cinema, making them appreciate it more, which can foster new sales. A typical illustration is the "Gangnam

[8] See http://techcrunch.com/2012/11/25/bittorrent-matt-mason-interview/.
[9] See http://en.wikipedia.org/wiki/Peer-to-peer_file_sharing.
[10] See for example http://grafodexia.blogspot.fr/2011/02/japanese-econometrics-study-filesharing.html.

style" video clip which attracted (albeit on YouTube rather than through P2P networks) many views, becoming a worldwide commercial success in 2012. The mechanisms at work here are similar to *seeding* strategies in marketing, where a good is offered to some users with the hope that some people within their circle of friends and acquaintances will be convinced to purchase that good. In the same vein, some apps are downloadable for free only for a short period, after which interested users must pay to get them; it is expected that this seeding will foster demand.

There are thus conflicting opinions on this business-loss issue, even though nowadays most industry experts believe that P2P has an adverse effect on the economy. This effect can be investigated through econometric analysis to establish the main causes. The audio sales were analyzed in [235] by looking at data containing over 10,000 album-weeks, to get as accurate an estimate as possible. In this example, the effect of file sharing on record sales could not be statistically proved on average, but it was shown that high-selling albums actually benefit from file sharing. The econometric model in [235] is as follows. It first rightfully states that it is not appropriate to use the relation

$$S_i = X_i\beta + \gamma D_i + \mu_i,$$

where S_i is the number of (observed) sales for an album i, X_i is a vector of album characteristics, D_i is the number of downloads, β is a vector of parameters to be inferred along with γ, which quantifies the effect of downloads, and μ_i is a random variable. Indeed, the number of downloads is likely to be correlated with unobservable and difficult-to-measure album characteristics that can create a bias in the estimations. It was rather suggested that one should consider a timed relation

$$S_{i,t} = X_i\beta + \gamma D_{i,t} + \sum_s w_s t^s + \nu_i + \mu_{i,t},$$

where ν_i is an album-dependent effect, t is for time in weeks, and the summation allows for a flexible time effect. Some tools are in addition used to better capture the time-varying album-specific heterogeneity in $\mu_{i,t}$. Using this type of analysis on large data, the parameter γ was estimated to be around zero, meaning that there was no sensitive effect due to downloads. In an often-cited reference [16] about music CD sales in Canada, still based on an econometric study, the author, Birgitte Andersen, even found that "P2P file sharing tends to increase rather than decrease music purchasing." However, in 2010 in [17], the same author came to the same conclusion as [235], claiming that "the Industry Canada data showed no association between the number of P2P files downloaded and CD album sales." We remark here that other models of the impact of P2P on demand may have different conclusions; our goal is not to defend one claim or the other, but we want to stress that analysis is here too a key issue if one wishes to go beyond *a priori* beliefs.

4.6 Economics of content delivery networks

Content delivery networks (CDNs) are distributed systems deployed in many data centers within the Internet, at edges of the network, in order to put data closer to end-users and offer better QoS and availability. There exist three categories of CDNs:

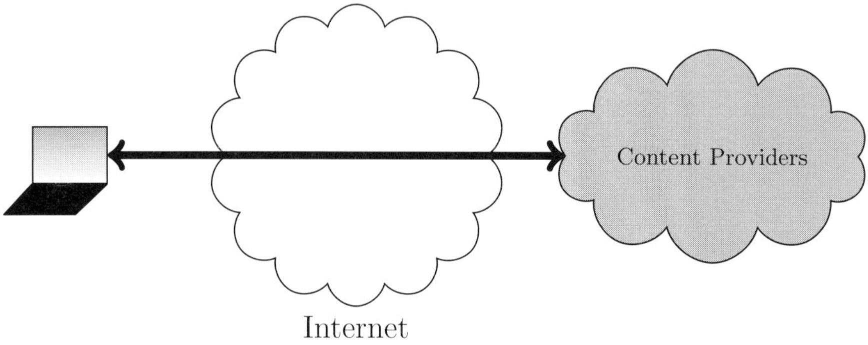

Figure 4.4 Communications without a CDN.

commercial/proprietary ones (the ones we are going to focus on), cooperative/academic ones, and P2P-based overlays, which are P2P networks working in the way described in the previous section. CDNs, especially commercial ones, use and select caches to serve users' requests. See [56, 96] for a general description of CDNs and more generally **overlay networks**.

The role of CDNs is gaining increasing importance as a means to deal with inter-domain relations, and their impact within the network neutrality debate may be of primary importance (see Chapter 7). In addition to providing better service than with direct access to the remote server, they allow one to avoid transit charges for traffic needing to reach end-users, if they are properly placed within the network. CDNs are thus paid by content or service providers for delivering their traffic to users and avoiding having to pay for those transit costs. All service and content providers can be interested in CDNs, but it is typically true for major ones such as YouTube, Dailymotion, Facebook, and Netflix (Netflix accounted for more than 30% of downstream traffic in US fixed networks in 2012). The challenge for CDNs is to cost-effectively distribute content on the Internet.

Figures 4.4 and 4.5 respectively illustrate communications without and with a CDN. It is easy to understand that a CDN with caches at the edge of the network limits communications within the Internet and improves QoS. As described in [56], popular content providers (CPs) have to pay ISPs an amount of the order of tens of thousands of dollars to host their sites; using CDNs may significantly reduce this cost. On the other hand, they imply more complex exchanges for the users to be redirected to caches, and smart mechanisms for the CDN to select and put into caches the (most) relevant content.

Among the main CDNs, we can mention Akamai, which managed about 60% of the market in 2012 according to experts, delivering between 15% and 20% of Internet traffic, with an estimated revenue of $1.1 billion in 2011. Its main competitors are Amazon CloudFront, CDNetworks, Limelight, etc. Even Internet and access network providers are now developing their own CDNs in order to reduce the backbone load (which is a way to avoid a costly infrastructure upgrade), which is mainly coming from the increase of streaming video traffic. The cost of caching must then be compared with the cost of transport within the network. A challenge in this area is that some ISPs might not be

Table 4.5 CDN77 pricing as of February 2013

Monthly traffic	USA and EU	Asia and Australia	South America
Under 30 TB	\$49/TB	\$125/TB	\$185/TB
31–100 TB	\$45/TB	\$120/TB	\$160/TB
101–400 TB	\$30/TB	\$100/TB	\$135/TB
401–1,000 TB	\$25/TB	\$85/TB	\$110/TB
Over 1,001 TB	\$19/TB	\$70/TB	\$95/TB

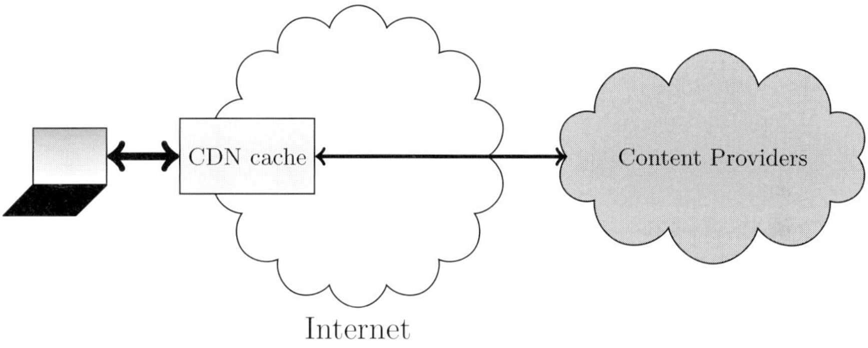

Figure 4.5 Communications with a CDN. Thanks to the cache, data from the content provider do not need to be downloaded each time a user makes a request.

willing to allow CDNs to place caches in their network, because this could limit their own benefits by reducing their transit revenues.

The pricing schemes used are not advertised for Akamai, but Amazon CloudFront pricing schemes are available at http://aws.amazon.com/cloudfront/pricing/. Another simple illustration is CDN77, with prices from February 2013 given in Table 4.5.[11] It can be seen that prices depend on the region, because costs do too. A decreasing price per unit is charged as the total traffic increases. This decreasing-unit-price behavior was shown to be the best practice in [137]. The model there considers a CDN and a time period of interest. The number X of requests to the CP during the period is unknown and treated as a random variable. If the CP had invested to process on average μ requests in a period, the utility of the CP would be

$$u(X, \mu) = v(X) - c(\mu) - k\ell(\mu, X),$$

where $v(X)$ is the valuation from X requests, $c(\mu)$ the cost for investing in resources to process μ requests, k the unit cost of a lost request, and $\ell(\mu, X)$ the number of lost requests. The goal of the CP is then to invest in $\mu = \mu^*$ maximizing the expected utility $\mathbb{E}[u(X, I)]$. This has to be compared with the situation when the CP delivers its traffic

[11] See http://www.cdn77.com/pricing.

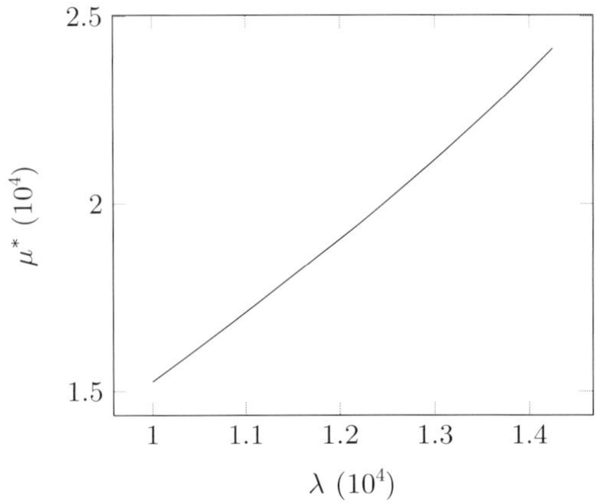

Figure 4.6 The optimal capacity investment μ^* for a content provider in terms of the mean arrival rate λ of requests.

through a CDN. Then its utility can be expressed as

$$u'(X) = v(X) + \tau(n)X - C_0 - p(X),$$

where $\tau(n)$ is the benefit for each request from using the improved QoS thanks to the CDN use, with n a parameter defining the size of the CDN (τ increasing with n), C_0 the (random) transaction cost to give the content to the CDN with cumulative distribution function H, and $p(X)$ the price charged to the CP by the CDN. If $\max_\mu \mathbb{E}[u(X, \mu)] > \mathbb{E}[u'(X)]$, the CP will prefer managing its traffic itself, whereas it will prefer using the CDN otherwise. In order to compute μ^*, the content server is modeled by an M/M/1/K queue with arrival rate $\lambda = \mathbb{E}[X]$ (see [287] for elements of queuing theory) and service rate μ (CDNs, on the other hand, use server farms and experience negligible queuing effects). Then the expected number of lost requests is [287]

$$\mathbb{E}\left[\ell(\mu, X)\right] = \frac{\lambda(1 - \lambda/\mu)(\lambda/\mu)^K}{1 - (\lambda/\mu)^{K+1}}.$$

Even considering the quadratic cost function $c(\mu) = a\mu - b\mu^2$ to get a concave cost, determining μ^* analytically is not tractable. But this can easily be done numerically. Figure 4.6 displays the optimal μ in terms of λ when $a = 3.46$ and $b = 0.000043$, for $K = 5$ and $k = 20$ (values taken from [137]). With our form of $c(\mu)$, in order to have an increasing cost function, we restrict μ to values in $(\lambda, a/(2b)]$.

A CP will subscribe to the CDN with a cost C_0 if $\mathbb{E}[u'(X)] > \mathbb{E}[u(X, \mu^*)]$, i.e., if

$$C_0 < \tau(n)\lambda + c(\mu^*) + k\mathbb{E}[\ell(X, \mu^*)] - \mathbb{E}[p(X)].$$

This occurs with probability $H(\tau(n)\lambda + c(\mu^*) + k\mathbb{E}[\ell(X, \mu^*)] - \mathbb{E}[p(X)])$. If $g(\lambda)$ is the mass of CPs with rate λ (C_0 being independent of the values of λ), the expected number of CPs seeking association with the CDN is

$$g(\lambda)H(\tau(n)\lambda + c(\mu^*) + k\mathbb{E}[\ell(X, \mu^*)] - \mathbb{E}[p(X)])$$

and the expected revenue of the CDN is

$$R = \int_{\lambda} g(\lambda)H(\tau(n)\lambda + c(\mu^*) + k\mathbb{E}[\ell(X, \mu^*)] - \mathbb{E}[p(X)])\mathbb{E}[p(X)]\mathrm{d}\lambda,$$

with $\mathbb{E}[p(X)] = \sum_{x=0}^{\infty} p(x)e^{-\lambda}\lambda^x/x!$. The goal of the CDN is thus to find the optimal $\mathbb{E}[p(X)]$ for each λ (since the distribution of X depends on λ), denoted by $p^*(\lambda)$. If h is the density associated with the cumulative distribution function H, the first-order condition gives the equation

$$p^*(\lambda) = \frac{H(\tau(n)\lambda + c(\mu^*) + k\mathbb{E}[\ell(X, \mu^*)] - p^*(\lambda))}{h(\tau(n)\lambda + c(\mu^*) + k\mathbb{E}[\ell(X, \mu^*)] - p^*(\lambda))}. \tag{4.11}$$

The "game" is as a last step to determine a price function $p(x)$ such that $p^*(\lambda) = \mathbb{E}[p(X)]$.

Example 40 Very briefly, considering the example in [137], assume C_0 to be uniformly distributed over [0, 1]. Thus (4.11) becomes $p^*(\lambda) = \tau(n)\lambda + c(\mu^*) + k\mathbb{E}[\ell(X, \mu^*)] - p^*(\lambda)$, i.e.,

$$\begin{aligned} p^*(\lambda) &= \frac{1}{2}\tau(n)\lambda + c(\mu^*) + k\mathbb{E}[\ell(X, \mu^*)] \\ &= \frac{1}{2}\left[\tau(n) + a\alpha + \frac{k(\alpha - 1)}{\alpha^{K+1} - 1}\right]\lambda - b\alpha^2\lambda^2/2, \end{aligned}$$

assuming also (to simplify) that $\mu^* = \alpha\lambda$ for a fixed constant α in order to get the second line (as an approximation of the almost-linear relationship observed in Figure 4.6). A price function with this expectation is (adding a term in the brackets)

$$p(x) = \frac{1}{2}\left[\tau(n) + a\alpha + b\alpha^2 + \frac{k(\alpha - 1)}{\alpha^{K+1} - 1}\right]x - b\alpha^2 x^2/2.$$

Note that this price function gives volume discounts to CPs, in accordance with most CDNs' policies (the double derivative with respect to x is negative), and a CDN of larger size can charge more (since the function is increasing with n).

---

5 Interactions among network service providers

5.1 Introduction

We focus in this chapter on network service providers, which are operators offering network capabilities to "others" (end-users, content providers, application/service providers, etc.). As evidenced in Chapter 1, the last decades have witnessed a multiplication of the number of such network service providers, proposing a variety of services. We have, as examples, wired access operators, wireless operators, transport operators, and bandwidth brokers, all of which now coexist in the increasingly complex telecommunications ecosystem. The relationships between those stakeholders can be of very different natures: one can indeed find customer–provider interactions, competition situations, cases with service providers as peers needing to collaborate, and quite often some complex combinations of those (e.g., the case of mobile virtual network operators (MVNOs), which rent their infrastructure from a wireless operator they will be competing with).

Among those relationships, the *competition* aspect has been exacerbated by regulators, with the purpose of benefiting users. Indeed, the regulators' objective is in general to ensure fair behavior of actors for the good of society. Their actions in particular aim at favoring new entrants in the market, especially at the access network level, by applying strong anti-trust policies, and by favoring user *churn* – i.e., users switching between providers. For example, in France, the latest 4G operator, Free, was supported by the regulatory authority forcing incumbent operators to lease some of their resource until Free develops its own infrastructure; this led to an overall decrease of access prices. At the European level, regulators strive to avoid collusion among the main operators, which may try to exclude rivals when setting standards; this was typically a concern in 2001 about the five major operators (Deutsche Telekom, France Télécom, Telecom Italia, Telefónica, and Vodafone). We remark, however, that "excessive competition" might lead to counterproductive outcomes, putting into question the survivability of the service providers (and, thus, of some services), due to phenomena such as price wars that we will analyze in this chapter.

For those reasons, competitive situations are of primary concern in this chapter. Sometimes the relations are more complex than just direct competition. In Section 5.2 we focus on wireless operators competing, not yet for users, but to acquire the right to use the radio spectrum – the so-called *spectrum licenses*. Such rights are classically sold through *auctions*, which were introduced in Chapter 2, but that we address here through a more detailed approach, introducing some specific models and also covering the practical implementation side. Section 5.3 is devoted to the competition that takes place after operating infrastructures and rights have been determined: access service providers then endeavor to attract users so as to maximize revenue, coping with the trade-off between the price per subscriber and the number of subscribers. Since the final decision – to subscribe or not, and to which provider – is in the hands of the customers, user behavior first has to be modeled; the section introduces several modeling possibilities. Using those models, which determine the demand levels of providers, we investigate the strategies played by competing providers in various contexts. First, we study price competition in wireless and wireline access networks, leading to two-stage games played on different time scales (user association on the smallest one, price competition among providers on the largest one), with the pricing strategies being based on the anticipated results of the smaller-time-scale game. We investigate also the possible strategies that providers can implement to prevent users from churning. Another level of game is then added, leading to three-stage games, where long-term decisions such as investments are taken on an even larger time scale. The section ends with high-level considerations that a regulator might have, regarding the decision to license the wireless spectrum or to leave it unlicensed, in a logic of welfare maximization. Section 5.4 focuses on the specific situation when some wireless access providers, MVNOs, use some of the network infrastructure of another provider (called a mobile network operator (MNO)), against a payment. That situation is unusual, since MVNOs are at the same time clients of their MNOs, renting resource from them, but also their competitors. Finally, Section 5.5 tackles the incentive issues that arise in networks where players are asked to relay the data flows of the others (such as inter-domain and wireless ad-hoc networks), while Section 5.6 considers the other special case of community networks (i.e., users offering a part of their access bandwidth to members of their community).

5.2 Auctions for wireless spectrum

5.2.1 Why use auctions?

The usage of wireless spectrum has long been regulated, to separate the different types of services (radio and television broadcasting, GPS, personal telecommunications, medical applications, governmental and military applications, ...). As the need for wireless spectrum increased, governments decided to *lease* its usage through paid **licenses**, giving to their owner the right to use a precise band of spectrum within a given geographic area, for a period of time (typically, ten years). The interest in applying charges is twofold. First, selling licenses yields revenues to governments, but also their prices match, to some extent, a guarantee that the spectrum will be used efficiently. Indeed, license buyers have to get some return on the license investments, and thus effectively use the spectrum to provide valuable services.

With the explosion of the demand for wireless spectrum to provide personal communications – starting with mobile telephony – and the enormous economic stakes for the private companies offering those services, spectrum has become an extremely scarce resource. For example, in 2011 mobile data traffic almost tripled, and may be again multiplied by 16 by 2016 according to current projections [109]. Deciding how much spectrum to allocate to each operator, and how much to charge for those licenses, has, then, become a very delicate matter. The problem is indeed the need to avoid setting prices too high, discouraging too many operators, while the opposite issue (setting prices too low) may incentivize operators to lease more spectrum than they actually need, leading to an inefficient use of spectrum (and less income for governments). The right price to set depends heavily on the value that those licenses have to operators, which information they are obviously reluctant to disclose; hence the need for more elaborate ways to lease the spectrum.

We are therefore here typically in a situation of mechanism design, as described in Chapter 2. The specific outcome targeted by governments is an efficient allocation of spectrum (i.e., spectrum being put to its most productive use), which cannot be computed without some input from those who create value out of spectrum (the operators). Hence the natural use of *auctions* to address that issue. Note that the revenue aspect of the mechanism is not evoked here, but it can certainly play a role in the decision for an auction scheme.

That approach started in 1993, when the US Congress authorized the Federal Communications Commission to assign licenses through auctions; about 100 auctions have been carried out since. Also in Europe, auctions have progressively been adopted. For example, for the allocation of 3G licenses in the late 1990s to early 2000s, several countries applied auctions – among which were Germany, the UK, Austria, and Belgium – while others – such as France, Sweden, Portugal, Spain, and Finland – preferred to use so-called **comparative hearings** (sometimes called "beauty contests," as evoked in Subsection 1.2.3). The latter scheme consists in setting fixed license prices, and selecting among the candidates according to some prespecified criteria that have been formulated as objectively as possible, such as the quality of the service to be provided to the public (including the areas of coverage), full-time participation in station operations by the owners, past broadcasting experience, and the technical efficiency of the frequency use. Interestingly, the fixed prices can be very low, and were even null (licenses allocated for free) in Sweden and Finland for 3G. But the trend is definitely towards auctions, and all European countries have actually applied auctions to allocate the first 4G licenses.

We remark that bids have to be chosen with care, as highlighted by the 3G spectrum auction carried out in 1999 in the USA: because of the dot-com bubble bids were very high, and buyers ended up defaulting on their payments after the 2001 telecom crash. The auction had to be re-run.

We review here the auction rules that have been applied, and their evolution as a response to the problems which were observed. But we first define the notions of *substitutes* and *complements*, which play key roles in studying these auctions.

Spectrum licenses: substitutes or complements?

Consider a country made up of two regions, and a spectrum band license to be sold for each region. For an operator wishing to have national coverage, it is mandatory to obtain both licenses: covering only one part of the country would not be acceptable, or would bring only small revenues. For that operator the two licenses are **complements**: the value of owning them both exceeds the sum of the values of owning each one separately.

Even if two licenses cover the same region, they can be complements if they are so small that the operator must get them both in order to achieve significant economies of scale. On the other hand, if only one license is enough for the operator to provide a satisfying service to his clients, then the operator can be interested in getting one license or the other, and his valuation for the pair would be below the sum of the individual valuations for each. In that case the licenses are said to be substitutes.

In practice, whether licenses are substitutes or complements depends largely on the definition of the licenses by the regulators [216]. For example, if one were faced with only national operators then it would be preferable to sell only country-wide licenses, which removes the complementarity between regional licenses. In most cases, it is assumed that licenses are substitutes. This is, for example, perfectly reasonable if all the licenses cover the same region and correspond to comparable spectrum bandwidths. In such cases buyers are indifferent to the specific *set* of licenses that they obtain, but rather focus on the *number* of such licenses.

5.2.2 Auction rules and evolution

Why not simply apply Vickrey–Clarke–Groves auctions?

As we saw in Chapter 2, a way to incentivize operators to truthfully reveal their valuation for the spectrum would be to use the VCG auction scheme. Concretely, each operator would be asked to declare in a bid how much they would be willing to pay for each possible subset of the whole license set. Then the auctioneer could compute an efficient resource allocation among operators on the basis of those bids, and charge each one his social opportunity cost.

From a theoretical perspective, VCG auctions therefore appear as the perfect tool in that context. However, their application in practice raises several issues [215, 250].

- The first problem concerns the computational complexity needed in order to implement the allocation and pricing rules. Indeed, denoting by L the number of licenses to sell in an auction, the bid dimension is then $2^L - 1$ for each bidder, since the bids have to be for each subset of the L licenses. In practice, often several hundred licenses are sold: thus the optimization problem to be solved in order to find an efficient allocation becomes untractable.
- The second issue is of the same order, but from the bidders' point of view. it becomes too complex to evaluate what to bid (even with the intention of being truthful), because of the extremely large number of combinations to consider.
- Finally, operators are deeply reluctant to disclose information about how highly they value the spectrum. While VCG rules require them to fully reveal that information, even if they end up without any license, other schemes, such as ascending auctions, allow them to limit the amount of data they disclose.

Those limitations finally prevent the use of VCG auctions to allocate spectrum licenses in practice. A simpler mechanism has to be implemented, where operators provide only limited information along the way. The "price" to be paid for that is then likely to be in terms of the efficiency of the final result.

Simultaneous ascending auctions

Simultaneous ascending auctions were originally proposed by Preston McAfee, Paul Milgrom, and Robert Wilson for the FCC spectrum auctions. They were first used in 1994, and have been extensively applied around the world since then.

The mechanism is quite simple, and relies on the following principles.

(i) All items (licenses) are sold *simultaneously*: to each license is assigned a price that evolves (it only increases, unless a bid is withdrawn) through successive *rounds* (of predefined duration).
(ii) At each round, bidders can bid (or not) for each license. At the end of the round, for each license the highest price – called the "standing high bid" – and the identity of the highest bidder are made public. That particular bid is kept for the next round and treated as the current license price, so that buyers other than the standing high bidder have to bid above that price.
(iii) No package bids are possible, i.e., if a bidder is interested in several licenses, he or she has to submit a separate bid for each of them.
(iv) The auction ends simultaneously for all items, when a round sees no new bid on any license. Licenses are then allocated to their standing high bidders, who pay their bid level.

Two additional rules are also imposed to guarantee that the auction finishes in a reasonable number of rounds.

- The **activity rule** prevents buyers from waiting too long to bid on a license. That rule is implemented through the notion of **eligibility**, which represents the rights for each potential buyer to bid (or not) on the set of licenses. It is expressed as a quantity (typically, the total population covered by the licenses you bid for is limited). The initial eligibility may differ among buyers; it is determined before the auction starts during a phase called the **qualification stage**, and is often associated with an initial deposit. Then, for each license, eligible buyers have to be the standing high bidder or submit valid (i.e., above the standing high bid) bids sufficiently often to keep that eligibility, otherwise the eligibility decreases so that they are not allowed to bid on as many licenses in the subsequent rounds. Mathematically, if a buyer's activity a is observed during round n when that buyer had an eligibility $e(n)$, then his or her eligibility in the next round is updated to $e(n+1) = \min(e(n), a/f_n)$, with f_n the activity requirement during round n (f_n being typically between 0.8 and 0.99; see [215] for details).
- To prevent prices from increasing too slowly, a **minimum price increment** is imposed. In general it consists in a given percentage of the standing high bid (between 5% and

20%), and can slightly change between rounds. Note that the standing high bidder does not have to increase his or her bid.

Those rules are also justified by game-theoretic arguments, since they help prevent some undesirable behaviors from self-interested bidders. The advantage of the minimum price increment rule is quite obvious: since winning bidders pay the amount indicated in their bid, they should look for the smallest price such that they remain the only interested bidder, and therefore increase bids by small steps. The rule then speeds up that price discovery process, at the cost of precision of the discovered price, and sometimes at the cost of efficiency. Indeed, consider two bidders 1 and 2 with respective valuations of 104 and 105 for one license, with bidder 1 being the standing high bidder with a bid of 100; if the minimum price increment exceeds 5% of the standing high bid then bidder 2 does not over-bid, and bidder 1 wins the license even though he or she is not the one with the highest valuation (the allocation is not efficient).

The activity rule, together with the minimum-bid-increment rule, guarantees that the auction will end in a reasonable time (since standing high bids increase multiplicatively with the number of rounds). Moreover, it prevents strategies such as **auction sniping**, which is observed in auction platforms like Ebay. Auction sniping consists in submitting one's bid just an instant before the auction end time (recall that Ebay auctions' ending times are known), so that other bidders cannot respond; that strategy offers the advantage of limiting the information disclosed to other bidders, while benefiting from the information revealed by them. Bidders implementing that strategy introduce a considerable bias in the price discovery which the ascending auction aims at, rendering the first rounds of the auction meaningless. Hence the strong interest of the activity rule. In simultaneous ascending auctions, the activity requirement increases on approaching the end of the auction. Low activity requirements in the first auction rounds (when there is most uncertainty about what can be obtainable) allow bidders to shift among packages while maintaining their eligibility.

Simultaneous selling allows bidders to adapt the set of licenses they bid on as the auction continues, and avoids many situations of regret that sequential selling can create, like having bought one license but being unable to obtain a second one at a reasonable price (in the complementary licenses case), or having bought an expensive license and observing that other licenses are sold later for lower prices (in the substitute case).

Example 41 (A simultaneous ascending auction scenario) At this point, let us illustrate the auction scheme through a (simplified) case of a simultaneous ascending auction scenario for two licenses A and B, involving three potential buyers. The valuations of the three bidders for their possible allocations are given in Table 5.1. We remark that the licenses are substitutes for bidders 1 and 2, whereas they are complements for bidder 3. Now assume that, at each round, the minimum bid increment is 10% of the standing high bid, and the activity requirement is $f_n = 0.8$. The auction starts with reserve prices 30 and 20 (standing high bids, with the auctioneer – bidder 0 – as the standing high bidder). Consider a particular trajectory for the first auction rounds depicted in Table 5.2.

Table 5.1 Valuations of the three bidders for the simultaneous ascending auction scenario of Example 41

	A	B	AB
Bidder 1	100	80	130
Bidder 2	90	60	130
Bidder 3	50	30	140

Table 5.2 The first five rounds of a simultaneous auction. Winning bids are boxed; for each round the standing high bid is indicated, with the identity of the high bidder in parentheses. The symbol • corresponds to the high bidder not over-bidding, and the symbol X is used when the eligibility constraint prevents bidding on more licenses.

Round	Bidder	Eligibility	Licence A	Licence B
			High bid = 30 (0), Min. increment = 3	High bid = 20 (0), Min. increment = 2
1	1	$e_1 = 2$	33	22
	2	$e_2 = 2$	[36]	23
	3	$e_3 = 2$	34	[26]
			High bid = 36 (2), Min. increment = 4	High bid = 26 (3), Min. increment = 3
2	1	$e_1 = 2$	40	29
	2	$e_2 = 2$	•	[30]
	3	$e_3 = 2$	[42]	•
			High bid = 42 (3), Min. increment = 4	High bid = 30 (2), Min. increment = 3
3	1	$e_1 = 2$	46	no bid
	2	$e_2 = 2$	[48]	•
	3	$e_3 = 2$	•	[33]
			High bid = 48 (2), Min. increment = 5	High bid = 33 (3), Min. increment = 3
4	1	$e_1 = 1.25$	53	X
	2	$e_2 = 2$	•	[36]
	3	$e_3 = 2$	[56]	•
			High bid = 56 (3), Min. increment = 6	High bid = 36 (2), Min. increment = 4
5	1	$e_1 = 1.25$	X	40
	2	$e_2 = 2$	[62]	•
	3	$e_3 = 2$	•	[40]

Table 5.3 The end of the auction scenario

Round	Bidder	Eligibility	Licence A	Licence B
			High bid = 62 (2), Min. increment = 6	High bid = 40 (3), Min. increment = 4
6	1	$e_1 = 1.25$	X	[44]
	2	$e_2 = 2$	•	no bid
	3	$e_3 = 2$	[68]	•
			High bid = 68 (3), Min. increment = 7	High bid = 44 (1), Min. increment = 4
7	1	$e_1 = 1.25$	X	•
	2	$e_2 = 1.25$	[76]	X
	3	$e_3 = 2$	•	[48]
			High bid = 76 (2), Min. increment = 8	High bid = 48 (3), Min. increment = 5
8	1	$e_1 = 1.25$	X	[53]
	2	$e_2 = 1.25$	•	X
	3	$e_3 = 2$	[86] **withdrawn**	•
			High bid = 76 (0), Min. increment = 8	High bid = 53 (1), Min. increment = 5
9	1	$e_1 = 1.25$	X	•
	2	$e_2 = 1.25$	[84]	X
	3	$e_3 = 2$	no bid	no bid
			High bid = 84 (2), Min. increment = 8	High bid = 53 (1), Min. increment = 5
10	1	$e_1 = 1.25$	X	•
	2	$e_2 = 1.25$	•	X
	3	$e_3 = 0$	X	X

In round 3, bidder 1 bids on only one license. Indeed, from the standing high bids and the minimum increments, bidding on both would mean paying at least $46 + 33 = 79$ for the package, hence a net benefit of $130 - 79 = 51$. On the other hand, bidding only on license A corresponds to a "current" higher benefit of $100 - 46 = 54$ (and also higher than if bidding on license B only). As a result, it is useless for bidder 1 to try to win both licenses, and he or she accepts the reduction in eligibility (here expressed as the number of licenses you can bid on) which is applied in the next round. Note, however, that bidder 1 can still bid on license B in the next rounds: only the *number* of new bids is affected by the eligibility constraints. Bidder 1 does so in round 5 for example, when license B becomes more interesting to him or her than is license A.

The same phenomenon occurs for bidder 2 in round 6 (see Table 5.3): it becomes more interesting to bid on just one license (here, A).

We remark that, in round 3, bidder 3 bids above his or her valuation for the sole license B. This is due to these two licenses being complements for that bidder: if bidder 3 obtains A, then he or she is willing to pay up to an extra 70 monetary units to also get B. Hence bidder 3 continues to bid on that license, hoping to win the auction for A. In round 4, bidder 3 also bids above his or her valuation for licence A: again, he or she is willing to pay up to 90 for that license, provided that also B can be obtained, and that the total price paid remains below 120. That phenomenon is among the drawbacks of simultaneous ascending auctions: bidders with complementary preferences have to guess whether it is worth continuing to participate.

At the end of round 8 (see Table 5.3), bidder 3 discovers that there is no hope of winning both items for a total below 120 (his or her valuation for the bundle), and hence ceases bidding on license B. But, being the standing high bidder on licence A with a price above his or her valuation, bidder 3 has to *withdraw* that bid. Indeed, bidder 3 faces the so-called **exposure problem**; i.e., bidder 3 may win license A despite being willing to pay for it only if B is obtained too, which is no longer an option. This is the reason why most simultaneous ascending auctions include a **bid withdrawal rule**. The withdrawal rule allows bidder 3 to remove a bid; the standing high bid is then lowered to the second highest bid, but the associated bidder is changed to the auctioneer (in case the second-highest bidder has shifted his or her focus towards other licenses in the meantime). There is a penalty for withdrawing a bid, which equals the loss to the auctioneer, i.e., the difference between the withdrawn bid and the final winning bid for the license.

No new bids are submitted during round 10, hence the auction ends: license A goes to bidder 2 for a price of 84, and license B goes to bidder 1 for 53. Bidder 3 has to pay a penalty of 2 for withdrawing his bid of round 8.

While the example above illustrates the main rules, of course some slight variations exist. For example the activity rule has some flexibility; i.e., in practice the eligibility is reduced only after a few rounds of inactivity (instead of one round of inactivity in the example) to prevent technical errors from yielding illegitimate eligibility reductions. More detailed descriptions can be found in [68, 70, 215], for the auctions applied in the USA.

Going deeper into the practical implementations of simultaneous ascending auctions, we think it is worth mentioning that most auctions are run online, through a web-based interface. Interestingly, the practice has evolved from bidders submitting numerical values for their bids to bidders simply clicking on one button (out of about ten) indicating by how many times the minimum increment they want to raise their bid. This has been implemented to prevent errors (adding or forgetting one digit could have dramatic consequences), but also mainly to prevent bidders from using the numerical bid values as a hidden communication channel (while no communication is allowed among bidders during the auction, the less significant bits in the numbers could indeed be used to give information to the other bidders, e.g., to say things like *"I'll stay out of New York, if you stay out of Los Angeles"* [69, 72]). That new bidding procedure has therefore been imposed. Note, however, that this increases the likeliness of having several high bidders in each round: in such cases the identity of the winning bidder is most often decided through a lottery (as was done during round 5 in Example 41).

The use of withdrawal has also been constrained, since it has been observed that some bidders used this rule strategically (as a signal to the other bidders, or to maintain eligibility without bidding on the licenses they are really interested in until the very end – the so-called parking strategy – see [70]).

Do simultaneous ascending auctions perform well?

From a theoretical point of view, simultaneous ascending auctions perform quite satisfactorily. The rules are quite simple, but it is difficult to highlight the best bidding strategies that operators should implement. However, a simple strategy "naturally" emerges. This corresponds to adopting a myopic behavior; i.e., bidding in each round as if it were the final one and as if no other bidder were able to submit new bids. Concretely, each bidder would consider the price of each license as equal to the minimum price he or she can expect (the high bid if he or she is the high bidder, or the high bid plus the minimum increment otherwise). Then the **straightforward bidding** strategy, defined in [215], consists in bidding with those prices, on a (maximal) set of licenses maximizing one's resulting utility. We remark that this strategy was implemented by bidder 1 in the scenario in Example 41 above. The two other bidders also used a strategy quite similar to it; they sometimes bid only slightly above the minimum price they could declare.

Given the constraints imposed by the auction rules, the straightforward bidding strategy appears to be a good candidate. Indeed, the activity rule prevents "wait and see" behaviors, so potential buyers have to constantly bid on sufficiently many licenses. Then, to avoid prices going too high, bidding the minimum allowed price seems reasonable. No better strategy clearly emerges from a careful game-theoretic analysis of the auction scheme, which was actually among the objectives of the auction designers. However, the enormous financial amounts involved strongly elicit the finding of better strategies like the ones cited above (using price declarations to exchange information among bidders, playing with bid withdrawals, . . .), leading to rule adjustments.

Some mathematical results regarding the performance of simultaneous ascending auctions and straightforward bidding have been shown by Milgrom [215].

- When all licenses are substitutes for a bidder, then bidding straightforwardly ensures a non-negative revenue (in any round, i.e., even if the auction were to be ended at a random round).
- If licenses are substitutes for all bidders, then when they all bid straightforwardly the auction leads to an efficient final assignment. More precisely, the total valuation of the assignment is close to the maximum reachable value, the distance being upper-bounded by the sum of single bid increments over all licenses.

That last result speaks in favor of choosing small minimum bid increments, to guarantee efficiency of the license assignment. But, on the other hand, this increases the duration (number of rounds) of the auction, hence the common practice of starting with large bid increments (to speed up the price discovery phase) and reducing them on approaching the end of the auction.

We remark that, in Example 41, bidders follow a strategy that is similar to straightforward bidding, and the assignment reached is actually the most efficient one, with a total assignment value of 170.

In terms of practical performance of simultaneous ascending auctions, some analysis on previously run auctions can be found in [70]. The main conclusions are the following, recalling that the actual valuations of operators are unobservable.

- The fact that hardly any license resales have been observed after the auctions were carried out suggests that licenses are efficiently allocated.
- Also, the revenues yielded by auctions have often significantly exceeded the values expected by governments and operators, which can stem from the scheme's property of progressively reducing the uncertainty on prices, and eliciting more aggressive bidding behavior.

But, as pointed out in [70], simultaneous ascending auctions are not robust against collusion among bidders. We illustrate that problem below.

Example 42 (Collusion in simultaneous ascending auctions) For the scenario of Example 41, the three bidders could agree to let license A go to bidder 2 and license B to bidder 1 (the efficient allocation), and to bid only on the license they are supposed to obtain. As a result, there would be new bids only during round 1 (bidder 2 bidding 33 on license A and bidder 1 bidding 22 on license B), so that the auction would end with those prices, the threat being to switch to the straightforward bidding strategy if one bidder deviates. Bidder 3 could be given a small payment to agree not to bid at all (his or her utility in the straightforward bidding scenario being 0), so that all bidders would be strictly better off with that agreement. (Or, alternatively, the overall gains with respect to the non-cooperative bidding could be shared among the three bidders according to the Shapley value.)

Here the efficiency of the assignment is not affected, but the revenue is significantly reduced, whence the willingness of bidders to communicate despite the interdiction (e.g., through the use of bid values as a hidden signal).

Another consequence of collusion is that it can harm efficiency as well, through a phenomenon called **demand reduction**.

Example 43 (Demand reduction in simultaneous ascending auctions) Consider again the scenario of Example 41, but with bidder 1 willing to pay 180 (instead of 130) to obtain both licenses. The efficient allocation would now consist in bidder 1 obtaining both licenses, but it can happen that an agreement is found among the bidders, giving license A to bidder 2 and license B to bidder 1. Indeed, with such an agreement (the same one as in Example 42), bidder 1 obtains a utility of $80 - 22 = 58$, while everybody bidding straightforwardly would lead to prices of around 90 and 60 for license A and B, respectively (because of bidder 2), and hence a utility of around $180 - 150 = 30$ for bidder 1. Bidder 1 is thus incentivized to accept the agreement and reduce his or her demand, leading to an inefficient assignment in terms of total value.

We remark that there does not need to be an explicit agreement: bidder 1 alone may anticipate that reducing his or her own demand to one license would lead to a better situation for him or her, even if the others bid selfishly. Indeed, if there were only one competitor interested in only one license, then the (unilateral) demand reduction prevents prices from rising above the initial values set by the auctioneer.

Fears of such collusions (or recognized cases) have occurred worldwide, e.g., 3G auctions in Thailand in 2012 that were refuted afterwards, and tacit collusion with bid signaling and retaliation in FCC auctions [69]. The problems of collusive bidding and demand reduction can be reduced by forbidding communication among bidders, and through a careful design of the items to be auctioned. Indeed, grouping several licenses into a single item prevents demand reduction, and thus favors efficiency when licenses are complements. In that case, it also protects against **hold-up strategies**, consisting in buying a license for a limited price (by convincing the other bidders of one's willingness to bid high, as illustrated in the previous examples), and then reselling it after the auction to an operator with complementary preferences. On the other hand, auctioning packages of licenses also hinders the appearance of smaller operators, e.g., those willing to offer regional services. Hence a **band plan** (form of the licenses auctioned) that includes at the same time small license blocks and larger blocks may be used, as was done in the main auctions in the USA in 2006 and 2008. Unfortunately, such band plans create other undesirable effects.

- During the auction process, a bidder can easily switch from a large block to a set of small blocks yielding equivalent rights, but the opposite is made difficult by the activity rule. Indeed, since it is unlikely that the bidder will be outbid on all small blocks, he or she may have to withdraw the corresponding bids (where he or she is still the high bidder) and face withdrawal penalties. This affects negatively the flexibility of bidders and the efficiency of the final assignment.
- The other disadvantage of auctioning large blocks is to reduce competition, since the potential candidates for them are few and in general well identified: other bidders may be deterred from bidding on such blocks, which can end up being sold for low prices.

The problems raised by simultaneous ascending auctions have been observed in implementations all over the world (see [70]), fostering discussions regarding other schemes.

5.2.3 Evolving from simultaneous ascending auctions . . .

The issues raised by the simultaneous ascending auction scheme speak in favor of mechanisms allowing **package bidding** (i.e., bidding on bundles of licenses), such as the combinatorial auctions introduced in Chapter 2.

Among those, a promising solution is the **combinatorial clock auction** mechanism [71], which tends towards more efficiency, using the simplification that many licenses are substitutes and can be sold as a single divisible good. The mechanism consists of two phases, and its main principles are the following.

- When they are substitutes (e.g., on close frequency values), licenses are grouped as a single product, so that bidders specify *quantities* they want to buy of each type of product instead of declaring prices for each license. This reduces considerably the number of different bundles to specify, and thus the complexity of maximizing the (declared) value of an allocation that is based on bids.
- During the first phase (called the *clock stage*), an iterative process is carried out: a unit price is assigned to each product (type of license), and bidders have to declare how much of each product they would buy at the current price values, building their own bundle with those prices. Prices of products for which there is excess demand are increased in the next round, and the clock phase ends when no product is experiencing excess demand. All bids submitted during all rounds in that phase are stored by the auctioneer.
- The second phase (called the *supplementary round*) consists in allowing all bidders to submit new bids; more precisely, they can increase the price values for the packages they have bid for in the clock stage.
- The mechanism involves an activity rule, whose principle is the same as in simultaneous ascending auctions (you have to bid on large packages to maintain the right to obtain large packages). Its mathematical expression is more complicated than in simultaneous ascending auctions, but the objective is the same, i.e., to favor price discovery by eliciting truthful bidding behavior. The activity rule limits the bids that can be made during the clock auctions, and also during the supplementary round.
- The whole set of collected bids (clock-stage bids and supplementary-round bids) is then treated by an optimizer to compute the license assignment and the prices to be paid. The license assignment is chosen so as to maximize efficiency (i.e., the sum of declared values by bidders), and prices equal the VCG auction prices (if they are within the core, as defined in Chapter 2), or the values in the core nearest to the VCG prices.

The interested reader is referred to [71] for details of the implementation of the combinatorial clock auction scheme. Recall that prices being in the core implies that no group of losing bidders is offering more to the auctioneer than the winning bidders pay. On the other hand, VCG prices aim at eliciting truthful bidding: when they are too low, such that they are not in the core, the pricing rule raises them to some minimal possible value satisfying the core constraint. The rules as a consequence aim at keeping prices low and focusing on efficiency (and incentive-compatibility) rather than revenue, the argument being that, due to the numerous economic spillovers, governments should always benefit in the long term from an efficient allocation of licenses.

Combinatorial clock auctions solve most of the problems of simultaneous ascending auctions (collusion, demand reduction, the exposure problem). They have been successfully applied for electricity and gas products in Europe, and more recently for spectrum in the UK [71]. But most countries still apply simultaneous ascending auctions.

5.2.4 . . . to incentive auctions

The next big spectrum auction in the USA will concern frequencies that had previously been used for television broadcasting. Thanks to the analog-to-digital transition, the spectrum needs of broadcasters have been dramatically reduced, so that some of the bandwidth initially reserved for television can be used for other purposes. This is the **digital dividend**, which the FCC intends to optimize through a global re-allocation of licenses in those frequencies among broadcasters, broadband operators, and unlicensed spectrum services. The proposed rules were released on October 2, 2012 [60], opening the discussion on several aspects. The term **incentive auction** was coined by the FCC, to insist on the benefits that current license holders (broadcasters) should get from relinquishing some of their rights.

The idea here is to use double-sided auctions (i.e., on the supply as well as demand sides), through a three-stage mechanism.

(i) Broadcasters are asked to "sell" some of their spectrum through a so-called **reverse auction**, i.e., an auction where the bidders are the sellers (bids contain prices that broadcasters are willing to receive). Three options are offered to broadcasters (for a monetary compensation):
 - simply relinquishing some of their usage rights;
 - releasing an ultra-high-frequency (UHF) television channel in exchange for a very-high-frequency (VHF) television channel (indeed, the FCC intends to re-allocate a maximum of UHF spectrum for broadband services (operator licenses and unlicensed usages)); and
 - relinquishing some of their usage rights, in exchange for a share of a television channel (with another licensee).

(ii) The relinquished spectrum and the spectrum remaining to broadcasters are then **repacked**, to form sufficiently large contiguous bands.

(iii) Those liberated bands are finally sold through another (more classical) auction, called a **forward auction**. The most likely candidate schemes are the simultaneous ascending auction and the combinatorial clock auction. The revenues of that auction are then partly used to pay broadcasters for having relinquished some of their spectrum rights.

The document [60] calls for comments from all participants, to define sufficiently clear and fair rules regarding all practical aspects of such a scheme, among which are

- the relative timing of the three stages (to be held sequentially or in parallel);
- the form of the repacking for technical aspects of reuse (block sizes of 5 MHz (like LTE) or 6 MHz (like current digital TV channels), paired (uplink plus downlink) blocks or not, specific location of the blocks, . . .); and

- the specific rules for each type of auction (bid format, package bidding allowed or not, use of reserve prices, simultaneous ascending auction or combinatorial clock auction, ...).

As of May 2013, the way those auctions will be run was thus still undecided. Among the objectives of the incentive auction is the fostering of innovation, by freeing up some significant amount of unlicensed spectrum (repacked spectrum that will not be auctioned and hence will become usable for free). This is motivated by the considerable economic impact of services based on unlicensed spectrum in the last decade (such as WiFi access); the regulators do not want the available spectrum resources to limit the development of such services or the appearance of new ones [109].

The first incentive auctions are expected to be run in 2014 [109]; the future will tell us whether they represent a real breakthrough in spectrum management.

5.3 Competition between access providers

Spectrum license auctions are a typical situation where wireless operators directly compete in their bids. But the most common understanding of competition is a bit less direct, since it involves the decisions made by *users*: access providers fight to attract users and make revenue. This section is devoted to the definition and mathematical analysis of some situations of that type.

To study competition for customers, we first need to model the decisions made by users regarding their choice of an access provider. We therefore summarize here the most commonly used mathematical assumptions for user behavior, illustrating them with some specific competition models.

5.3.1 Association models based on user utility

A first way to model the decision of users is to consider them as self-interested actors willing to maximize their individual utility. More precisely, for given provider choices of the other users and given provider strategies (e.g., prices), each user has a utility associated with connecting to each provider, and is expected to select the provider yielding the highest utility.

Example 44 (A price-war situation) Consider a very simple situation, where two access providers, 1 and 2, compete for users, offering services of identical quality, and no congestion effects occur. Competition is through prices only (no "reputation" effect is considered), and we denote by p_i the subscription price set by provider i, $i = 1, 2$. Since no congestion occurs, the overall utility experienced by a user u is independent of the choices made by the others: it can be modeled as the difference between the user's willingness to pay, V_u, for the service (which is the same for both providers) and the price actually paid. As a result, all users u such that $V_u > \min(p_1, p_2)$ will simply select the cheapest provider, with the other users preferring not to subscribe to either. (What happens when prices are equal has to be specified in the model, since users are then indifferent between the two providers; we assume here that users split equally among providers.) Then, when the user willingness-to-pay distribution is continuous, it is easy to see that there is no equilibrium with a provider making revenue in the pricing game played by providers. Indeed, take a situation (p_1, p_2) with provider 1 making some positive revenue: this implies $p_2 \geq p_1 > 0$. But provider 2 could strictly improve its revenue by setting its price just below p_1, hence attracting all users and making a strictly positive revenue. But exchanging the roles of providers, provider 1 could then do the same, this leading to null prices. Such a situation is called a **price war**, and, even though it is desirable for users in the short run, it can jeopardize the survivability of the providers and be harmful for society in the long run.

That price-competition setting (with users selecting the cheapest provider) is also known as **Bertrand competition**: prices decrease until the price reaches the marginal cost of a provider (with that provider then leaving the market). In the example above, the marginal cost was taken to be 0.

Wardrop equilibria

When utilities depend on the provider selected by the other players (possibly through congestion effects), we then have a game played among users that has to be studied to investigate the competition among providers. We provide below an example with non-atomic users, so that the user equilibrium notion is the Wardrop equilibrium.

As an illustration, let us consider the case of competition among a set $\mathcal{N}$ of providers implementing loss-based pricing, i.e., a pricing model studied in Subsection 3.5.2 (see in particular Figure 3.4), where all packets sent are charged so that the perceived price per successfully transmitted packet increases when losses occur. We denote by C_i the transmission capacity (in packets per time slot) of provider i, and assume that losses occur when demand exceeds capacity on a network, in which case lost packets are randomly chosen (more exactly, there is an equal loss probability for all packets sent to a network). Consequently, the perceived price $\bar{p}_i$ per successful packet with provider i is

$$\bar{p}_i = p_i \max(1, d_i/C_i), \tag{5.1}$$

where p_i is the price charged per packet sent by provider i and d_i is the total demand (number of packets sent per time slot) of that provider.

Users are assumed to be infinitesimal, and selfishly select the cheapest provider in terms of perceived price when they have a choice. Finally, demand is assumed elastic; i.e., the aggregated number of packets that all users wish to send (successfully) is a strictly decreasing function D of the perceived price. The function D is assumed to be continuous and such that $D(0) > \sum_{i \in \mathcal{N}} C_i$, and can stem from a distribution of valuations for the service among users and/or from individual elastic demands. We will treat two specific scenarios in terms of network topology.

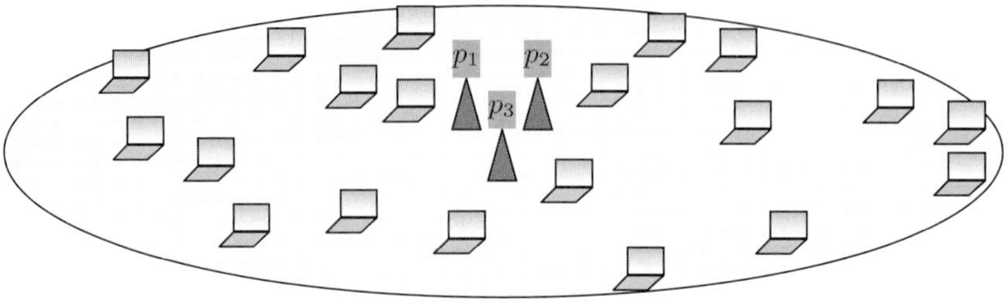

Figure 5.1 A network topology with all providers competing for the same set of users.

Common coverage area. First consider that all users can reach all providers, as depicted in Figure 5.1. Then, if we denote by d_i the total demand of provider i, the Wardrop equilibrium conditions can be summarized by the following system of equations:

$$\begin{aligned} \forall i \in \mathcal{N}, \qquad & \bar{p}_i = p_i \max\left(1, \frac{d_i}{C_i}\right), \\ \forall i \in \mathcal{N}, \qquad & \bar{p}_i > \min_{j \in \mathcal{N}} \bar{p}_j \Rightarrow d_i = 0, \\ & \sum_{i \in \mathcal{N}} d_i = D(\min_{ij \in \mathcal{N}} \bar{p}_j). \end{aligned} \tag{5.2}$$

The first relation expresses the perceived price for each provider, according to (5.1); the second one comes from the Wardrop condition (only the cheapest options are selected), and the last equation links total demand to the (common) perceived price of all chosen providers.

The interaction situation can reasonably be analyzed using backward induction, by assuming that providers set their price so as to maximize revenue, *anticipating* the outcome of the non-atomic game played by users: even if they are playing first, providers make use of what the Wardrop equilibrium would be for any price profile $\boldsymbol{p} = (p_1, \ldots, p_{|\mathcal{N}|})$. The following set of results can then be established (see [194] for proofs).

(i) For any price vector $\boldsymbol{p} = (p_i)_{i \in \mathcal{N}}$ there exists a Wardrop equilibrium, and the corresponding perceived price of each provider is unique. This is illustrated in Figure 5.2, where the minimum perceived price (when users select the cheapest options) and the inverse demand function when the aggregated demand varies are plotted. A Wardrop equilibrium corresponds to an intersection point of those curves. If several providers set the same price $\bar{p}$ and $D(\bar{p}) < \sum_{j:p_j \leq \bar{p}} C_j$ (the case of Figure 5.2(b)), then users are indifferent between those providers, and any spreading of the demand $D(\bar{p}) - \sum_{j:p_j < \bar{p}} C_j$ among those providers such that none of them gets congested is a Wardrop equilibrium.

(ii) If, between two outcomes leading to the same revenue, each provider prefers the one for which it manages smaller demands, then when the demand elasticity $(p/D(p))\mathrm{d}D(p)/\mathrm{d}p$ is below -1 there is a unique Nash equilibrium of the price-competition game: this equilibrium corresponds to all providers setting the same

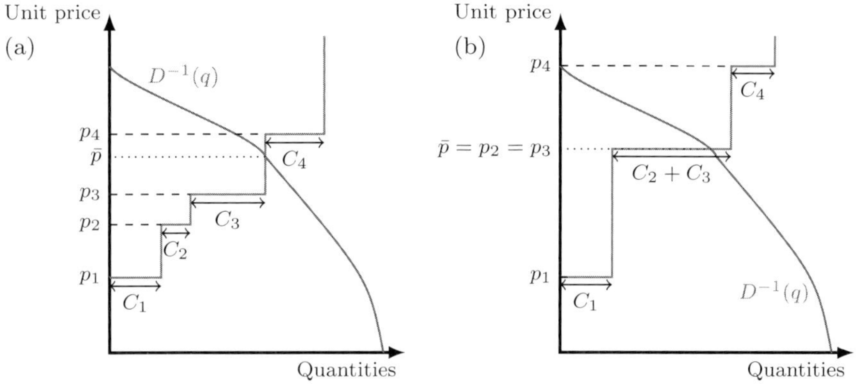

Figure 5.2 The common perceived price on all chosen providers when the total quantity served is given by the stair-step function. At a Wardrop equilibrium, that perceived price is $\bar{p} = \min_{i \in \mathcal{N}} \bar{p}_i$ (intersection with the inverse demand curve). (a) Unique Wardrop equilibrium. (b) Infinite number of Wardrop equilibria.

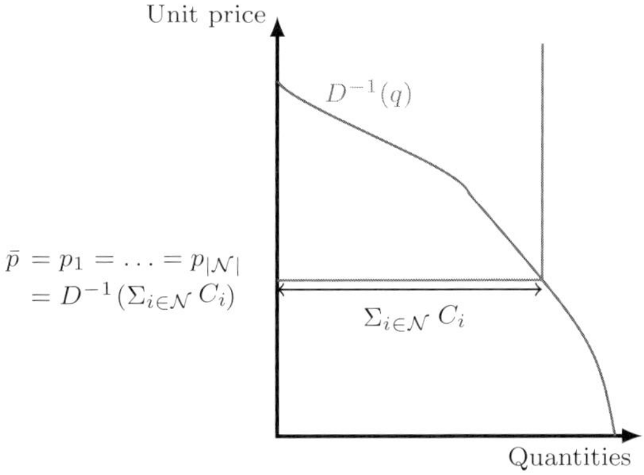

Figure 5.3 Wardrop equilibrium at the Nash-equilibrium price vector, i.e., $p_i = p^* = D^{-1}\left(\sum_{j \in \mathcal{N} C_j}\right)$.

price $p^* := D^{-1}\left(\sum_{i \in \mathcal{N}} C_i\right)$ [194], where D^{-1} is the generalized inverse of D, i.e., $D^{-1}(y) := \min\{x \geq 0 : D(x) \leq y\}$.

(iii) The Wardrop equilibrium for that price vector $\boldsymbol{p}^* = (p^*, \ldots, p^*)$ is unique and such that $d_i = C_i$ for each provider $i \in \mathcal{I}$. In other terms, all providers have their capacity fully used, but none is congested (see Figure 5.3). This situation is actually the one which maximizes the social welfare: the price of anarchy equals 1.

Different coverage areas. Now we treat a simple case where the providers' coverage areas differ. Consider two wireless access points owned by competing operators, 1 and 2, offering service over a given area, as depicted in Figure 5.4. Provider 2 operates in a subdomain of provider 1, due to use of a limited-range technology, e.g., a WiFi cell versus an LTE one. The zone covered by both providers (hence, where they actually

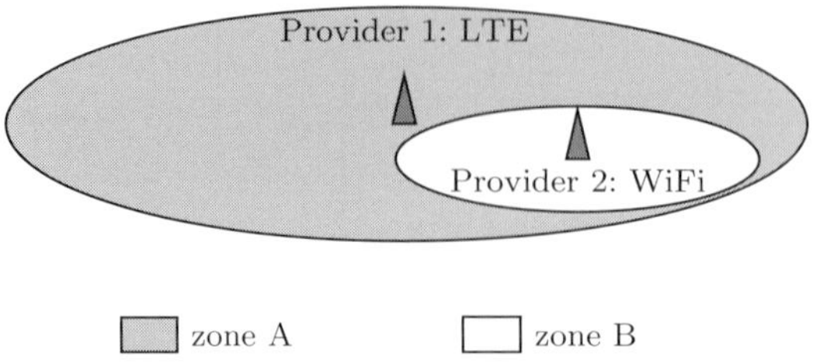

Figure 5.4 The competition situation: providers and topology [190].

compete) is denoted by B, while provider 1 acts as a monopoly in zone A. Even though the number of providers has been reduced to two, this framework considerably complicates the analysis with respect to the case of a common coverage area treated just above. However, we think the results provide interesting insights into the reasons why different prices can be observed.

Assume that a proportion α of the users is within the coverage area of provider 2 (zone B), for some α, $0 < \alpha < 1$ (a proportion $1 - \alpha$ of users will then be in zone A). An outcome from user choices is characterized by demands d_{A} (in zone A, hence for provider 1), $d_{1,\mathrm{B}}$ (aggregated demand from users in zone B choosing provider 1), and d_2 (aggregated demand from users in zone B choosing provider 2). Such an outcome corresponds to an equilibrium when the following system is satisfied:

$$\bar{p}_1 = p_1 \max\left(1, \frac{d_{1,\mathrm{A}} + d_{1,\mathrm{B}}}{C_1}\right), \tag{5.3}$$

$$\bar{p}_2 = p_2 \max\left(1, \frac{d_2}{C_2}\right), \tag{5.4}$$

$$(1-\alpha)D(\bar{p}_1) = d_{1,\mathrm{A}} \min\left(1, \frac{C_1}{d_{1,\mathrm{A}} + d_{1,\mathrm{B}}}\right), \tag{5.5}$$

$$\alpha D(\min(\bar{p}_1, \bar{p}_2)) = d_{1,\mathrm{B}} \min\left(1, \frac{C_1}{d_{1,\mathrm{A}} + d_{1,\mathrm{B}}}\right) + d_2 \min(1, C_2/d_2), \tag{5.6}$$

$$\bar{p}_1 > \bar{p}_2 \Rightarrow d_{1,\mathrm{B}} = 0, \tag{5.7}$$

$$\bar{p}_1 < \bar{p}_2 \Rightarrow d_2 = 0. \tag{5.8}$$

Relations (5.3) and (5.4) give the perceived price on each provider, as a function of the price charged, capacity, and demand of that provider. Equations (5.5) and (5.6) express the demand relations in zones A and B (in terms of successful packets per time slot), respectively: the left-hand sides correspond to demands as functions of the perceived price in each zone, and on the right-hand sides are the actual successful packet rates in each zone (computed from the sent packet rates). We remark that the perceived price in zone A is always $\bar{p}_1$ (users have no choice), while it is $\min(\bar{p}_1, \bar{p}_2)$ in zone B (users select the lowest perceived price). Finally, (5.7) and (5.8) reflect Wardrop's first principle in zone B (where users can choose): only the cheapest provider(s) can get demand.

Again, the reasoning at the pricing level is done by backward induction. A rigorous analysis for that model is carried out in [190]; we summarize here the main results.

- For every price profile (p_1, p_2) with strictly positive prices, there exists a (not necessarily unique) Wardrop equilibrium, and the corresponding perceived prices $(\bar{p}_1, \bar{p}_2)$ are unique.
- If the demand elasticity $(p/D(p))\mathrm{d}D(p)/\mathrm{d}p$ is strictly smaller than -1 for all $p > 0$, then the pricing game played by providers has an infinity of Nash equilibria of the form $\{p_1 \in (0, p_1^*], p_2 \in (0, p_2^*]\}$, where

  if $C_1/(1-\alpha) \leq C_2/\alpha$,

$$p_1^* = D^{-1}\left(\frac{C_1}{1-\alpha}\right) \geq p_2^* = D^{-1}\left(\frac{C_2}{\alpha}\right); \tag{5.9}$$

  and if $C_1/(1-\alpha) > C_2/\alpha$,

$$p_1^* = p_2^* = p^* = D^{-1}(C_1 + C_2), \tag{5.10}$$

  where D^{-1} is again the generalized inverse of D. Despite there being infinitely many Nash equilibrium prices, all yield the same perceived prices, $\bar{p}_i = p_i^*$, and the same revenues to providers, $R_i^* = p_i^* C_i$, $i = 1, 2$.

Let us take a step back here. Recall that the condition on elasticity implies that $pD(p)$ decreases with p, so providers should try to set low prices in order to maximize revenue. However, there is no benefit from lowering the price too much, since providers can only "sell" their capacity (extra packets being lost, which affects the perceived price and reduces demand), so a natural objective is to sell all of one's capacity. The case when $C_1/(1-\alpha) \leq C_2/\alpha$ corresponds to a *capacity per user* offered by provider 2 alone in zone B larger than the one provider 1 can offer in zone A only. Hence provider 2 can sell all of his capacity C_2 in zone B for a perceived price $p_2^* = D^{-1}(C_2/\alpha)$, a perceived price that provider 1 can never beat: assume $\bar{p}_1 < p_2^*$, then the demand for successful traffic in zone A is $(1-\alpha)D(\bar{p}_1) > (1-\alpha)D(p_2^*) \geq C_1$, which provider 1 cannot fulfill. As a result, provider 1 cannot compete in zone B and leaves it to provider 2. Provider 1 then generates revenue by selling its capacity in zone A only, where provider 1 has a monopoly. On the other hand, when $C_1/(1-\alpha) > C_2/\alpha$, the two providers compete in zone B, with each of them getting some demand.

Depending on the Nash equilibrium selected by providers, the social welfare can vary from 0 (when prices tend to 0) to its maximal possible value (when $p_i = p_i^*$ for $i = 1, 2$). The latter case is quite likely to be chosen by providers (e.g., if we add some small traffic management costs). Since it leads to a socially optimal situation, the price of stability of the whole game, defined in Subsection 2.3.1 as the best-case ratio comparing the social welfare for the non-cooperative outcome with the optimal value, equals 1. Similarly, since low prices also correspond to an equilibrium, the price of anarchy is unbounded.

It is worth noticing that, in the example above, we have a situation where a two-level game involving self-interested users and providers can lead to a socially optimal outcome, i.e., the price of stability equals 1. Because of the Wardrop conditions, a price war occurs when no provider experiences congestion; but the price war stops because of congestion effects: decreasing one's charging price further then does not bring any revenue improvement. Some more elaborate models of competition based on Wardrop user behavior can be found in [3, 144].

Utility with random components

In the previous (Wardrop) model, all users had the same comparative perception of the alternatives (here, the providers to select). As we saw previously, this can create threshold effects (all demand switching from one provider to the other after a slight price change, as in Example 44) and price wars.

Since such phenomena are not commonly observed in reality, one can look for improvements in the model leading to more continuity in the aggregated user behavior. A way to do this is to assume that not all users are exactly the same: while some are only looking for the cheapest option, others will consider it worth switching providers only if the price difference is large enough. Furthermore, some users can be willing to stay with the most expensive provider, even without clear evidence of a better quality of service: this can explain why incumbent providers often remain slightly more expensive than their competitors, without losing too much market share. Such effects are due to rather subjective aspects, such as perception of the overall service quality (e.g., through the reputation of the provider) and attachment to the provider's brand (that can be affected by advertising).

The number of aspects other than price that can affect users' choices, and their somehow intangible nature, make it difficult to encompass each of them in a model. Therefore, a natural way to proceed is to *aggregate* all these unknown effects into a single value for each user and each provider that represents the non-monetary benefit (or cost) that this particular user associates with this provider. Then we can still assume that each user makes a utility-maximizing choice, but now users are heterogeneous, meaning that they do not necessarily all prefer the same provider(s), even without congestion effects.

Such an approach is typical of **discrete-choice** models, which are frequently used in econometrics [34]. To be more specific, the utility for a user u making the choice i (from a discrete set of options) is supposed to equal $v_i + \kappa_{u,i}$: the term v_i encompasses the objective aspects of the option (e.g., price and QoS) and is the same for all users, while $\kappa_{u,i}$ is an unobserved user-specific value that is treated on the global level as a random variable. In most cases, it is assumed that the variables $(\kappa_{u,i})_{u,i}$ are all independent, and that for each option i the variables $(\kappa_{u,i})_u$ are identically distributed, so that the probability distributions of $\kappa_{u,i}$ for each option i completely characterize the model, and the subscript u can be omitted. Then, from those distributions, one can compute the *probability* that a user selects option i for each i; when the population is sufficiently large this corresponds to the proportion of users making that choice.

Let us come back to the case of competing access providers implementing a flat-rate pricing scheme, and neglect congestion effects. The subscription price of provider i is denoted by p_i, and each user has to decide which provider to subscribe to, if any. We suppose here that the average value for an ISP i is affected by the price it charges, but also by some average reputation factor $x_i > 0$, through the standard logarithmic relation

$$v_i = \alpha \log\left(\frac{x_i}{p_i}\right), \tag{5.11}$$

where $\alpha > 0$ is a sensitivity parameter. The logarithmic functional originally stems from psychophysics (the relationship between the magnitude of a physical stimulus and its perceived intensity is often logarithmic), and has recently been observed also in the context of telecommunications [253]. We remark that a null price p_i yields an infinite value v_i, so a free option will always be preferred over one for which there is a charge. The average reputation factor can be interpreted as follows: if $x_1 = 1.2$ and $x_2 = 1$ for example, then on average customers are willing to pay 20% more with provider 1 than with provider 2 (the reputation compensates for the larger price). Finally, we have to model the utility of the "no-provider" option, which we label by the index 0: we take the same form as for the other options, i.e., we denote by v_0 the average (negative) value of not having Internet access, and still consider a random part κ_0 in the individual utility of option 0.

Following the literature on discrete-choice models, we assume that the user-specific random variables κ_i follow a Gumbel distribution of mean 0; i.e., their distribution satisfies $\mathbb{P}[\kappa_i \leq y] = \exp(-\exp(-y/\beta - \gamma))$, where $\gamma \approx 0.5772$ is Euler's constant, and we take $\beta = 1$.

It can then be proved under that assumption of a Gumbel distribution (see [34]) that the probability of a user choosing the option $i \in \{0, 1, 2, \ldots, n\}$ (n being the number of competing providers) equals $\exp(v_i)/(\sum_{j=0}^{n} \exp(v_j))$. Defining $p_0 := \exp(-v_0/\alpha)$ as some "equivalent price" of option 0, and using our expression (5.11) for v_i, that probability can be expressed in terms of prices as

$$\sigma_i(\boldsymbol{p}) := \frac{(x_i/p_i)^\alpha}{\sum_{j=0}^{n}(x_j/p_j)^\alpha}, \tag{5.12}$$

with $\boldsymbol{p} = (p_0, p_1, \ldots, p_n)$ and $x_0 := 1$. We assume here that the number of users is large, so that the probability σ_i corresponds to the market share of provider i for $i = 1, \ldots, n$.

We can observe the effect of the sensitivity parameter α: when α tends to infinity, users focus on the cheapest option(s) (the one(s) with the smallest p_i/x_i). On the other hand, α going to 0 leads to a uniform spreading of users among all alternatives. We remark also that we have chosen the same sensitivity parameter α for all options for simplicity, but the study could also be carried out with different α_is: in (5.12) the α exponent of x_j/p_j just has to be replaced by α_j for each j.

Now let us study the price competition between two providers, labeled 1 and 2. Assuming without loss of generality that the total mass of users is 1, the revenue of provider i equals

$$R_i(p_0, p_1, p_2) = p_i\sigma_i(p_0, p_1, p_2) = \frac{p_i(x_i/p_i)^\alpha}{(x_1/p_1)^\alpha + (x_2/p_2)^\alpha + 1/p_0^\alpha}. \tag{5.13}$$

Let us immediately treat the particular case of null prices: if a provider sets a null price, then it makes no revenue but also attracts all users unless the competitor sets a null price, hence preventing that competitor from making revenue. Thus null prices $(p_1, p_2) = (0, 0)$ constitute a Nash equilibrium, albeit a peculiar one: as soon as one provider sets a strictly positive price, it is strictly better to also fix a strictly positive price than to set a null price. Hence providers should try to avoid that equilibrium, when possible.

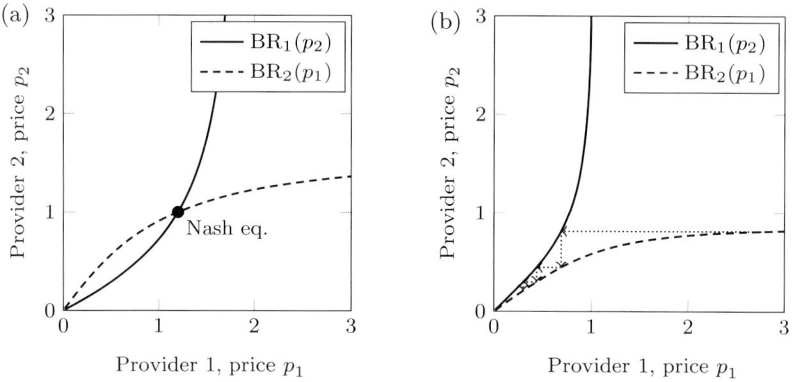

Figure 5.5 Best-response functions for the price competition, with $p_0 = 1$, $x_1 = 1.2$, and $x_2 = 1$. (a) $\alpha = 1.5$. (b) $\alpha = 2.5$.

We now look for equilibria different from (0, 0). An equilibrium price profile (p_1, p_2) is such that each provider plays a best response to the price of the opponent; we therefore compute here the best-response functions.

From (5.13), we obtain, for all strictly positive prices (p_1, p_2),

$$\frac{\partial R_i}{\partial p_i} = \sigma_i(1 - \alpha(1 - \sigma_i)),$$

with σ_i given in (5.12). Since σ_i is a probability and is strictly positive when prices are non-zero, the case when $\alpha \leq 1$ leads to infinite best-response prices (i.e., maximizing R_i given p_j, $j \neq i$). This somehow corresponds to a small price elasticity of demand: a small α means that user choices are not much affected by prices, hence raising prices improves revenues since only a small proportion of users will switch.

From now on, we make the more realistic assumption that $\alpha > 1$. For $i = 1, 2$ and $j \neq i$, the maximization of R_i in terms of p_i for a fixed $p_j > 0$ leads to $1 - \alpha(1 - \sigma_i) = 0$: indeed, $\partial R_i / \partial p_i$ has the same sign as $1 - \alpha(1 - \sigma_i)$, which is decreasing in p_i from (5.12). This yields

$$\sigma_i = \frac{(x_i/p_i)^\alpha}{(x_i/p_i)^\alpha + (x_j/p_j)^\alpha + 1/p_0^\alpha} = 1 - 1/\alpha,$$

$$(x_i/p_i)^\alpha + (x_j/p_j)^\alpha + 1/p_0^\alpha = \frac{\alpha (x_i/p_i)^\alpha}{\alpha - 1},$$

$$(x_i/p_i)^\alpha \frac{1}{\alpha - 1} = (x_j/p_j)^\alpha + 1/p_0^\alpha.$$

The best-response function of provider i is then

$$\mathrm{BR}_i(p_j) = x_i \left((\alpha - 1) \left(\left(\frac{x_j}{p_j} \right)^\alpha + \left(\frac{1}{p_0} \right)^\alpha \right) \right)^{-1/\alpha}. \tag{5.14}$$

An example of those best-response functions is plotted in Figure 5.5 for two different values of the sensitivity parameter α. We remark that for $\alpha > 2$ there is no equilibrium with strictly positive prices. Indeed, such an equilibrium would correspond to

$\sigma_1 = \sigma_2 = 1 - 1/\alpha$, which is possible only if $\alpha \leq 2$ (recall that $\sigma_1 + \sigma_2$ is the probability of users selecting no provider, thus $\sigma_1 + \sigma_2 \leq 1$). When $\alpha \in (1, 2]$, solving the system $p_1 = \mathrm{BR}_1(p_2)$, $p_2 = \mathrm{BR}_2(p_1)$ yields a unique solution (as illustrated in Figure 5.5(a)). That solution (p_1^*, p_2^*) is such that $\sigma_1 = \sigma_2 = 1 - \alpha$, hence $x_1/p_1^* = x_2/p_2^*$, which gives

$$p_i^* = x_i p_0 \left(\frac{2-\alpha}{\alpha-1}\right)^{-1/\alpha}, \qquad i = 1, 2.$$

We find here again the condition that $\alpha \in (1, 2]$. Interestingly, we also observe that the equilibrium prices are proportional to the "reputation" factor x_i, i.e., $p_1^*/p_2^* = x_1/x_2$. Hence the relative operator prices directly reflect their relative reputations. Similarly, since the equilibrium market shares are equal ($\sigma_1 = \sigma_2 = 1 - 1/\alpha$), the providers' revenues are also proportional to their reputation factors: $R_i = x_i \sigma_i = x_i(1 - 1/\alpha)$.

Finally, if $\alpha > 2$ we remark that $\mathrm{BR}_i(p_j) < p_j x_i / x_j$, so that successive best responses lead to a strictly decreasing sequence of ratios p_i/x_i, and eventually to null prices as illustrated in Figure 5.5(b).

The discrete-choice model presented above is developed in [66] to analyze the relations between ISPs in terms of traffic exchange. Interestingly, the formula (5.12) is also considered in [50] for the particular case $\alpha = 1$ and $x_i = 1, \forall i$, but with no mathematical interpretation at the user-utility level: the authors' main argument for the formula involves some customer stickiness (or, interchangeably, inertia or loyalty), which represents the users' reluctance to switch providers if the price difference is small. That kind of approach – modeling directly the *aggregated* behavior of users without user-utility expressions – provides a different way to study the relations between service providers, which will be developed below.

5.3.2 Aggregated demand models

Equation (5.12) is an example of an aggregated demand function: it quantifies the overall demand that a provider can expect, as a function of the strategies of all stakeholders (here, all providers). While that particular demand function can be justified by reasoning on user utilities, such a justification is not mandatory, and it often happens that a demand function is chosen because it satisfies the main trends expected from aggregated user behavior. We now present some possibilities.

Linear demand

The most typical example – and most frequently used model – is the **linear demand** model [9, 95, 292]: for a set $\mathcal{N} = \{1, \ldots, n\}$ of providers competing on price (and/or quality, but we present here just the case of prices), the demand level d_i of provider $i \in \mathcal{N}$ with the price vector $\boldsymbol{p} = (p_i)_{i \in \mathcal{N}}$ is postulated to be given by

$$d_i(\boldsymbol{p}) = d_{i,0} - \alpha_i p_i + \sum_{j \neq i} \beta_{ij} p_j, \tag{5.15}$$

where $d_{i,0}$, α_i, and $(\beta_{ij})_{j \in \mathcal{N} \setminus \{i\}}$ are strictly positive parameters. The parameter α_i represents the direct effect of price on demand (the higher my price, the smaller my demand), while the parameters β_{ij} model the indirect effect of the competitors' prices (the higher their price, the larger my demand). Some additional assumptions are usually made:

- $\alpha_i > \sum_{j \neq i} \beta_{ij}$ for all i, to represent the fact that all prices increasing by the same value should lead to a demand decrease for each provider; and/or
- $\alpha_i > \sum_{j \neq i} \beta_{ji}$ for all i, to represent the fact that a provider increasing the price it charges cannot lead to an increase in total demand.

One of the advantages of that model is its simplicity: the global demand vector $\boldsymbol{d} = [d_1, d_2, \ldots, d_n]^{\mathrm{T}}$ can be written as the result of a simple linear operation

$$\boldsymbol{d}(\boldsymbol{p}) = \boldsymbol{d}_0 - \boldsymbol{A}\boldsymbol{p},$$

where $\boldsymbol{d}_0$ and $\boldsymbol{p}$ are column vectors, and $\boldsymbol{A}$ is the matrix such that, for $i, j \in \mathcal{N}$,

$$A_{ij} = \begin{cases} \alpha_i & \text{if } i = j \\ -\beta_{ij} & \text{otherwise.} \end{cases}$$

That simple model often leads to tractable derivations, as in the special case studied in Example 19 (Chapter 2): providers' revenues are indeed quadratic functions of prices, which are easy to maximize. We now briefly treat the case of a duopoly.

Example 45 (Competition between two providers with linear demands) Consider two providers (named 1 and 2) whose demands are given by (5.15). For any price $p_j \geq 0$ set by the competitor, there is a unique revenue-maximizing price for provider $i \neq j$, i.e., the best response is

$$\mathrm{BR}_i(p_j) = \arg\max_{p_i} p_i\big(d_{i,0} - \alpha_i p_i + \beta_{ij} p_j\big) = \frac{d_{i,0} + \beta_{ij} p_j}{2\alpha_i}.$$

Then solving the system $p_1 = \mathrm{BR}_1(p_2)$, $p_2 = \mathrm{BR}_2(p_1)$ gives a unique solution when $4\alpha_1\alpha_2 > \beta_{12}\beta_{21}$ (which is usually the case). That unique Nash equilibrium is characterized by

$$p_i = \frac{2\alpha_j d_{i,0} + \beta_{ij} d_{j,0}}{4\alpha_1\alpha_2 - \beta_{12}\beta_{21}},$$

for $i, j = 1, 2$ and $j \neq i$. The corresponding revenues are

$$R_i = \alpha_i \left(\frac{2\alpha_j d_{i,0} + \beta_{ij} d_{j,0}}{4\alpha_1\alpha_2 - \beta_{12}\beta_{21}} \right)^2.$$

It can be checked that, when $\alpha_1 = \alpha_2 = \alpha$ and $\beta_{12} = \beta_{21} = \beta$, the result is the one we obtained in Chapter 2 (Example 19).

Such an analysis encompasses the main effects of the competition; namely, an incumbent operator can be expected to have a larger $d_{i,0}$ and a smaller α_i than its competitor, which leads to larger prices. Similarly, the indirect price effects are reflected according to intuition: a larger β_{ij} (propensity of users to subscribe to i when p_j increases) induces provider i to raise his price.

However, the simplicity of the linear demand model comes with different types of drawbacks. The most obvious one is that the expression (5.15) can give negative values, which is an unrealistic assumption. This problem is typically solved by taking only the positive part in (5.15); however, that solution raises new issues since (5.15) then has to be considered with care, e.g., when differentiating revenues.

Another more subtle issue also arises when one tries to investigate the outcome of cooperation between two providers. Indeed, consider one of them increasing its price so that (5.15) gives a negative value (the demand for that provider is then set to 0): the problem is that, according to (5.15), the demand for the other providers goes on increasing. As a result, a way to reach an infinite revenue would be to have one provider setting an infinitely large price, and the other one benefiting from the infinite demand that results; this is clearly an unsatisfactory feature of the model. That drawback of the linear demand model is more complicated to deal with: while it sounds reasonable to have some indirect effect of prices (through the β_{ij} parameters), that effect represents the fact that some users switch providers when prices vary, which cannot occur when the provider raising its price has no demand. Ideally, the indirect effect from a provider should stop as soon as the demand for this provider becomes zero, but this is not easy to translate into a direct static formula giving the demand vector as a function of the price vector $\boldsymbol{p}$. A nice attempt can be found in [12], where the reasoning is done by having all prices increasing at the same speed until $\boldsymbol{p}$ is reached, with the indirect effects from one provider stopping when its demand hits 0. However, the analytical simplicity of (5.15) seems to be lost for good ... Non-linearity seems a relevant other option, but it also generally complicates analytical derivations; a few examples follow.

Attraction models

The other main stream of aggregated demand models generalizes the approach of user utilities with random parts, as we saw before in this chapter. The principle is to assume that the market shares of the providers are proportional to their *attractiveness*. Mathematically and following the notations of [9], the attractiveness a_i of a provider i is a function of strategic choices such as the price p_i and quality q_i, and the market share of provider i is given by

$$\sigma_i = \frac{a_i(p_i, q_i)}{\sum_{j \in \mathcal{N}} a_j(p_j, q_j) + a_0},$$

where a_0 represents the attractiveness of the outside option (i.e., the no provider choice).

Such demand models are called **attraction models**; when providers play on price only and $a_i(p_i) = (x_i/p_i)^{\alpha}$ we have the demand expression (5.12) found previously. Similarly, with $a_i(p_i) = 1/p_i$ we find the "stickiness-based" function used in [50]. Note actually that the general case can also be interpreted from a random-utility model as before, with user utilities of the form $U_{u,i} = \log a_i + \kappa_{u,i}$, where $\kappa_{u,i}$ is a user-specific random variable following a Gumbel distribution.

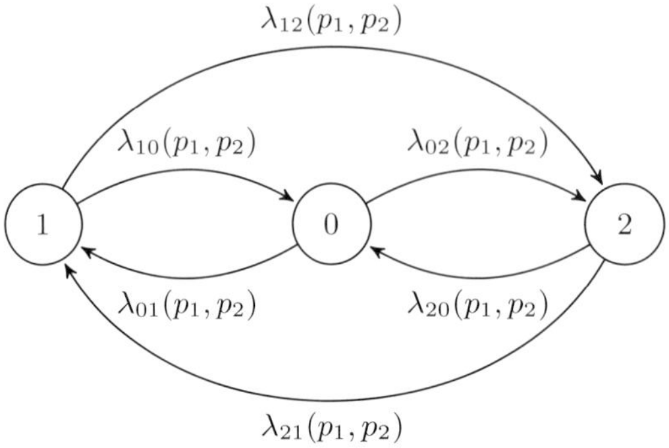

Figure 5.6 A continuous-time Markov-chain model for the switching behavior of a user. State $i = 1, 2$ corresponds to the user being with provider i, while state 0 means that the user chooses to have no provider.

Dynamic models

The telecommunications ecosystem is an extremely fast-changing environment. A salient example is the mobile market, through the phenomenon called **churn** (users switching providers): yearly migration rates can often reach 25% [301]. In that context, it makes sense to take that dynamicity into account in the economic model. Furthermore, as evoked previously, the number of factors capable of affecting user choices and the large number of users involved speak in favor of considering *random* models. We therefore present here an analysis where user behavior over time (more precisely, the choice of the operator to which he or she subscribes) is modeled as a random process. More precisely, and to keep things simple, we consider a *finite-state continuous-time Markov chain* [234], as is done in [181]:

- the situation of the user is represented through a *state* (state i if the user is with provider i, state 0 if the user forgoes the service);
- after some exponentially distributed random time depending only on his or her current state i and the destination state j (through the parameter, or transition rate, λ_{ij}), the user switches providers (or goes to having none) and finds himself or herself in state j.

Let us consider a specific case, where two providers compete by playing on price only, price being indeed one of the most relevant determinants of churn [46]. Then the churning rates are assumed to depend on the price vector $\boldsymbol{p} = (p_1, p_2)$, with p_i the price set by provider i. We therefore model the churning behavior of a user with the Markov chain displayed in Figure 5.6. Using standard Markov-chain analysis, we can easily prove the existence and uniqueness of a *steady-state probability distribution*, i.e., a vector $\boldsymbol{\pi} = (\pi_i)_{i=0,\dots,2}$ of probabilities such that the proportion of time the user spends in state $i = 0, 1, 2$ tends to π_i as time passes. A sufficient condition is that $\lambda_{ij}(p_1, p_2) > 0$ for all i, j. Interestingly, when all users are assumed to follow independently the behavior described in Figure 5.6, π_i also represents the stationary proportion of the population that is in state i (the market share of provider i, for $i = 1, 2$). The steady-state probabilities are simply obtained by solving the system

$$\forall i, \quad \pi_i \sum_{j \neq i} \lambda_{ij} = \sum_{j \neq i} \pi_j \lambda_{ji},$$
$$\sum_i \pi_i = 1.$$

Those steady-state probabilities then directly give the average per-user revenue that a provider $i = 1, 2$ can expect: that revenue simply equals $R_i := p_i \pi_i$. Since the probability vector $\boldsymbol{\pi}$ depends on the price vector $\boldsymbol{p} = (p_1, p_2)$, the revenue (utility) of each provider depends on both prices; hence there is a non-cooperative game on prices.

Since providers are assumed to be able to anticipate the user behavior resulting from a given price vector, the game can be thought of as a Stackelberg game (see Chapter 2), with providers as leaders and the user as the follower (who adapts to the leaders' actions through his or her transition rates, the outcome being the steady-state probabilities). As in previous competition models, the trade-off that providers face is between the revenue per customer and the "number" of customers (more precisely, the probability of being selected by the customer, which decreases with the price). Maximizing one's revenue in terms of one's price for any value of the competitor's price yields the best-response correspondence, and a Nash equilibrium is an intersection point of the best-response curves, as developed in Chapter 2.

For this model, we cannot prove in general the existence or uniqueness of a Nash equilibrium. When transition rates are simple functions of prices, we may find the form of the Nash equilibria analytically (see [181]); but we choose rather to perform a numerical study here, with some realistic transition rates.

Estimating transition rates

Let us briefly have a look at the literature on churn models. In contrast to our approach, most existing work considers *discrete* time, i.e., the authors estimate transition probabilities (within a period, say a year) instead of rates. Therefore, some conversion to continuous time must be applied. The most widespread model to represent churn probabilities is the so-called logit model [155, 252], which involves a linear function of a number of churn determinants. The probability that a user churns during the next period is taken to be of the form $\mathbb{P}_{\text{churn}} = 1/(1 + \exp(-\sum_{i=1}^{n} \beta_i X_i))$, where X_i, $i = 1, \dots, n$ are the explanatory variables (churn determinants) and β_i, $i = 1, \dots, n$, are weights representing the relative importance of those determinants.

Since we focus on price in our model, the effect of the other churn determinants can be grouped into some term γ, yielding $\mathbb{P}_{\text{churn}} = 1/(1 + \gamma \exp(-\beta p_1/p_2))$ for the probability that the user switches from provider 1 to provider 2 within a particular period. We remark that we take the *ratio* of (losing to winning provider) prices as the churn determinant, this being consistent with the model studied previously, where we claimed that users are sensitive to *relative* (rather than absolute) variations of quantities: see, for example, how (5.12) can be rewritten in terms of price ratios.

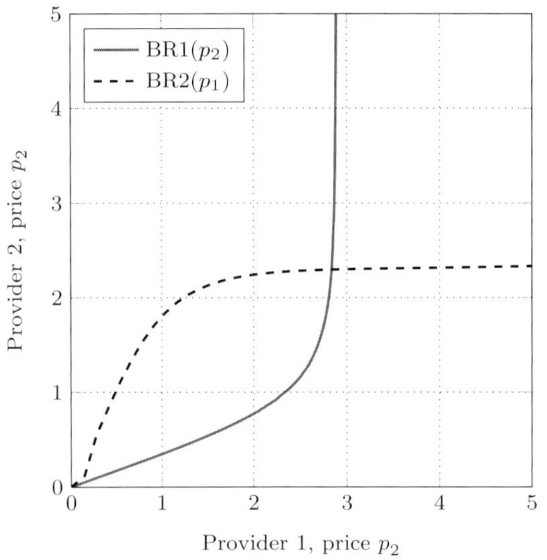

Figure 5.7 The best-response prices of both providers.

For our continuous-time transition rates, we slightly modify the discrete-time expressions to the form

$$\lambda_{ij}(p_i, p_j) = \frac{1}{\gamma_i e^{-\beta p_i/p_j}} = \frac{1}{\gamma_i} e^{\beta p_i/p_j}, \tag{5.16}$$

keeping similar main interpretations and effects: the influence of the ratio of prices, price sensitivity β, overall likeliness γ_i of staying in state i regardless of prices (QoS, reputation, ...). Choosing provider-specific values of γ_i allows one to introduce asymmetry among providers.

As regards the outside option (the choice of having no provider), represented by state 0 in our model, we assume (as in a previous model in this chapter) that being in state 0 corresponds to a *cost* p_0, reflecting the inconvenience arising from the absence of service. Consequently, all transition rates from state $i \in \{0, 1, 2\}$ to state $j \in \{0, 1, 2\} \setminus \{i\}$ are supposed to be given by (5.16).

Let us now analyze the model for those transition rates, with the following parameter values [181]: $p_0 = 1$, $\beta = 0.5$, $\gamma_1 = 2$, $\gamma_2 = 1$, and $\gamma_0 = 1$. Note that the heterogeneity is in favor of provider 1: independently of prices, users tend to stay longer with that provider. The best-response curves of both providers are plotted in Figure 5.7, to highlight the Nash equilibria of the game. Two equilibria are found, namely (0, 0) and $\boldsymbol{p}^* \approx (2.84, 2.29)$. As before, the case of null prices is not satisfying, since it brings no revenue to the providers, and, moreover, it is not stable: if one provider deviates slightly from that situation by setting a strictly positive price, then successive best replies lead to the other (stable) Nash equilibrium $\boldsymbol{p}^*$. We will therefore focus on that equilibrium in the following, when it exists. We also notice that $\boldsymbol{p}^* = (p_1^*, p_2^*)$ reflects the asymmetry due to $\gamma_1 > \gamma_2$: provider 1 takes advantage of having a better reputation (or the user

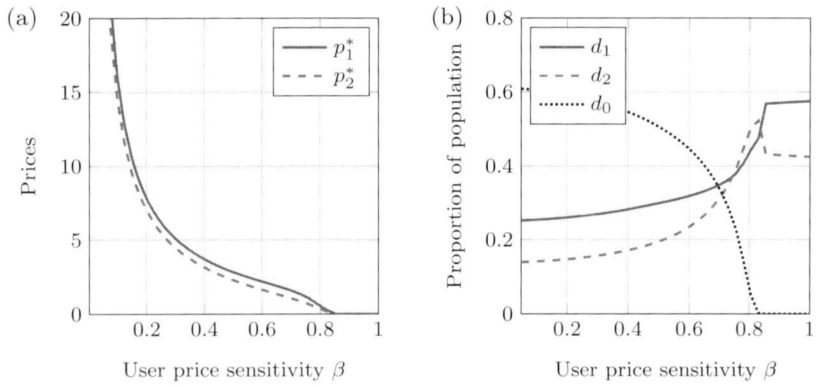

Figure 5.8 The influence of the price sensitivity factor β [181]. (a) Nash-equilibrium prices. (b) The steady-state distribution of users.

having a larger reluctance to change provider) to set a price higher than that charged by the competitor.

Recall that p_0 corresponds to the cost perceived by the user for not having any service. In rapidly evolving sectors such as telecommunications, that cost is very likely to evolve (e.g., Internet access and cellular telephony are now almost mandatory for many users: hence there are very high no-service costs, whereas this was not the case in the infancy of those technologies). We can investigate the effect of that cost p_0 on the outcomes of the game, assuming that the variations of p_0 take place on a larger time scale than the user behavior, so the Nash equilibrium of the pricing game is computed as explained before. As could be expected from the form of the transition rates (i.e., in terms of price *ratios*), numerical computations show that the Nash-equilibrium prices (p_1^*, p_2^*) are proportional to p_0, so that if the parameters other than p_0 remain the same we have $(p_1^*, p_2^*) = p_0 \cdot (2.84, 2.29)$. Consequently, if providers are aware of an increase in p_0 (an increase in value of the service), they should raise their prices correspondingly. Another consequence is that the resulting proportions of users in each state (the market share of each provider) remain unchanged if providers follow that policy. If they do not, then their market share naturally increases.

Now let us have a look at the impact of the parameter β (users' sensitivities to price differences). We expect providers to decrease their prices when β increases. Nash-equilibrium prices (p_1^*, p_2^*) and the resulting distribution of users among the three states when β varies are displayed in Figure 5.8. We observe (similarly to the attraction model summarized by relation (5.12)) that, above a given threshold (around 0.85 here), user price sensitivity is such that providers engage in a price war and prices tend to 0. Below that threshold, prices decrease when the sensitivity β increases, as expected, which results in more users selecting one of the providers (see Figure 5.8(b)).

Another aspect of the model that is worth considering is the asymmetry among providers, which is reflected by the likeliness γ_i to remain with provider i. In Figure 5.9 we keep $\gamma_2 = 1$, and vary γ_1 from 1 (symmetric providers) to 5 (a strong advantage for provider 1). The Nash-equilibrium p_1^* increases as expected, with provider 1 taking

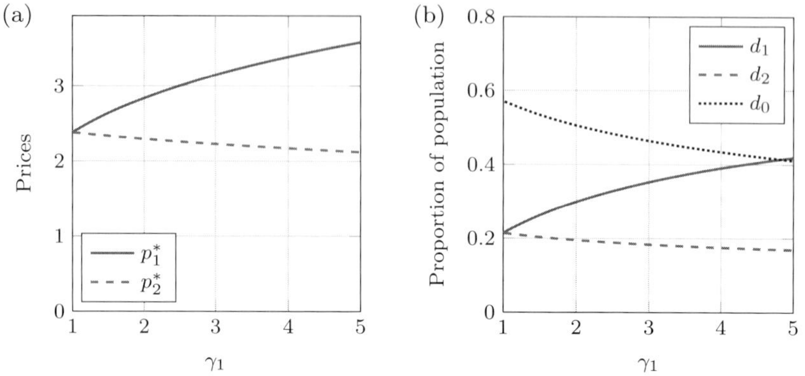

Figure 5.9 The influence of asymmetry among providers. (a) Nash-equilibrium prices. (b) The steady-state distribution of users.

advantage of a higher reputation to set larger prices. The effect on price p_2^* and on market shares was less obvious; for the parameters chosen we observe that the better reputation of provider 1 induces a price reduction by the competitor, but also a reduction in the competitor's market share.

Building more developed dynamic models

As we claimed before, one of the main challenges faced by mobile operators is retaining customers. This may imply the use of unfair practices such as delaying the migration process [196, 229], which can be encompassed in a dynamic model: one can add a state i', representing the fact that the user has expressed the will to leave provider i but is being delayed. From that state, three events can occur:

(i) the provider finally releases the client (after some time depending on the retention policy of the provider);
(ii) the regulator intervenes (possibly when alerted by the client) to release the client, and imposes a sanction fee on the provider;
(iii) the client decides to give up on the idea of leaving the provider.

The first two alternatives lead to the user leaving the provider, but in the second case an additional fee is paid by the provider. Also, notice that, while the client is in state i', he or she continues to pay the subscription price to provider i. From the point of view of provider i, the trade-off in the retention policy is therefore between the gain in terms of subscription revenues due to retention (from users staying in state i' for a while, and, in some cases, deciding to stay with provider i), and the sanction fees imposed by the regulator. If the sanction fee is not too large and users are quite patient, retention policies can be beneficial to providers, which can explain the increasing number of sanctions imposed by regulators to stimulate competition in the mobile market in the 2000s.

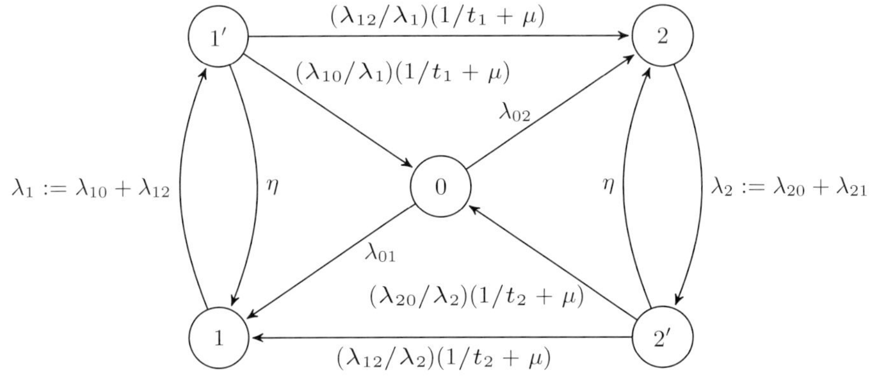

Figure 5.10 A dynamic model for user behavior with retention by providers.

To keep the model simple, one can assume that:

- users want to change providers (from i to j, with possibly i or j equal to 0) after some random time, exponentially distributed with parameter λ_{ij};
- the retention duration by provider $i = 1, 2$ is exponentially distributed with mean t_i;
- users alert the regulator after some time modeled as an exponentially distributed random variable with parameter μ; or
- refrain from changing providers after some time following an exponential distribution with parameter η.

We remark that the assumptions of exponentially distributed times can be relaxed, at the cost of an increase in the state space. For example, retention times are often constant: this can be quite well approximated by a hypoexponential distribution of $m = 3$ or 4 stages (i.e., the retention time of provider i being then modeled as the sum of m exponentially distributed variables with individual expected value t_i/m).

When all durations are modeled by exponentially distributed random variables, the user behavior can be modeled by the Markov chain drawn in Figure 5.10. It is noticeable that the state $1'$ does not contain all the information about the user: the model "loses" the state (0 or 2) intended by the user when he or she took the decision to leave provider 1. However, this is not a problem for the study, since that information is irrelevant for the metrics we compute: we could similarly consider that users decide to leave provider 1 after some time (exponentially distributed with rate λ_1), and choose whether to go to provider 2 or to the no-provider state only when released by provider 1. The same reasoning holds for provider 2 and state $2'$.

To analyze such a model, we first notice that the Markov chain has a unique steady-state distribution, and denote by π_s the steady-state probability of state s. Then it is easy to express the (per user and per time unit) revenue R_1 of provider 1 in terms of those probabilities: that revenue equals

$$R_1 = \underbrace{p_1(\pi_1+\pi_2)}_{\text{income}} - s_0 \underbrace{\pi_2\mu}_{\text{sanction frequency}} ,$$

where p_1 is in monetary units per time unit and s_0 (the sanction price) is in monetary units. Considering the counterpart expression for provider 2, one can study the non-cooperative game played on prices and retention times, after selecting the form of the transition rates in terms of those strategic variables.

5.3.3 Providers competing in multiple-time-scale decision games

In practice, operators have a lot of decisions to make, involving different time horizons. Let us consider a wireless access provider for example: the basic resources needed to provide some service are spectrum (that in most cases has to be bought through auctions, as seen previously in this chapter), access points, and a backbone network.

Wireless licenses are typically sold for a decade-long period; similarly setting access points involves some long-term estate-leasing agreements and some significant investments in equipment (we also include backhaul links – to ensure the connections between access points and the backbone – in this time category). As regards backbone connectivity, the decision can be seen either as a long-term one (when the operator builds or upgrades its own backbone network) or a shorter-term one (when the operator outsources that part to another company).

Decisions on applicable policies are more flexible: operators can change – within the regulatory constraints – their prices, and/or the way they deal with QoS provisioning (e.g., by setting priority classes) on shorter time scales.

Finally, users base their decisions (to subscribe or not, and to which provider) on all those aspects: pricing policy, services provided (and their QoS), and areas of coverage. As we saw before, the market of mobile users is very volatile, and regulation also exacerbates competition by favoring churn among providers. This therefore shall be the smallest-time-scale decision level.

Given that description, how should a provider bid in the spectrum auction, i.e., what should the valuation for the different license bundles be? Also, what strategy should be set in terms of access points or backbone transmissions? And finally, what QoS and pricing policies should one implement?

All those decisions also have to take into account the *competitive* context, i.e., providers are playing a non-cooperative multi-level game. Of course, there is a lot of uncertainty when making long-term decisions: about the acceptance of new services in the next ten years, or even about the nature of the services that might appear . . . however, those decisions can be based on the backward-induction reasoning highlighted in Chapter 2, where at each level the utility functions considered are given by the (Bayesian) equilibrium – if any – of the level below. The two last time scales of the game (price/QoS strategies and the association game for users) are the ones we considered in the previous subsection, but now the outcomes from those interactions are assumed to be anticipated by providers making their higher-level investment decisions.

We summarize here a quite simple model developed in [195], adding a third level on infrastructure and license investments to the price-game level and the user-decisions level. The model relies on the Wardrop-equilibrium notion for the user level, and then on the Nash-equilibrium notion for the pricing level (intermediate time scale) and for the investment level (largest time scale). The initial question addressed from the point of view of the providers takes the form "which technologies are worth investing in?", but answering it implies answering the following series of questions, descending lower and lower in the game levels.

- What technologies will my competitors propose?
- Given the set of proposed technologies, what prices will be set by each provider?
- For a given price vector and set of technologies offered, what are the user distribution and the corresponding revenues for each provider?

The capacity aspects are not treated here, since it is assumed that the transmission capacity is fixed for each technology. However, it is worth mentioning that, in other works, capacity investments and pricing decisions are taken simultaneously, still considering some Wardrop equilibrium for the subsequent choices of users [144].

Let us denote by $\mathcal{T}$ the set of technologies that are likely to be proposed to users (3G, 4G, WiFi, ADSL, ...), and by $\mathcal{N}$ the set of providers. The largest-time-scale decision that each provider $i \in \mathcal{N}$ takes therefore consists in choosing a subset $\mathcal{T}_i \subset \mathcal{T}$ of technologies to operate, by balancing costs and expected revenues.

In terms of prices at the intermediate level/game, we assume a technology-transparent scheme; i.e., each provider i will offer a given price p_i (say, per month) to grant access to users on any of the available technologies in $\mathcal{T}_i$. This type of offer is becoming ever more common, and tends to be a simplification from the users' point of view, in the context of a multiplication of mobile devices and access technologies: users do not need (and in most cases, do not want) to know which specific technology is used for what purpose and with what device; they just want their services to be available. The choice of a specific technology may be made by the device (remaining transparent to the user), or by the user; this does not affect our model as soon as the device tries to select the best option. Owing to congestion effects, not all users will select the same provider or technology: two subscribers of a provider will pay the same price but may be directed to two different technologies (WiFi and LTE for example, because one becomes more congested otherwise), in order to lead to a better price/QoS tradeoff for all of them. This is reflected by the Wardrop-equilibrium conditions, implying that all chosen (operator, technology) pairs will have the same total cost, also below the total cost of the other options. That total cost, which we also call the *perceived price* and denote by $\bar{p}_{i,t}$, is modeled as the sum of the subscription price p_i and a monetary-equivalent QoS-based cost $\ell_{i,t}$ for each operator i and technology $t \in \mathcal{T}_i$ depending on the load on that technology. To be more specific, we distinguish between technologies with *licensed spectrum* (such as 3G) and technologies with *shared spectrum* (such as WiFi): in the former case, congestion arises solely from the demand level $d_{i,t}$ on the corresponding (operator, technology) pair (i, t), while in the latter case interference comes from all users on the same spectrum (technology t); hence the congestion depends on $\sum_{j \in \mathcal{N}: t \in \mathcal{T}_j} d_{j,t}$:

$$\forall i \in \mathcal{N}, t \in \mathcal{T}_i,$$

$$\bar{p}_{i,t} = \begin{cases} p_i + \ell_{i,t}(d_{i,t}) & \text{if } t \text{ is licensed} \\ p_i + \ell_t(\sum_{j \in \mathcal{N}: t \in \mathcal{T}_j} d_{j,t}) & \text{if } t \text{ is a shared-spectrum technology,} \end{cases} \quad (5.17)$$

Table 5.4 The net benefits matrix when provider 1 is positioned on WiFi only, while provider 2 is already positioned on WiFi and 3G (and hence has smaller 3G license and infrastructure costs) [195]. Values are in hundreds of euros per month and per cell, the symbol "–" indicates that the pricing game has no Nash equilibrium with strictly positive prices.

	Provider 2							
Provider 1	∅	3G	4G	3G, 4G	WiFi	WiFi, 3G	WiFi, 4G	WiFi, 3G, 4G
∅	0;0	0;19	0;26	0;38	0;22	0;37	0;41	0;**48**
3G	14;0	12;17	11;22	8;31	12;20	9;32	8;35	6;**41**
4G	26;0	22;15	20;20	17;29	22;19	19;30	17;33	**14**;**37**
3G, 4G	32;0	27;13	24;17	18;23	**27**;17	**21**;25	**18**;27	12;**29**
WiFi	22;0	20;17	19;22	17;**32**	0;0	–	–	–
WiFi, 3G	32;0	27;14	25;19	20;**26**	–	–	–	–
WiFi, 4G	41;0	35;13	**33**;17	**27**;**23**	–	–	–	–
WiFi, 3G, 4G	**43**;0	**36**;11	32;14	24;**17**	–	–	–	

where $\ell_{i,t}$ and ℓ_t are continuous and strictly increasing functions reflecting the congestion cost of the technology t.

Finally, all users prefer the cheapest option, whose total price will be denoted by $\bar{p} := \min_{i\in\mathcal{N},t\in\mathcal{T}_i} \bar{p}_{i,t}$, but the aggregated demand level is naturally assumed to decrease with $\bar{p}$, through some continuous demand function.

Using very general results on non-atomic routing games [1], we can establish the existence of a Wardrop equilibrium for the lowest game level (users choosing a provider and a technology), and the uniqueness of the perceived price for each option (i, t). The uniqueness of the Wardrop equilibrium (i.e., of all values $(d_{i,t})_{i\in\mathcal{N},t\in\mathcal{T}_i}$) is guaranteed only if we fix a rule regarding demand distribution on each shared-spectrum technology among several providers with the same price; for example, an even distribution can be assumed.

The study of the pricing game among operators at the intermediate level (using the above Wardrop equilibrium) is more involved, and in the general case the existence of a Nash equilibrium – or its uniqueness – cannot be proved. However the numerical study carried out in [195] (with two competing providers, linearly decreasing demand, and QoS-related costs given by the delay in an M/M/1 queue) exhibited unique Nash equilibria with positive prices for the pricing game with most technology choices $(\mathcal{T}_1, \mathcal{T}_2) \in \mathcal{T} \times \mathcal{T}$, actually for all cases where the shared-spectrum technology is proposed by only one provider. In the other cases, a price war occurs to attract all users of the shared-spectrum technology, leading prices to zero.

Plugging those equilibrium points into a matrix summarizing the total net benefit of each provider (including subscription revenues and license and infrastructure costs) for each combination $(\mathcal{T}_1, \mathcal{T}_2)$ of technologies then provides us with a two-player game in normal form as defined in Chapter 2 for the choice of technologies on the largest time scale: we can indeed obtain results as indicated in Table 5.4, where the terms in the payoff matrix are of the form $U_1; U_2$, the value being in bold when the corresponding strategy is among the provider's best responses to the competitor's choice. The data used for this particular scenario are aimed at reflecting the situation in a country like France (see [195] for more detailed justifications of the particular values considered).

- The country considered is made of about 10,000 cells (a cell being understood at the 3G–4G level: to provide WiFi coverage more access points are needed due to the limited coverage radius).
- Both providers are already positioned on WiFi, but in addition provider 2 operates 3G while provider 1 does not.
- WiFi coverage is assumed to be at no cost for providers, while 3G incurs license costs estimated at €541 per month and per cell – considering a total license price of €649M valid for ten years. Provider 2 has no 3G infrastructure costs since it already owns one; the infrastructure costs for provider 1 are estimated at €1,167 per cell and per month (infrastructure costs of 1.4B€, for ten years).
- For 4G, license and infrastructure costs are estimated at €649M and €340M, respectively, making a total monthly cost per cell of €824.
- The aggregated demand per cell is assumed to be linear, of the form $D(\bar{p}) = [300 - 3\bar{p}]^+$, where $\bar{p}$ is the cheapest perceived price (given by (5.17)) among all available choices of provider and technology.
- All congestion-sensitive cost functions are taken to be of the form $\ell_t(x) = 1/(c_t - x) - 1/c_t$, i.e., the average waiting time in an M/M/1 queue. For 3G and 4G we take $c_t = 28$ Mb/s and $c_t = 40$ Mb/s, respectively, and recall that these technologies are licensed (i.e., if technology t is operated by provider i, congestion comes only from users selecting this provider and this technology; hence the cost depends only on $x = d_{i,t}$). For WiFi, the shared-spectrum technology, we assume $c_t = 25$ Mb/s.

We observe that the game has two Nash equilibria, where 4G is always implemented by both providers, while 3G is offered only by the provider already positioned on that technology, and WiFi is proposed only by one operator. Interestingly, that matrix highlights the effect that some regulatory measures can have on the outcome of the game: appropriately setting the license prices, for example (that act as additive constants for strategies involving each technology), means that the technological investment game can lead to a totally different Nash equilibrium. It is then up to the regulator to decide which outcome should be favored (for example, the one maximizing social welfare), and to set license prices accordingly. In the present case, reducing the 3G license price for provider 1 may lead to both providers offering that technology, and an improved social welfare. Such a reasoning implies considering one additional game level, that of the regulator fixing license prices as the leader in a Stackelberg game.

5.3.4 To license or not to license resources?

The previous example illustrates the importance of regulation (through license prices) on the technologies implemented by operators. But the regulator can take even more drastic decisions, by deciding *whether* to license some part of the spectrum, or to leave it unlicensed and accessible for free by operators. The tremendous success of WiFi, and the economic growth it allowed, speaks in favor of developing license-free spectrum

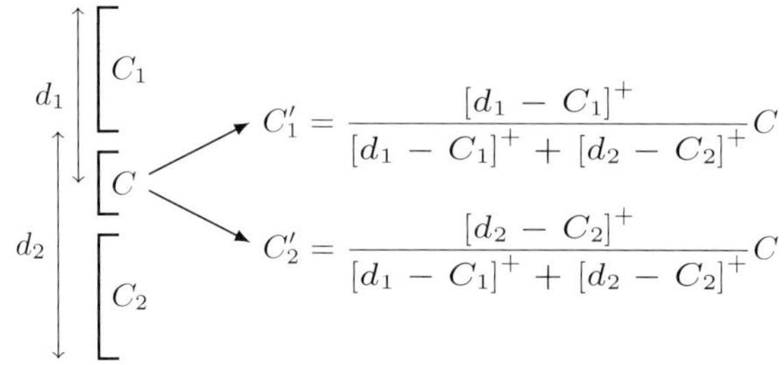

Figure 5.11 Sharing of the unlicensed spectrum capacity C [188]: d_i is the demand of provider i and $x^+ := \max(x, 0)$.

usage: this is among the objectives of the upcoming incentive auction in the USA, as we have already mentioned. Moreover, with licenses some spectrum bands may be underused while others are congested; sharing the capacity for a more efficient use of the scarce resource may be beneficial to society.

Let us consider a slightly different model from the ones of the previous subsection, mixing licensed and unlicensed spectrum usage, which was introduced in [188]. More precisely, consider two providers, each provider $i = 1, 2$ owning some licensed bandwidth to serve up to a demand C_i. In addition, some part C of the spectrum is unlicensed, and can be used by operators when their demand exceeds the capacity of their licensed bands. We assume that this unlicensed spectrum is shared among providers, proportionally to their excess demand as illustrated in Figure 5.11.

Let us take a pricing model we have already used, with users being charged for all sent packets and thus perceiving a congestion-sensitive cost per packet sent successfully (see (5.1)). One can then analyze the price competition among providers: the best-response correspondences for the values $C_1 = 1.2$, $C_2 = 2.4$, and $C = 0.4$, and user demand function $D(p) = [10 - 3p]^+$ were actually plotted in Chapter 2 (Figure 2.3), with prices as actions, and highlight a unique Nash equilibrium with non-zero prices [188]. Also, the impact of the amount of unregulated spectrum can be studied; the main welfare metrics when the proportion of the unlicensed spectrum increases are plotted in Figure 5.12.

It appears that unlicensed spectrum favors users at the expense of providers since the shared spectrum exacerbates price competition (lowering one's price increases one's demand and thus one's share of the unlicensed capacity). The overall effect of unlicensing in terms of social welfare is negative, but remember that the model is quite specific and does not encompass some WiFi-like services such as free Internet access, which are being proposed ever more widely.

5.4 Client but competitor: the (unsustainable?) situation of MVNOs

To stimulate competition among wireless access providers, and possibly foster the creation of new services, governments endeavored in the early 1990s to promote the entry of new operators into the mobile market, even if they did not own all (or any of)

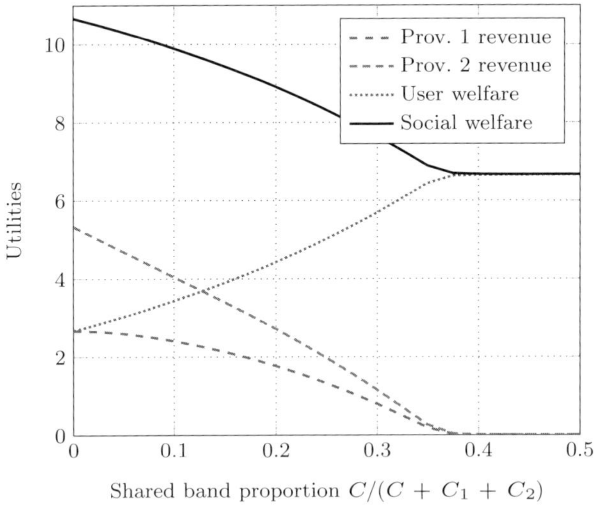

Figure 5.12 Utilities at Nash equilibrium when the share of unlicensed spectrum varies, with the ratio C_2/C_1 being kept constant at 2 (Prov., provider).

the necessary resources (license, access points, backhaul links, backbone network). The way to avoid that barrier to entry is to have those new entrants use some of the resources of the operators already in place – called mobile network operators (MNOs). Hence the name **MVNO**s, for mobile virtual network operators. Of course, MNOs get paid by MVNOs for the resources they use. Among notable MVNO successes, we can mention Virgin Mobile in the UK. As of October 2012, 634 active MVNOs are in operation worldwide.

Wireless spectrum being the bottleneck resource in general, we focus on that specific resource. Spectrum leasing can take three different forms. First, the MNO and MVNO can share the same radio spectrum without any differentiation. Alternatively, MVNO users may have lower-priority access than MNO users. Finally, there can be a strict separation among spectrum bands, with the MNO transferring to the MVNO the rights for a part of the licensed spectrum [86]. For that last option, one can rely on long-term contracts to determine the amount and price of spectrum leased to the MVNO, or one may prefer frequent re-computation [274].

The MNO/MVNO paradigm creates some quite novel situations, with MVNOs being the *client* of their host MNO, but also *competing* with MNOs (and other MVNOs) to attract customers. Therefore, the incentives for MNOs to agree to host one or several MVNOs must be considered with great care, inducing potentially a need for intervention from regulatory bodies. We review in this section some of the approaches to that problem.

5.4.1 Exploiting secondary markets

The possibility for MNOs to lease their extra spectrum and the associated revenue gain are studied in [7, 224], in special cases where the MVNO is not seen as a competitor whose success may harm the MNO. We do not develop those aspects here, but rather focus on competition effects since they jeopardize the viability of MNO–MVNO partnerships.

5.4.2 Can MNO–MVNO associations survive? Observations from different countries

It is interesting to note that the integration of MVNOs, which was initiated in the last decade, has not always been successful: while several successful entries have taken place in the USA and Europe, also failures were observed, notably in Asia. As a successful example, the market share of MVNOs in the French residential market grew from 6% in 2009 to about 13% in 2012 (according to the French regulatory authority ARCEP, http://www.arcep.fr/, January 2013). The main determinant highlighted in [275] is the structure of the market which the MVNO enters: when several stakeholders contribute to the different levels of service provision (operators, application providers, content providers, . . .), then the chance that an MVNO will succeed is increased, with respect to a vertically integrated market where most service levels are controlled by the same entity. Such general observations seem rather intuitive:

- in a market driven by a vertically integrated incumbent, users switching to the MVNO for the access part of the service chain still rely on the incumbent for the other service levels, which reduces the gain from switching (there is no way to totally leave the incumbent);
- the competition on the access level can be biased by the incumbent, which can easily decrease its access prices to beat the MVNO and compensate for revenues lost through the other service levels.

Another point worth noticing here is that government incentives to host MVNOs can be put in place as early as during the license-allocation process. This was the case in France in 2011 for 4G license auctions, where bids from operators committing themselves to hosting MVNOs were multiplied by a factor of up to 5/3 for the bid ranking. Three providers out of the four competing bidders took that option; just one declined [21]. Such strong incentives drastically reduce the MNOs' degrees of freedom with respect to MVNOs, since the commitments involve very specific MVNO hosting conditions (amount of capacity leased, maximum charging price, . . .). They constitute a way to ensure the viability of the MNO–MVNO model; however, we will not consider such constraints in the next subsection, focusing rather on the market mechanisms that can enter into play in less-regulated settings.

5.4.3 Can MNO–MVNO associations survive? Theoretical approaches

A nice economic study on the case when one MNO has the possibility to lease some of its spectrum to an MVNO can be found in [32], the focus being on the conditions under which an MNO has an interest in coexisting with a (competing) MVNO.

More precisely, the model considers an MNO that can produce resources (transmission capacity) with a fixed unit cost, and sell to users for a given unit price, to be fixed so as to maximize revenue. There is also the possibility of selling resources to an MVNO, at a discounted price. The model is quite involved mathematically; we do not detail it here, but rather draw the main conclusions from the analysis in [32], in terms of necessary conditions for the MNO–MVNO relationship to be mutually beneficial.

(i) The MVNO must benefit from a strong "brand effect," i.e., it must be able to attract users even if its price and quality are no better than those of the MNO. This holds when the MVNO is well known in sectors outside telecommunications and can take advantage of its reputation.
(ii) To limit competition with the MNO, the service offered by the MVNO must be sufficiently differentiated from that of the MNO. In other terms, MNO and MVNO services must not be too close substitutes, otherwise the demand for the MNO will suffer too much from the MVNO's entrance.
(iii) Finally, the discount offered to MVNOs must be sufficiently large for the "resource reselling" activity of the MVNO to be profitable. This is likely to be the case when several MNOs compete (on prices) to attract MVNOs.

Those facts are quite intuitive; however, having them as the outcomes of a mathematical model allows one to weigh the trade-offs among the adverse effects, and to check for real cases whether the explicit conditions for MNO–MVNO partnerships to be viable (mainly involving demand-elasticity conditions) are satisfied. Some similar results were obtained in [124, 166, 167], with a quite different model involving Wardrop-equilibrium modeling for the user behavior, instead of aggregated demand functions.

A different model is considered in [73], with the advantage of encompassing two types of MNOs, namely one incumbent and a more recent entrant into the market (called a *follower*). The difference between them is that the incumbent is less likely to suffer from the entry of the MVNO, due to having a larger *brand loyalty* (or faithfulness) from users, meaning that users are more reluctant to leave the incumbent than they would be to leave another MNO. In that context, using a model where market shares of operators depend on their prices, the authors of [73] show that

- the MNO–incumbent partnership can be beneficial for both parties, at the expense of the (non-incumbent) MNO;
- an MVNO is better off associating with an incumbent rather than with another MNO.

This is in accordance with what is observed in practice.

Multi-stage (Stackelberg) game models are considered in [84, 147], where (competing) MNOs are seen as leaders and MVNOs as followers. The study of such games is performed using the backward induction method of Chapter 2, and exhibits interesting results summarized below.

- If the number of MNOs increases, so does the incentive to agree to host an MVNO (since the negative effect of having a new competitor is "diluted" among all MNOs).
- There can be situations where each MNO prefers hosting an MVNO rather than seeing it hosted by a competitor, but where all MNOs would prefer that the MVNO be without a host. In such cases, MNOs could reach an agreement to refuse to contract

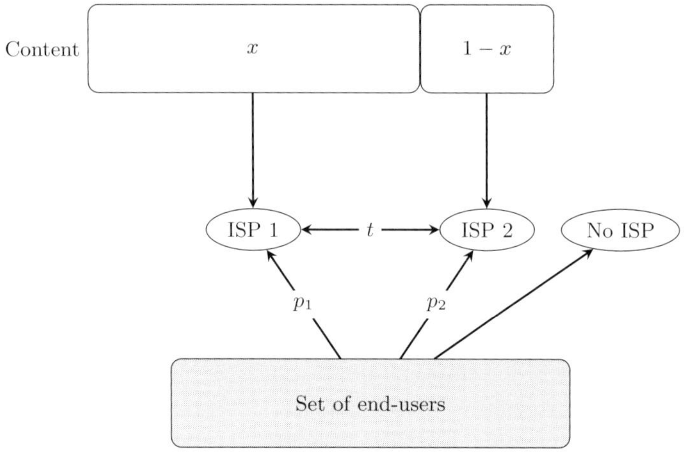

Figure 5.13 A situation with content hosted by two competing ISPs.

with MVNOs (hence the need for a regulatory intervention to enable the entry of MVNOs).

5.5 The economics of interconnection

As its name indicates, the Internet involves a large number of interconnected networks. Those networks are owned by different entities that have to collaborate in order to provide worldwide connectivity to their users.

5.5.1 An example

Customers may subscribe to an ISP different from the one providing access to the content provider they are interested in, as illustrated in Figure 5.13, featuring two ISPs that compete to attract end-users and also host different portions of the overall content. Users pay a flat-rate fee p_i to get access from provider i, and t represents a unit price that ISPs can charge each other for transferring traffic. In the model, a proportion x of the overall content is hosted by ISP 1, while ISP 2 hosts the remaining part $1 - x$. Figure 5.13 is, of course, a considerable simplification of the actual relations among actors: the current Internet counts much more than two ISPs, there are some providers contracting only with content providers and not with end-users, there might need to be some transit providers (see Figure 1.6) between the ISPs' networks – which get paid only for transferring data, and also content delivery networks (CDNs) can play an important role in the distribution of content to end-users. However, we believe this simple model captures some of the important features of inter-ISP relationships. In particular, if end-users subscribing to ISP 1 want to access content hosted by ISP 2, the download traffic has to be transferred by ISP 2 to ISP 1. There are actually mixed interests with regard to the treatment of that traffic: for ISP 2 it represents some extra load due to ISP 1's subscribers, hence ISP 2 can expect to be paid by ISP 1 for it. On the other hand, forwarding that traffic to all interested users can be part of the contract that ISP 2 has with its content provider clients. We will nevertheless favor the former interpretation, i.e., with ISPs selling to their end-users access to data (all of it, in practice), and thus being responsible for rewarding other actors who relay their subscribers' traffic. This is reasonable when considering free content, but the problem can be seen the other way around when paid content is involved: in that case some subscribers of ISP 1 accessing some content hosted by ISP 2 would pay the content provider, who would use some of those revenues to pay ISP 2 for providing access to all users. In that context, ISP 1 may argue that it should obtain a share of those revenues, due to its contribution.

Let us therefore consider here free content, and the responsibility for granting access being held by the end-users' ISP. Note that the model focuses on relations with end-users; the revenues from content providers and the possible competition to attract them are neglected here. This could be studied as a higher-level game, by considering that content providers select their ISP on a larger time scale; or alternatively, one could consider ISP–CP interactions on the same time scale as ISP–user ones, as is done in [162], i.e., consider the whole two-sided market where ISPs compete to attract both content providers and users. This will be discussed in Chapter 7. We assume here instead that ISP–CP settlements are given and fixed.

In practice, ISPs may have **peering** relationships: they forward all content without any charge (hence $t = 0$ in Figure 5.13). However, such arrangements hold within certain limits, in particular in terms of relative amounts of traffic to transfer in each direction. Indeed, if one ISP serves traffic for a competitor much more than it uses the reciprocal service, the peering agreement can be questioned, but this may induce complaints about the sustainability of the full connectivity principle of the Internet, such as in the argument of Cogent against France Telecom.[1] In the model here we consider three possible alternatives.

- Free peering: all traffic is forwarded for free ($t = 0$).
- No transfer: this is the extreme scenario when ISPs finally refuse to transfer traffic from their content providers to the competitor's network. As a result we have two separated networks, clients of each ISP having access only to the content hosted by that ISP. Such a case is unlikely to occur in practice, but can be a threat, as highlighted in the above Cogent–France Telecom argument.
- Paid transit: all traffic is forwarded, but each ISP pays a price per volume of traffic coming from the other ISP. We will consider several ways of setting the unit price t (which could be imposed by a regulator, or settled through negotiation).

We consider a multi-level game, where the transit price t (if any) is set first, then a pricing game among ISPs is played, and finally users select their ISP. As was done previously on several occasions, we use backward induction to analyze that situation.

In terms of user behavior, we consider the discrete-choice model developed earlier in this chapter (i.e., the utilities of users include a price-dependent part and a random

[1] See http://ec.europa.eu/competition/ecn/brief/04_2012/fr_cogent.pdf.

part), with the average value v_i among users for selecting option $i \in \{0, 1, 2\}$ (0 being the no-ISP choice) depending only on the price per unit of reachable content. More specifically, we consider that, if a proportion x_i of the available content is reachable by subscribing to ISP $i \in \{1, 2\}$, then the price per unit of content is x_i/p_i, and we choose the average value $v_i = \alpha \log(x_i/p_i)$, with $\alpha > 0$ a sensitivity parameter. Similarly to what we did previously with that model (just above (5.12)), we can also introduce a perceived cost p_0 for not having access to any content, so that the average value of the no-ISP option equals $v_0 = \alpha \log(1/p_0)$. Assuming as before that the random parts of utilities follow a Gumbel distribution, the proportion of users choosing each option is given by the same formula as (5.12):

$$\sigma_i(p_0, p_1, p_2) := \frac{(x_i/p_i)^\alpha}{(x_1/p_1)^\alpha + (x_2/p_2)^\alpha + 1/p_0^\alpha}. \tag{5.18}$$

In this simplified model, we consider utilities of providers to stem only from user subscription revenues, plus monetary exchanges in the case of paid transit. In the latter case, given our assumption that the subscriber's ISP is responsible for guaranteeing access to all content and pays a unit price t to the ISP hosting the content, the overall utilities of the ISPs are

$$U_1 = p_1\sigma_1 + t\underbrace{(x\sigma_2 - (1-x)\sigma_1)}_{:=\Delta_{1,2}}, \tag{5.19}$$

$$U_2 = p_2\sigma_2 - t\underbrace{(x\sigma_2 - (1-x)\sigma_1)}_{:=\Delta_{1,2}}, \tag{5.20}$$

where $\Delta_{1,2}$ is the differential amount of traffic transferred from ISP 1 to ISP 2. Indeed, ISP 1 relays $x\sigma_2$ units of traffic for ISP 2's customers, and ISP 2 relays $(1-x)\sigma_1$ units of traffic for ISP 1's customers, assuming that the interests of users for content are independent of their ISP, and that the expected amount of traffic for each user is normalized to 1 (with x hosted by ISP 1 and $(1-x)$ by ISP 2).

When utilities come only from subscriptions (i.e., in the free-peering and no-transfer cases), the price competition can be studied exactly as on pages 182–185, just with a different interpretation of the parameters x_i (reputations in (5.12), and proportions of reachable content in (5.18)). In the peering, all content is reachable from any ISP, thus $x_1 = x_2 = 1$, whereas $x_1 = x$ and $x_2 = 1 - x$ when ISPs refuse to forward traffic. The results obtained previously for the price competition can therefore be directly used here.

Finally, the case of paid transit is more difficult to handle analytically. By analyzing the properties of the best-response functions $\mathrm{BR}_i^t(\cdot)$ (the best-response functions of ISP $i = 1, 2$ when the transit price is t, which are single-valued, continuous, lower-bounded by some $m_i > 0$ and upper-bounded by some $M_i < \infty$), we can show that the application

$$\begin{aligned} g : [m_1, M_1] \times [m_2, M_2] &\mapsto [m_1, M_1] \times [m_2, M_2] \\ (p_1, p_2) &\to (\mathrm{BR}_1^t(p_2), \mathrm{BR}_2^t(p_1)) \end{aligned}$$

is continuous over the compact convex set $[m_1, M_1] \times [m_2, M_2]$. Applying the Brouwer fixed-point theorem, g has a fixed point that constitutes a Nash equilibrium with strictly

Table 5.5 Outcomes to expect from the price competition game (stable equilibria)

	$\alpha \leq 1$	$1 < \alpha \leq 2$	$\alpha > 2$
Free peering	No equilibrium (prices tend to infinity)	Nash equilibrium $p_1 = p_2 = p_0 \left(\frac{2-\alpha}{\alpha-1}\right)^{1/\alpha}$ $U_1 = U_2 := U^{\text{peering}} = p_0(2-\alpha)^{1/\alpha}(\alpha-1)^{1-1/\alpha}\alpha$	Nash equilibrium $p_1 = p_2 = 0$ (price war) $U_1 = U_2 = 0$
No transfer	No equilibrium (prices tend to infinity)	Nash equilibrium $p_1 = x p_0 \left(\frac{2-\alpha}{\alpha-1}\right)^{1/\alpha}$ $p_2 = (1-x) p_0 \left(\frac{2-\alpha}{\alpha-1}\right)^{1/\alpha}$ $U_1 = x U^{\text{peering}}$ $U_2 = (1-x) U^{\text{peering}}$	Nash equilibrium $p_1 = p_2 = 0$ (price war) $U_1 = U_2 = 0$
Paid transit, $t > 0$	No equilibrium (prices tend to infinity)	Existence of an equilibrium, strictly positive prices $U_1 > 0, U_2 > 0$	Existence of an equilibrium, strictly positive prices $U_1 > 0, U_2 > 0$

positive prices in $[m_1, M_1] \times [m_2, M_2]$. Moreover, we can establish that each ISP has a strictly positive revenue at that equilibrium. We refer the interested reader to [66] for a detailed proof.

The results of the price competition in the three cases are summarized in Table 5.5. A fact worth mentioning is that the peering situation Pareto-dominates the "no-transfer" situations: all stakeholders are worse off in the latter case, with the ISPs seeing their revenue multiplied by their share of hosted content and the users perceiving the same overall value $v_1 = v_2 = \alpha \log(x_i/p_i) = \log((\alpha - 1)/(2 - \alpha))$. From this model, a disconnection is actually not preferred by both ISPs, whatever the proportion of content they host.

Interestingly also, we observe that paid transit prevents prices from dropping to 0 when users are very price-sensitive ($\alpha > 2$). Indeed, engaging in a price war to attract most users comes with the extra cost of paying the transit fees to the competitor for those users' content that the competitor hosts. Therefore, there are situations such that the paid transit relation is preferred by both ISPs, which seems a counter-intuitive idea at first sight.

Setting the transit fee

Let us focus on the paid transit case, and investigate how the transit price t should be fixed. A first idea can be drawn from Figure 5.14, where the utilities of ISPs and the users' welfare are plotted. The latter is defined as the cumulated (i.e., averaged, since the total mass of users is assumed to be 1) gain for users with respect to the no-ISP

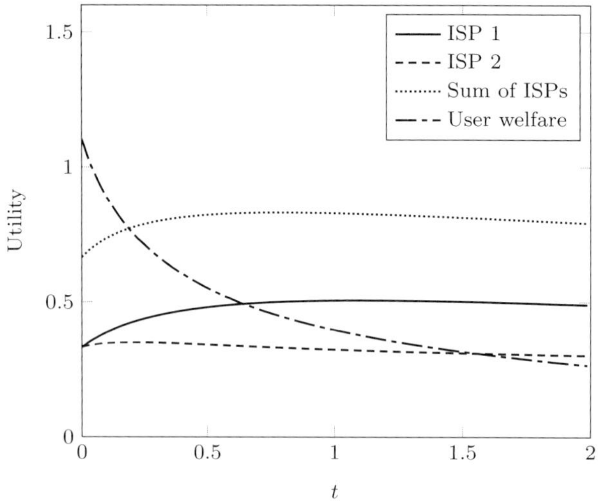

Figure 5.14 Utilities of the stakeholders at Nash-equilibrium prices, as a function of t, with $x = 0.8$ and $\alpha = 1.5$.

choice, and is given by

$$\begin{aligned}\mathrm{UW} &= \mathbb{E}\left[\max(0, v_1 - v_0, v_2 - v_0)\right] \\ &= \mathbb{E}\left[\underbrace{\max(0, \alpha\log(x_1p_0/p_1) + \kappa_1 - \kappa_0, \alpha\log(x_2p_0/p_2) + \kappa_2 - \kappa_0)}_{:=Z}\right].\end{aligned}$$

But for $z > 0$ we have

$$\begin{aligned}\mathbb{P}(Z \le z) = \mathbb{P}\Big(&\{\alpha\log(p_0) + \kappa_0 > \alpha\log(x_1/p_1) - z + \kappa_1\} \\ &\cap \{\alpha\log(p_0) + \kappa_0 > \alpha\log(x_2/p_2) - z + \kappa_2\}\Big),\end{aligned}$$

i.e., $\mathbb{P}(Z \le z)$ is the probability that a user selects no ISP if each ISP $i = 1, 2$ sets a subscription price $p_i e^{z/\alpha}$. Thus

$$\mathbb{P}(Z \le z) = \frac{(1/p_0)^\alpha}{e^{-z}((x_1/p_1)^\alpha + (x_2/p_2)^\alpha) + 1/p_0^\alpha}$$

from (5.12). Then we obtain

$$\mathrm{UW} = \mathbb{E}[Z] = \int_{z=0}^{+\infty} \mathbb{P}[Z > z]\mathrm{d}z = \int_{z=0}^{+\infty} \frac{ke^{-z}}{ke^{-z} + 1}\,\mathrm{d}z,$$

with $k := (x_1p_0/p_1)^\alpha + (x_2p_0/p_2)^\alpha$. This yields

$$\mathrm{UW} = \log(k + 1) = \log\left(1 + \left(\frac{x_1p_0}{p_1}\right)^\alpha + \left(\frac{x_2p_0}{p_2}\right)^\alpha\right). \tag{5.21}$$

Figure 5.14 confirms that paid transit ($t > 0$) can be preferred over free peering ($t = 0$) by providers, and it is here the case even with $\alpha < 2$. On the other hand, if one wants

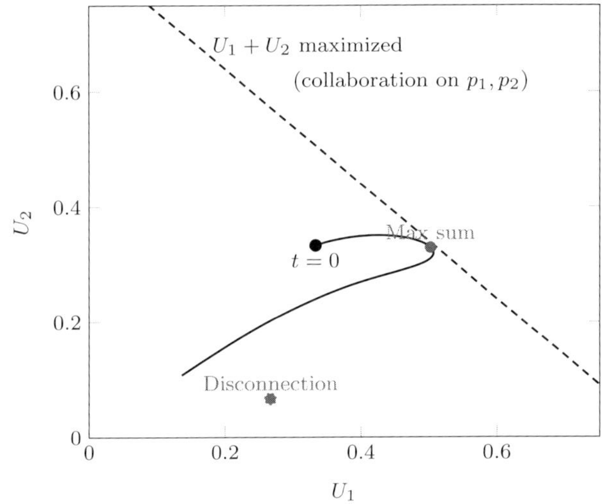

Figure 5.15 ISP utilities at Nash-equilibrium prices, when the transit price t varies (continuous curve). The parameter values are $\alpha = 1.5$, $x = 0.8$, and $p_0 = 1$.

to maximize user welfare then free peering should be enforced; this fact can actually be proved analytically (see [66]).

Another option to choose the transit price can be to assume that the two ISPs cooperate, and thus set the transit price to maximize the sum of their revenues (some sharing of the gains, possibly using the Shapley value, being performed afterwards). Finally, the situation where the two ISPs collaborate completely (i.e., also setting their subscription prices to maximize their aggregated revenue) is of interest: in that case transit prices are irrelevant, and the ISPs get a larger cumulated revenue than in any competitive situation (the analysis can be found in [66]).

The choice of t from the point of view of providers is better highlighted in Figure 5.15, where the revenues of both ISPs are displayed in all three cases: free peering ($t = 0$), paid transit (with a continuum of values for the transit price t), and disconnection. The collaborative case, where providers agree on subscription prices to maximize the sum of their revenues, is also plotted.

We remark that, if ISPs compete on subscription prices but fix the transit price to maximize the sum of their revenue (corresponding to the point "Max sum" in Figure 5.15), their aggregated revenue is almost as high as in the fully cooperative case (collaboration on subscription prices).

Finally, it can be reasonable to consider that ISPs remain competitors to attract users, but can *negotiate* the value of the transit price t; i.e., they can engage in a discussion to reach an agreement about that value. Here again, the negotiation process can be viewed as a game. That type of game has not been developed within the game-theory part in Chapter 2, so the interested reader is referred to [230, 243]. We shall mention only the solution concept generally considered as the expected outcome in that context, which is called the **Nash bargaining solution**. That solution involves a particular outcome, called

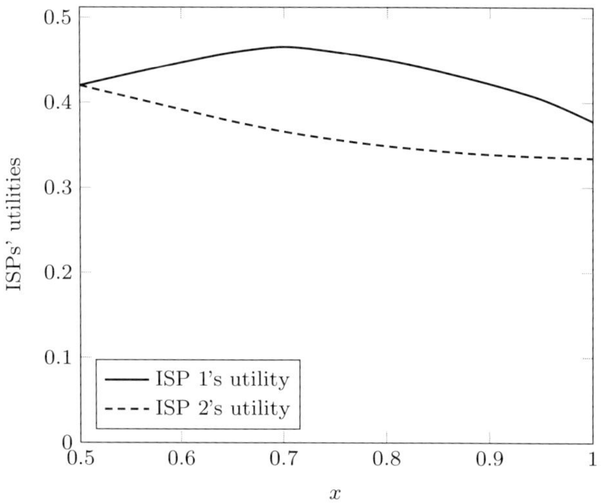

Figure 5.16 A comparison of each ISP's utility after the bargaining process with disconnection and enforced peering threats, when x varies, with $\alpha = 1.5$.

the *threat*, that describes the situation in the case in which no agreement is reached. For a given threat situation, yielding utilities U_1^{threat} and U_2^{threat}, the Nash bargaining solution is defined as a situation maximizing the product

$$(U_1 - U_1^{\text{threat}}) \times (U_2 - U_2^{\text{threat}}).$$

Here, the reasonable threat situation is free peering: we could otherwise imagine that not reaching an agreement would lead to a disconnection between ISPs, but, as pointed out previously, that outcome is Pareto-dominated by the free-peering case, and is therefore an unsatisfactory outcome for both ISPs. As soon as $0 < x < 1$, both ISPs are strictly better off peering for free (see Table 5.5).

Finally, the effect of the amount of content hosted on the provider utilities with a negotiated transit price is shown in Figure 5.16. It is worth remarking that owning more content does not necessarily lead to a larger revenue: here ISP 1 has a maximal revenue when it hosts about 70% of the whole content.

5.5.2 The problem of incentivizing intermediate entities

One of the main problems with interconnected networks lies in the incentives to relay traffic. In the example of the previous subsection, traffic from the content provider to the end-user may have to transit through another network, owned by an entity having no contract with the content provider or the user: why should the owner of that network use his resources for that traffic? We will call networks needing the relaying of intermediate entities **multi-hop networks**; those include inter-domain as well as ad-hoc networks.

One might argue that in the case when the interacting entities are network operators, some reciprocity agreements can be found (leading to traffic peering), or some price for that service can be determined. This is perfectly true, and is applied in practice: many operators peer with each other when directly connected, while there exist other operators only selling transit services for a given unit price. Thus the market seems to find a solution by itself in those cases.

However, if we want to apply incentive schemes on the level of a much finer granularity, several problems occur. For example, we might want to apply auctions to make the most efficient use of the network resources, with those providing transmission capacities declaring how much they would like to be paid, and those needing transmission capacities declaring how much they would be willing to pay. In that case, we are faced with the limits of double-sided auction schemes pointed out in Chapter 2, in the sense that we cannot get simultaneously the four properties below:

- incentive-compatibility (i.e., no stakeholder having any incentive to misreport their valuation),
- efficiency (i.e., social welfare being maximized),
- individual rationality (i.e., all stakeholders being guaranteed to get a non-negative utility by participating), and
- weak budget balance (i.e., the system not needing monetary expenses for it to work).

Recalling the revelation principle (stating that incentive-compatibility comes at no extra cost), this means that we cannot even design a scheme leading to an efficient use of the resource and incentivizing all users to participate without injecting money. Since that option is often off the table, individual rationality and more frequently efficiency has to be given up, but there are many proposals that involve relaxing a property so that it is "almost satisfied" (see [286] for a survey on this).

It therefore appears that the problem of finding incentives to relay traffic for others has simple solutions when taken at a large-granularity level (through peering/transit agreements), but the problem becomes much more involved when one is dealing with it on a per-request basis. Unfortunately, such a granularity has to be considered in networks where agents have only short-duration interactions so that no long-term reciprocity stance can be secured. This is in particular the case with ad-hoc networks, which are very volatile due to user mobility, intermittent availability, and limited battery life.

5.5.3 Some proposals for ad-hoc networks or multi-hop cellular networks

A **multi-hop cellular network** is a cellular network (i.e., one with access points connected to a backbone) in which a user node may need to rely on other user nodes forwarding its traffic to reach an access point (in the uplink; for the downlink the user may also need other nodes to forward the data from the access point). So the problematic issue is very similar to that in ad-hoc networks, where data must be relayed by intermediate nodes until the recipient is reached: only here either the recipient or the sender is an access point. In both cases, mobile nodes are users of the network service, but also act as service providers when relaying traffic for the others.

A nice proposal worth mentioning is introduced in [15], where the scarce resource considered is energy (no bandwidth limitations are considered). What is interesting is that an auction mechanism is defined, so that, for a given transmission request,

- the *route* to be taken by the data (which will be the lowest-cost route),
- the *power* to be used by each node in that route (the minimal power needed to reach the next node in the route), and
- the *reward* (in monetary equivalents) for each intermediate node

are computed at the same time. The route discovery and power discovery are performed through the broadcasting of a "route-request" message that is incremented by each node receiving it. More precisely, on receiving such a message, a node adds its identity, and declares what unit price it asks for its energy expenditure, and with what power it retransmits that route-request message. Also, a last action performed before transmitting the route-request message is to compute the minimum power that could be used *from the previous node*: this is done using the power declared by the previous node and the power with which the message was received. In the retransmitted route-request message, the node replaces the power declared by the previous one with that minimum power. Therefore, the message received by a node contains the necessary information to compute the aggregated cost of a candidate path: the node can just discard messages costing more than previously treated ones. To avoid the multiplication of request messages, one can let each node accumulate all such messages for the same end-to-end request during a time period, and then select the cheapest-path message to enrich and forward. To be a bit more specific, the route request packets have the following form:

$$\text{Broadcasted by node } 0 : \langle \texttt{REQUEST}, s_{0,n}, 0, n, P_0^{\text{emit}}, c_0 \rangle$$
$$\text{Broadcasted by node } i : \langle \texttt{REQUEST}, s_{0,n}, 0, n, P_{0,1}^{\min}, c_0, 1, P_{1,2}^{\min}, c_1, \ldots, i, P_i^{\text{emit}}, c_i \rangle,$$

where 0 (i) is the node originating the request (forwarding it), n is the destination node, and $s_{0,n}$ is a sequence number aimed at distinguishing different requests. Each node appends its declared unit cost for energy c_i and the power P_i^{emit} with which it transmitted the message. Upon receiving those messages, node i computes the power $P_{i-1,i}^{\min}$ needed for the last hop until i and the associated cost of the suggested path $0, 1, \ldots, i$ as $\sum_{k=0}^{i-1} c_k P_{k,k+1}$, to select the cheapest candidate.

Finally, the route and the rewards are computed using VCG rules, ensuring that all node have an interest in being truthful on all fields of the route-request message they send, and do not have any incentive to modify it or discard it. The resulting routes are therefore efficient (i.e., cost-minimizing), the main problem being – as pointed out previously – that the scheme is not budget-balanced, i.e., rewards exceed the values that communicating nodes are willing to pay. The authors of [15] claim that this can be solved by taxing all users uniformly (a criticizable view), and provide upper bounds for the amount of money that has to be injected to have the mechanism work. Finally, another drawback of VCG auctions (vulnerability to collusion) is also pointed out. In [92] a similar approach is proposed, and it is suggested that the cost of the mechanism (due to the rewards exceeding the willingness to pay of the sender) be borne by the receiver, which is also criticizable (especially if the receiver did not request those data).

A fluid model is considered in [75], where intermediate nodes earn amounts of money according to the *power* but also the *transmission capacity* they devote to flows other than theirs. The model describes – through a differential equation – how the credit balance of each node evolves according to its contributions and service usage, stressing that a lower balance limits the rate with which that node can send its own data. The authors prove the global stability of that system (i.e., credit balances do not tend to 0 or to infinity, and traffic relaying occurs), and show how the position of a node affects its transmission possibilities: nodes with a central position can earn a lot from relaying traffic and then use that credit to send data, while nodes on the periphery earn money only through a "restoring force" in the model (redistributing partly the wealth of all nodes) and can transmit only with small throughputs.

A problem with those proposals (and in general with ad-hoc networks) is the management of the prices and rewards: in a context where communicating is difficult and decentralized schemes have to be used, how can we guarantee any exchange of money, or account balance? In many cases one has to make the extra assumption that there is a trusted entity managing the accounts of all nodes in the network, a solution that is not very satisfactory in decentralized networks. The same assumption is made in [311], where it is assumed that all nodes regularly inform that entity (called the Credit Clearance Service) in order to claim their rewards. Only in [311] a different pricing rule is applied, that ensures weak budget balance (the price paid by the sender exceeds the sum of the rewards); to avoid node budgets going down to 0 the excess could be redistributed among all nodes.

An intermediate approach is considered in [48, 49], where no central payment-managing entity is needed, but it is assumed that each node has a tamper-proof security module storing the account balance (in a "currency" specific to the service, called *nuglets*) of the node locally and updating it on the basis of paid transmissions and rewards received. The routing decisions are assumed to be taken beforehand, and the authors consider two possibilities for the charging.

- The *packet purse model*: each packet is sent with a certain budget (of nuglets), and intermediate nodes withdraw a given number of nuglets (typically fixed to 1) when transferring the packet.
- The *packet trade model*: each packet is sold by the sender to the first intermediate node, which then sells it (with a markup) to the next one, until the destination is reached. Then the destination node has to pay the price requested by the last intermediary.

The incentive problems here (about the reward values, the strategies consisting in sending packets several times to get more rewards, the counterfeiting of nuglets...) are all assumed to be solved by the use of the tamper-proof module, which technically prevents nodes from deviating from the protocol: those aspects are thus treated technically instead of through economic tools, reducing the strategy space of players.

Let us finally mention two alternative ways to proceed, not involving any kind of currency exchange. In [98, 99] a non-cooperative game among nodes is considered, where the strategy of each node is its *probability* of forwarding a packet for the others. Then the strategic decision is made depending on the service offered by the other nodes (i.e., their forwarding strategies) and on the cost of relaying traffic. Some strategies like "always cooperate," "always defect," and "tit-for-tat" (respond in proportion to the perceived service) can then be considered. Interestingly, the authors take into account the *dependences* among nodes, which reflect whether the forwarding strategy of a node affects the perceived service of another. This allows one to explain why some nodes can forward no traffic at all, but will not be punished for that since the relays they want to use do not perceive this strategy. A more centralized approach is taken in [205], where a mechanism that is based on reputation is proposed. The mechanism considered relies on the existence of an entity centralizing reports about the good or bad behaviors of all nodes (then also, malicious reports have to be detected), and on the use of those reputations to affect the routing mechanisms so as to avoid bad relays. While such a mechanism offers an improvement in performance when the proportion of non-cooperating nodes is given, it does not provide any actual incentive to relay traffic since non-cooperating nodes are not sanctioned in any way; they are even saved from expending energy by being removed from relay routes.

In summary, there are so far no perfectly satisfying economy-based solutions to incentivize mobile terminals to collaborate within an ad-hoc network. While in the military context such networks can be used without any problem because cooperation can be assumed, the lack of proper incentives is probably the main reason for the poor development of ad-hoc networks in civilian environments.

5.6 The economics of community networks

A **community network** is an access network offering roaming possibilities, which are based on collaboration among its own customers. More precisely, the idea is that (some of the) customers have Internet access – typically through ADSL – and open it – typically through WiFi – to the other customers, thereby offering them Internet access while away from home. The problematic issue here is quite similar to the one in ad-hoc networks: each user would like to benefit from the service (Internet access with roaming), but most are reluctant to share their own Internet access, mostly for security reasons, but also because it may reduce their own throughput. However, here there exist sustainable business models, due to the notion of community but most importantly to the fact that *monitoring* the proper behavior of the members is technically doable, and even quite easy in practice since deciding whether to share your access or not is classically done online (i.e., the community network managing entity is informed directly).

A typical example of a worldwide community network is Fon,[2] which is separated from the choice of an ISP: anyone having Internet access at a given location can enter the Fon community by acquiring a Fon router (there are also ISPs that are Fon partners, whose "boxes" already include the Fon router's capacities). The deal is that they are then allowed to connect to the Internet through any Fon WiFi router of the community.

[2] www.fon.com.

People not sharing their access can also connect through Fon routers, but have to pay for an access pass to do so.

We present here the main aspects of a mathematical model introduced in [199, 200] involving a revenue-focused community network manager, when users also have the possibility to get mobile Internet access through a "classical" wireless provider operating on licensed spectrum. Assuming that no congestion effect occurs in any of those networks, the only externalities among user choices are in terms of coverage: the greater the number of subscribers of the community network, the larger its coverage and hence its perceived quality for users. To keep things simple, the quality of the cellular (licensed) operator is normalized to 1, and the quality of the community network is supposed to be equal to the *proportion* n_{c} of the overall user population subscribing to the community network, thus being less than or equal to 1. Then each user u perceiving quality q and paying price p is assumed to have a linear utility function of the form

$$U_u = a_u q - p,$$

with a_u a user-specific parameter representing the sensitivity to quality. We consider a non-atomic game, with a very large number of players u for whom the values over the population are assumed to be uniformly distributed on an interval $[\alpha, \beta]$. If the user connects to no network at all, we assume $U_u = 0$.

Let us consider fixed subscription prices p_ℓ and p_{c} for the license-based provider and the community network. Without loss of generality we can assume that $p_\ell \geq p_{\mathrm{c}}$, since with $p_\ell < p_{\mathrm{c}}$ all users would prefer the license-based network to the community one. It is then interesting to investigate the user behavior over time. Indeed, each user shall choose the strategy (subscribing to the license-based provider, to the community network, or to none) maximizing his or her utility, but, since externalities are difficult to predict (users in general do not perform a complex game-theoretic analysis to try to anticipate the reactions of the others; actually in most cases they do not have the necessary information to do so), it is reasonable to assume that users react to the quality levels that they *observe*. More specifically, we discretize time and denote by $n_{\mathrm{c}}(t)$ the proportion of users subscribing to the community network at time t. Then at time $t+1$ each user u decides (recall that we assumed q to be 1 for the licensed-based operator and $n_{\mathrm{c}}(t)$ for the community network)

- to join the community network if $a_u n_{\mathrm{c}}(t) - p_{\mathrm{c}} > \max(0, a_u - p_\ell)$, i.e.,

$$a_u \in \left(\frac{p_{\mathrm{c}}}{n_{\mathrm{c}}(t)}, \frac{p_\ell - p_{\mathrm{c}}}{1 - n_{\mathrm{c}}(t)} \right); \tag{5.22}$$

- to join the license-based operator if $a_u - p_\ell \geq \max(0, a_u n_{\mathrm{c}}(t) - p_{\mathrm{c}})$, i.e.,

$$a_u \geq \max \left(p_\ell, \frac{p_\ell - p_{\mathrm{c}}}{1 - n_{\mathrm{c}}(t)} \right);$$

- not to join any network if $0 > \max(a_u - p_\ell, a_u n_{\mathrm{c}}(t) - p_{\mathrm{c}})$, i.e.,

$$a_u < \min \left(p_\ell, \frac{p_{\mathrm{c}}}{n_{\mathrm{c}}(t)} \right).$$

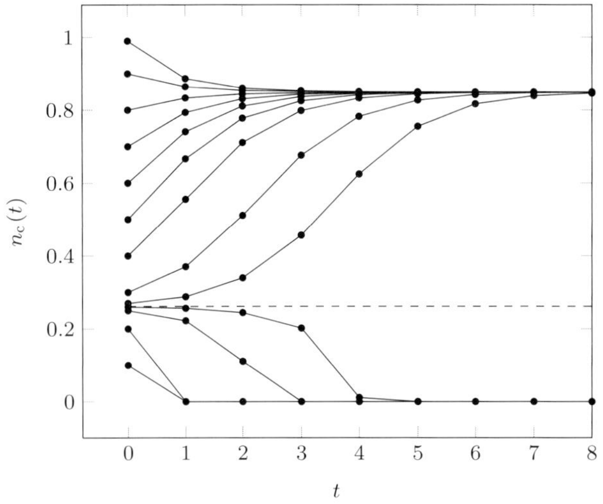

Figure 5.17 The evolution of the proportion $n_c(t)$ of users subscribing to the community network, for different starting conditions $n_c(0)$. The parameter values are $p_\ell = 10$ and $p_c = 2$, with the user sensitivity parameter a_u uniformly distributed over $[\alpha, \beta] = [1, 10]$.

We can notice from (5.22) that the interval of values for a_u for which users select the community network increases with $n_c(t)$, reflecting the externality effect: the more users on that network, the more attractive it becomes. Therefore the evolution of $n_c(\cdot)$ over time is monotonic, and depends on the initial quality offered (that equals $n_c(0)$): if $n_c(1) - n_c(0) > 0$ then the community network expands (possibly stopping its expansion after some time); otherwise it shrinks (also possibly ending at a non-zero value of n_c). In particular, the community network needs a minimal initial value $n_c(0)$ to have any hope of developing: indeed, from (5.22) we need at least $n_c(0) > p_c/\beta$ to have $n_c(1) > 0$ (although it does not suffice to have $n_c(1) > n_c(0)$, that condition being a bit more complex to express). Initiating such a process to reach the necessary critical mass is called **seeding**; this can be performed by providing the good (here, the access to the community network) for free or with a discounted price at the beginning of the offer, with the hope that this will attract more clients.

Summarizing, thanks to an integration of the density of a_u under the condition (5.22), the evolution of $n_c(t)$ over time is given by the recursive relation

$$n_c(t+1) = \frac{1}{\beta - \alpha}\left[\min\left(\beta, \frac{p_\ell - p_c}{1 - n_c(t)}\right) - \max\left(\alpha, \frac{p_c}{n_c(t)}\right)\right]^+, \qquad (5.23)$$

with $x^+ = \max(0, x)$. That behavior is plotted in Figure 5.17. The (monotonic, as expected) convergence occurs quite fast, in fewer than eight time periods even for values close to the seeding threshold (that equals 0.2616) above which the community network develops and below which it fails. The state reached after convergence can be interpreted as a Nash equilibrium of the game played among users: no user wishes to change choices.

Now we can consider the game played by network owners, who can be assumed to anticipate this convergence when fixing prices, hence creating a multi-level game. From the point of view of the community network aiming to optimize revenue, the focus will be on the revenues for the stationary situation: the price has to be low enough to attract users for one possible outcome of the user behavior, and the seeding policy $n_c(0)$ has to be high enough to lead to that specific outcome. Let us neglect here the cost of seeding, and concentrate on the pricing strategy: at a user equilibrium we must have, from (5.23),

$$n_c = \frac{1}{\beta - \alpha}\left[\min\left(\beta, \frac{p_\ell - p_c}{1 - n_c}\right) - \max\left(\alpha, \frac{p_c}{n_c}\right)\right]^+.$$

That expression can be used to compute the price p_c maximizing the revenue $p_c n_c$ of the community network for any given p_ℓ (with the appropriate seeding), in other words the best response of the community network to the price of the license-based operator.

Similarly, the corresponding proportion of users selecting the license-based provider is given by

$$n_\ell = \frac{1}{\beta - \alpha}\left[\beta - \max\left(\alpha, p_\ell, \frac{p_\ell - p_c}{1 - n_c}\right)\right]^+,$$

so that the best response to p_c can be computed by that provider.

The game is analyzed in [199]; it is in particular established that a Nash equilibrium in the pricing game between the two networks exists if and only if $\beta \geq \frac{3}{2}\alpha$, the equilibrium prices and revenues being then expressed analytically. The model is enriched a bit in [200], where the quality of the community network is given by $\min(1, \lambda n_c)$ for some parameter λ that can represent the density of the population (for a dense population, there is no need to have all users sharing their access to have a good coverage). Similar results are proven in that case for the pricing game.

Finally, let us mention that the models of user preferences can be complemented as well: while here users were sensitive only to the quality offered by the other members (which increases with the size of the community), it is also possible to model the nuisance from sharing one's access (which also increases with the community size). This approach is considered in [5], where the utility of a community network user is of the form

$$U_u = a_u q - p - cm,$$

with m the load offered by the user to the other roaming users and c the unit cost (possibly discounted by a reward for providing service to the community) perceived for that load, the other variables having the same interpretations as before. The user-specific parameter a_u can be interpreted as the propensity to roam of user u, in the same vein as for the previous model.

In that case, the adoption process is more complex, since there is no monotonicity. For example, users with a small a_u ("residential" users) may be willing to subscribe if the community is small (or mainly constituted of residential users): in that case the service m to offer to roaming users – and its associated cost – remains small. But, as more residential users join the community, the coverage of the community network improves, attracting roaming users (i.e., with a large a_u) but also increasing the load m borne by residential users, who may leave and in turn induce roaming users to leave because of poor coverage. Therefore, the result of the adoption dynamics depending on the service price and the seeding strategy is more difficult to predict (see [5]).

6 Interactions among content or application service providers

6.1 Introduction

While the previous chapter discussed competition between network service providers, which could be access network providers, or transit providers needing (or required) to cooperate to deliver traffic to its destination, the present chapter focuses on the competition at the content and application service provider level. We are going to see that the models can be formulated in a very general way, hence being closely related to what we described in the previous chapter for access network providers. Indeed, the main driver for customer choices is the price, but some notions of quality (of service) and reputation, among others, can or need to be dealt with too, leading to similar models for customers' choice of service provider (following the Wardrop principle, or some discrete-choice or stickiness models). Those choices will be briefly recalled in the next section. We thus have multi-level Stackelberg games with providers playing first on price and sometimes at an even higher-level content/service investment, anticipating the reaction of users to any strategy profile in terms of consumption.

The categories of content and application service providers which we have in mind are mainly (but are not limited to) (i) content providers, such as news web sites, for example, competing on the relevance and quality of content, design of the site, awareness, and attractiveness through advertising; (ii) online shops with similar characteristics; (iii) content delivery networks (CDNs), which have to attract content providers in terms of price too, but also in terms of the QoS for users through investments in strategically located capacities; (iv) similarly cloud service providers (even though we highlighted in Chapter 4 that for users a comparison of cloud services is complex due to the different service formulations, which are probably intended to allow providers to avoid fierce competition); (v) service applications such as search engines competing for keyword searches through relevance of the results, leading to more revenues from sponsored links (we will describe security providers as an illustration); (vi) application stores, albeit to a lesser extent, because for some operating systems the choice of store is limited; and (vii) architects/device builders. As said above, most of those competitive contexts can be encompassed in a general Stackelberg game, which will be briefly recalled in the next section, so we will afterwards focus on specific and arbitrarily chosen cases, by injecting some characteristic modeling properties into the model.

It is worth mentioning that competition at the content/application service level rose with the so-called dot-com bubble in the late 1990s, when Internet-based services or applications boomed and were financed as soon as "e-something" or "something dot com" was mentioned, a funding that was sometimes without any obvious viability. The dot-com crash occurred in 2000–2001, slowing the development of applications, bringing it back to a more realistic rate. Many businesses did not survive the crisis, and most experienced a dramatic collapse of their stock value on the market. It is interesting to note that the goal of companies at their start was (and still is for new entrants) to develop their base of customers as much as possible in order to get a chance to survive, even if this is done at the expense of constant financial losses in the first years; typical examples are Google and Amazon. As a maybe unexpected consequence of the crash, some surviving companies managed to get a dominant position in their market, leading to some areas becoming less competitive; Google in the advertisement and search engine markets is the immediate illustration. This type of dominant position and difficulty of emergence of competitors has to be kept in mind.

6.2 Competition at the content level

This section discusses general models of competition, but also (arbitrary) specific situations as illustrations.

6.2.1 General models

With full generality, the (abstract) models developed in Section 5.3 are also applicable to analyze competition at the application/content level. Basically, we can represent the competition by two-level games where the following apply.

(i) On the largest time scale, providers compete on the quality and design of their content/application and their price (if any).
(ii) On the smallest time scale, users select their provider given the strategy profile of providers. Note that they can even choose several providers, depending on the type of consumption, by splitting their usage between the providers. This is typically what happens for readers of news web sites, or often also for search-engine users.

The games are here too solved by backward induction, with the providers *anticipating* the reaction of users for a given profile of strategies on the largest time scale, making use of the subsequent reaction of users.

Price was considered as the (or one of the) key decision variable(s) for network providers in the previous chapter, since it is a parameter that has a major impact on users. This may still be the case for some specific service providers, such as cloud service providers, CDNs (in this case users are specific content providers themselves), e-commerce, content providers such as paid news sites, online video services, etc. The models of previous chapters can then be *directly* applied. However, there are situations where applications or content are free, and even where free and paid services compete; hence other parameters or variable decisions have to be emphasized. A few examples of such parameters follow.

(i) Content innovation and rejuvenation (or updating) for traditional web sites. This can be seen as a function of the level of financial investment made by the content provider towards activity on content. The utility (revenue) is then also a function of the investment level, as a function of the number of visits (through advertisement, but not necessarily only that), which is itself dependent on the profile of investments for all providers, and hence a game.
(ii) Web-site design and the amount of advertising incorporated. The more ads displayed, the larger the potential revenue, but displaying too many of them may deter users from consulting the content site. This has to be balanced, in relation to the choices made by competitors.

The models to analyze such a competition may thus again follow one of the three following approaches for customers' choices (as a summary of Section 5.3 adapted to the current context).

- The Wardrop principle, when customers are assumed non-atomic (briefly, the total demand can be split continuously, individual demands being negligible) and such that customers choose the provider yielding the smallest *perceived cost*, or, equivalently, the largest utility. The Wardrop principle thus says that at equilibrium all *used* providers have the same perceived cost, otherwise some customers would have an interest in switching providers, and that unused ones have a higher perceived cost (for the same reason). Here, cost can again be based on price, for paid services as mentioned above, and on QoS (or congestion effects), such that having more users implies a degraded service – the most relevant use case of this type of equilibrium notion. But other notions are usually involved, such as quality of content in addition to price, or instead of it in the case of free services. On top of that game, a game is played on the strategies of providers, making use of the subsequent Wardrop equilibrium. This type of model is of interest when congestion effects occur (a mass of users consuming the service creates a loss of utility for others); otherwise it ends up with only one provider with a positive demand.
- Discrete-choice models such that the valuation of a customer u for a provider, say i, is

$$V_i = \sum_{j=1}^{k} v_{i,k} + \kappa_{u,i},$$

where $v_{i,k}$ represents the measured objective value of a characteristic/attribute k at provider i, and $\kappa_{u,i}$ is an unobserved user-specific value that is treated on the global level as a random variable. In Section 5.3, a single attribute was considered, related to price, but more can be considered (by analyzing more the at-first-sight purely random aspects), as predefined functions of a corresponding decision variable such as reputation, content investments, design of the application, advertisement level, etc. Customers will select the provider maximizing their utilities; hence the proportion of customers choosing i is $\mathbb{P}[V_i = \max_j V_j]$. If the $\kappa_{u,i}$ are independent and identically distributed and follow a Gumbel distribution, we recall that we end up with the standard logit model

$$\mathbb{P}[V_i = \max_j V_j] = \frac{\exp\left(\sum_{j=1}^{k} v_{i,k}\right)}{\sum_{\ell} \exp\left(\sum_{j=1}^{k} v_{\ell,k}\right)},$$

which is also a standard stickiness model. Here again, providers may play on decision variables that impact the various $v_{i,k}$.

- Aggregated demand models, for instance, if a decision vector $\boldsymbol{s} = (s_i)_{i\in\mathcal{N}}$ is given for a set $\mathcal{N}$ of providers, with s_i the decision variable (or potentially vector) of provider i, the demand at i is given by

$$d_i(\boldsymbol{s}) = d_{i,0} - \alpha_i f(s_i) + \sum_{j\neq i} \beta_{ij} f(s_j),$$

where $d_{i,0}$, α_i and $(\beta_{ij})_{j\in\mathcal{N}\setminus\{i\}}$ are strictly positive parameters and f is a non-negative function. The provider game is here too solved similarly to what is described in Section 5.3. The decision variable can be the price, as in Section 5.3, with f the identity function (typical of paid applications/content), or again content/QoS investment.

The analysis can then be performed just like in the previous chapter.

6.2.2 Online TV competition

Let us focus on online TV as an illustration of existing competition. The amount of TV spending keeps increasing as a percentage of the gross domestic product (GDP), but there is competition among broadcasting, cable TV, satellite TV, and now Internet TV, whose market share is increasing. In all those categories, paid TV is also actually of increasing proportion, this being a reason for the increase in spending. As illustrations of the development of online TV, note that (TV) programs can now be found online, for example on iTunes, at low prices, namely for less than \$2. Also, Netflix, a USA-based company, started video streaming in 2007, including TV shows. A competitor, Hulu, started its business in 2008. Hulu was created by TV channels Fox and NBC, joined later by ABC. We remark that, although they are efficient in the USA, Hulu and Netflix are still not available in European countries, probably due to concerns from copyright owners.

To better understand the main decision variables of Hulu and Netflix, we can wonder what makes the difference between those two online movie and TV show providers. Price is not the main issue, since they charge an equivalent amount (around US\$8 a month for unlimited usage). But, on the other hand, the selection of programs, streaming QoS, and means of service are the main differentiators. Hulu is often considered to be better in terms of TV shows, while Netflix's catalog of movies is often believed to be larger. Streaming rates are also different, and may be a key component for customers (Netflix being usually thought to be better), but encoding differences have to be incorporated into the equation. Finally, not all programs are always directly available if one is not going through a website, which can create a difference; this is the type of thing we encompass in the terminology "means of service." Also included in this component are the choice of technologies to access online TV, which can be through smart TVs, through any portable device or computer, through set-top boxes, or through games consoles. All those aspects can be integrated into the decision variables/differentiators of a model.

Similarly, the competition can be further analyzed by incorporating into the game pay-per-view providers such as iTunes, Amazon, and the numerous regular TV channels whose business model is based on advertisement. We could also think of free TV network portals such as Viacom, whose online ads can be more efficiently targeted, as discussed for search engines and ad banners in Chapter 4.

Anticompetitive behaviors can be expected from traditional TV channels and movie makers refusing to have their programs presented by some online service providers, and choosing to supply them only to their own **aggregator** (an aggregator is one website concentrating content from multiple copyright owners). This content starvation could lead to some non-dominant actors running out of business, and aggregation can be analyzed similarly to bundling (see Section 3.6).

6.2.3 An illustrative model of competition among free CPs with advertising

To give a mathematical model of a specific situation, we present here a simplified version of the one in [63].

Consider some content, which could be a video sequence, a movie, or a TV show. This content is controlled (or offered) by a set $\mathcal{I}$ of different *free* CPs, which can play with the amount of advertising included on their webpage. CPs have different qualities *experienced* by users, depending on the number of clicks needed to reach the content, such as when ads are superimposed on the content and have to be clicked to be removed, or when there is a delay before accessing the content if a video ad is displayed first. The level A_i of advertising at the CP i is thus a *nuisance* for users, but provides additional revenue through clicks for the CP. We assume that CP i earns A_i each time its content is accessed, and that the quality of experience (QoE) V_i that the user has with CP $i \in \mathcal{I}$ is of the form

$$V_i = Q_i - A_i, \tag{6.1}$$

where $Q_i > 0$ is the intrinsic quality of the content of CP i (i.e., the quality that the user would experience if there were no advertising). To simplify the notation, we will define and use $a_i := A_i/Q_i$ as the *relative* amount of advertising introduced by CP i, such that $a_i \in [0, 1]$ because the amount of advertising will be reduced to ensure non-negative utility in (6.1). We also assume, to simplify the model in [63], a stickiness model such that the proportion C_i of users choosing CP i is proportional to the valuation associated with that CP:

$$C_i = \frac{V_i}{\sum_{j\in\mathcal{I}} V_j} = \frac{Q_i(1-a_i)}{\sum_{j\in\mathcal{I}} Q_j(1-a_j)}. \tag{6.2}$$

The expected revenue per unit of time for a content provider $i \in \mathcal{I}$, through advertising, is

$$R_i = C_i A_i = C_i Q_i a_i, \tag{6.3}$$

where it is assumed to be linear in the amount of displayed advertising. Each CP's revenue thus depends on its strategic choice a_i, but also on the amount of advertising $(a_j)_{j\in\mathcal{I}\neq i}$, which the other CPs fix through the market share C_i.

For this game between CPs on the level of advertising, we can establish the existence of a (non-trivial) Nash equilibrium. There exists a Nash equilibrium $a^{\mathrm{NE}} \in (0,1)^{|\mathcal{I}|}$, with $a_i > 1/2\ \forall i$.

To prove this, note first that setting $a_i = 1$ is a dominated strategy for CP i, and strictly dominated when at least one opponent j sets $a_j < 1$. The case when $a_i = 1$ for all i is therefore an unlikely situation that we disregard.

We thus assume that at least one CP $j \neq i$ sets $a_j < 1$. R_i in (6.3) is then a continuous function of a_i with $R_i = 0$ when $a_i = 0$ and when $a_i = 1$ and such that $\partial^2 R_i/\partial a_i^2 < 0$ (i.e., R_i is strictly concave) because it is of the same sign as $a_i Q_i - (Q_i + \sum_{j\in\mathcal{I}\setminus\{i\}} Q_j(1-a_j))$, thus having a unique maximum that is in $(0,1)$ as a best response to $(a_j)_{j\in\mathcal{I}\setminus\{i\}}$. This best response is the solution of

$$\frac{\partial R_i}{\partial a_i} = \frac{\partial}{\partial a_i}\left[a_i Q_i \frac{Q_i(1-a_i)}{\sum_{j\in\mathcal{I}} Q_j(1-a_j)}\right] = 0,$$

which is equivalent to solving the following quadratic equation in a_i:

$$a_i^2 - 2\left[1 + \sum_{j\in\mathcal{I}\setminus\{i\}} \frac{Q_j}{Q_i}(1-a_j)\right] a_i + 1 + \sum_{j\in\mathcal{I}\setminus\{i\}} \frac{Q_j}{Q_i}(1-a_j) = 0. \qquad (6.4)$$

The only solution in $(0, 1)$ is

$$a_i^{\mathrm{br}} = 1 + \psi_i - \sqrt{(1+\psi_i)^2 - (1+\psi_i)} = 1 + \psi_i - \sqrt{\psi_i(1+\psi_i)}, \qquad (6.5)$$

with (still in terms of $(a_j)_{j\in\mathcal{I}\setminus\{i\}}$)

$$\psi_i := \sum_{j\in\mathcal{I}\setminus\{i\}} \frac{Q_j}{Q_i}(1-a_j). \qquad (6.6)$$

Next, it can be seen (see [63] for details) that there exists an ϵ small enough that we have the best-response correspondence from $[0, 1-\epsilon]^{|\mathcal{I}|}$ to itself

$$G : \begin{cases} [0, 1-\epsilon]^{|\mathcal{I}|} \mapsto [0, 1-\epsilon]^{|\mathcal{I}|} \\ \qquad a \mapsto (a_i^{\mathrm{br}})_{i\in\mathcal{I}}. \end{cases}$$

Applying the Brouwer fixed-point theorem, it has a fixed point that constitutes a Nash equilibrium with strategies $a_i \in [0, 1)$.

The fact that $a_i > 1/2$ comes from the strict decreasingness on positive values of $f(x) := 1 + x - \sqrt{x(1+x)}$, a function considered so that $a_i^{\mathrm{br}} = f(\psi_i)$, and from $\lim_{x\to\infty} f(x) = 1/2$.

Example 46 Looking more closely at the simpler symmetric situation with $Q_i = Q$ $\forall i \in \mathcal{I}$, we get the expression

$$R_i = a_i \frac{1 - a_i}{\sum_{j \in \mathcal{I}}(1 - a_j)}. \tag{6.7}$$

The revenue optimization leads to

$$(1 - 2a_i) \sum_{j \in \mathcal{I}} (1 - a_j) + a_i(1 - a_i) = 0, \tag{6.8}$$

giving, for any $i, k \in \mathcal{I}$,

$$(1 - 2a_i) \sum_{j \in \mathcal{I}} (1 - a_j) + a_i(1 - a_i) = (1 - 2a_k) \sum_{j \in \mathcal{I}} (1 - a_j) + a_k(1 - a_k).$$

Stated otherwise, it gives

$$(a_k - a_i) \left(a_k + a_i - 1 + 2 \sum_{j \in \mathcal{I}} (1 - a_j) \right) = 0.$$

From our general result, $a_j > 1/2$ for all $j \in \mathcal{I}$ at a Nash equilibrium, so that the right-hand side of the above expression is strictly positive, giving $a_i = a_k$. Nash equilibria are necessarily symmetric, of the form $a_i = a$ for all $i \in \mathcal{I}$. On plugging that condition into (6.8), we obtain a unique equilibrium, with

$$a_i^{\mathrm{NE}} = \frac{n}{2n - 1} \qquad \forall i \in \mathcal{I}, \tag{6.9}$$

where n is the total number of CPs, i.e., $n := |\mathcal{I}|$, yielding the corresponding revenue

$$R_i = a \frac{1 - a}{n(1 - a)} = \frac{1}{2n - 1}.$$

From this expression, we can remark that the more competition (that is, as n increases), the less advertising at each CP at equilibrium, with an asymptotic value of $1/2$. Moreover, the sum of revenues $R = \sum_{i \in \mathcal{I}} R_i = n/(2n - 1)$ is also decreasing up to an asymptotic value of $1/2$.

Example 47 Consider now the case of an asymmetric duopoly. The best-response functions (6.5) become

$$\begin{aligned} a_1 &= 1 - \frac{Q_2}{Q_1}(1 \quad a_2) \left[\sqrt{1 + \frac{Q_1}{Q_2} \frac{1}{1 - a_2}} - 1 \right], \\ a_2 &= 1 - \frac{Q_1}{Q_2}(1 - a_1) \left[\sqrt{1 + \frac{Q_2}{Q_1} \frac{1}{1 - a_1}} - 1 \right]. \end{aligned} \tag{6.10}$$

We draw in Figure 6.1 an example of best-response functions (here symmetric, to check that the results are in accordance with the analytical ones from the previous example).

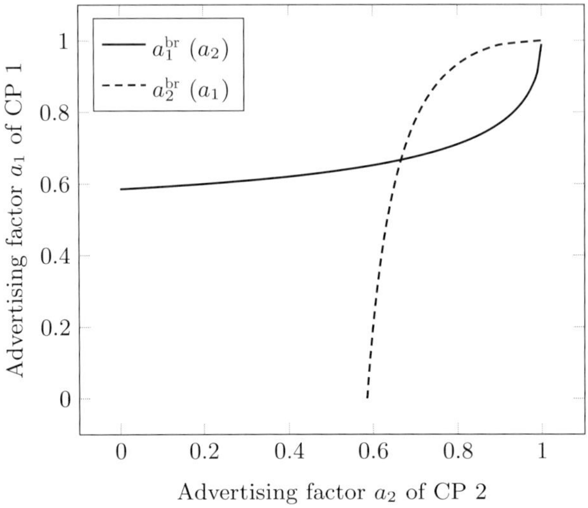

Figure 6.1 Best-response functions in the duopoly case for the competition between free CPs.

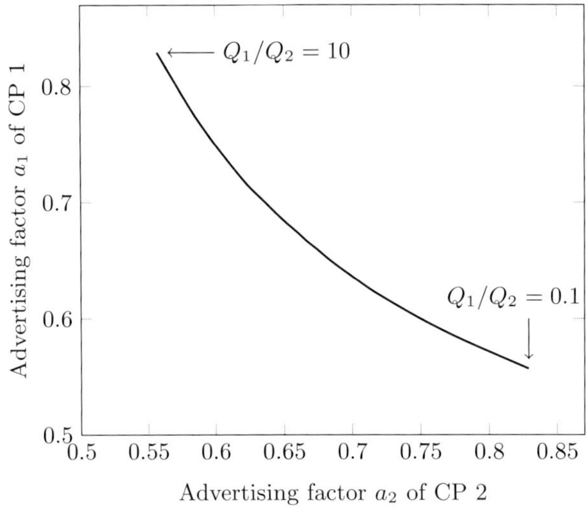

Figure 6.2 Nash-equilibrium points in a duopoly for various ratios Q_1/Q_2.

We can check that there are two Nash equilibria (the points where the curves intersect), first the unlikely dominated situation $a_1 = a_2 = 1$, which we disregard, but also another point with $a_i > 1/2\ \forall i$ (since, in the symmetric case here, we get $a_i = 2/3$ in accordance with the previous results).

Figure 6.2 investigates the evolution of that Nash equilibrium point for a range of ratios Q_1/Q_2 between the qualities at CPs. It is interesting to note here that a content provider with higher intrinsic quality can increase its advertising load.

For more numerical results on this type of model, the reader is advised to refer to [63].

6.3 A specific case: competition between search engines

To illustrate the consequences of competition at the service level, we are arbitrarily going to focus in this section on competition between search engines. The economics of search engines (through sponsored links) has been extensively studied in Section 4.2. Our aim is thus to see what influence competition exerts on the decisions of the various actors (advertisers, search engines, and even users).

Also of interest would be to investigate whether a search engine with a dominant position – typically Google's situation – has an interest in introducing barriers to entry for new competitors. If this happens at the expense of users, regulation procedures would then have to be considered. Any strongly dominant position may indeed lead to the possibility of agreements (potentially exclusive) of the dominant actor with some advertisers or content providers. For example, Google has been accused of favoring some content because of the additional revenue it can bring, such as YouTube content because YouTube is owned by Google and use of YouTube may yield further advertising revenue. As a consequence, competitors could be ranked lower in the search output than they "should" be, leading to their being less visited. This issue of **search neutrality** will be discussed more in the next chapter. The impact of such a dominant position has already been observed, and fought, in the past, when Microsoft imposed its Internet Explorer browser with its operating system. As other exclusive arrangements, Google is also now having default search toolbars in installed software provided by Adobe, on web browsers, and with the Android operating system. While Google is by far the most used search engine, the competition is trying to gain market share, the main competitors being Yahoo! and Bing, or Baidu in China. The issues raised in this paragraph are discussed, for example, in [93].

Two close models of competition between search engines, analyzing their ad allocation rules and pricing strategies, have been developed and studied in [191, 193] (for other search-engine competition models, the interested reader can also look at [24, 93, 119, 170, 171]). The models consist in a two-level game such that

- on the largest time scale search engines decide which allocation rule to implement, between revenue-based and bid-based; and
- on the shortest time scale advertisers decide how to split their advertising budget between the two search engines, depending on the return on investment this will bring to them.

The models are thus a bit more complicated than those in Section 4.2 for the allocation rule analysis because it is a multi-level game, but also because it involves budget constraints. To simplify the analysis (the goal is to highlight the effect of competition), consider a single keyword, composed by users at rate λ (i.e., representing the average number of searches per unit of time), and two search engines (SE), labeled 1 and 2, in competition for advertisers, and assume that each SE offers a *single* advertisement slot. Search engines are differentiated among other things by their market share: a proportion α $(1-\alpha)$ of searches are performed through SE 1 (SE 2). We consider k advertisers in competition for the slots, each advertiser having a budget b to spend and a valuation v per click for the considered keyword. The values of b and v are assumed to be randomly picked respectively from the cumulative distribution functions $G(b)$ and $F(v)$, identically and independently distributed between advertisers. The CTR of an advertiser at an SE is assumed to be separable (as was commonly done in Section 4.2), and thus is given by the product of the CTR of the search engine, q_1 and q_2 for SE 1 and SE 2, respectively, and the CTR c_i of the considered advertiser i. Advertisers are assumed to be truthful (which is not a strong assumption since a GSP auction for a single slot will be applied at each SE); i.e., the bid b_i of advertiser i is equal to his or her valuation v_i; the goal of advertisers is rather to choose which proportion of their budget b_i to spend on which engine. Let β_i denote the proportion of the advertising budget submitted to SE 1 by advertiser i, while $1-\beta_i$ is submitted to SE 2.

In order to express the utilities of advertisers, and later on of search engines, we need to define the probability $w_j(v_i)$ that advertiser i wins on SE j when his or her bid is v_i and the average price $\mathbb{E}[R_j|v]$ paid on SE j *given that the advertiser has won the auction when bidding* v, depending on the random information described above on the knowledge of competing advertisers. The rates at which advertiser i submits his or her bids to SE 1 and SE 2 are then

$$\lambda_i^{(1)} = \min\left(\frac{\beta_i b_i}{q_1 c_i \mathbb{E}[R_1|v_i] w_1(v_i)}, \alpha\lambda\right),$$

$$\lambda_i^{(2)} = \min\left(\frac{(1-\beta_i) b_i}{q_2 c_i \mathbb{E}[R_2|v_i] w_2(v_i)}, (1-\alpha)\lambda\right).$$

We have a minimum in the expressions because the number of potential requests before reaching the budget can be larger than the actual number of searches. In other words, it may be possible that the budget at one (or both) engine(s) is not reached. From this budget allocation, advertiser i can compute the probability $p_i^{(j)}$ to bid at SE j, as $p_i^{(1)} = \lambda_i^{(1)}/(\alpha\lambda)$ and $p_i^{(2)} = \lambda_i^{(2)}/((1-\alpha)\lambda)$. From those elements, for any strategy profile $\beta = (\beta_1, \ldots, \beta_k)$ of advertisers, the utility of advertiser i is the gains $v_i - \mathbb{E}[R_j|v_i]$ from clicks at both search engines when bids are submitted and the auction won:

$$U_i(\beta) = q_1 c_i w_1(v_i)\lambda_i^{(1)}(v_i - \mathbb{E}[R_1|v_i]) + q_2 c_i w_2(v_i)\lambda_i^{(2)}(v_i - \mathbb{E}[R_2|v_i]). \qquad (6.11)$$

We can then consider two possible situations regarding the strategic behavior of advertisers.

(i) Following [193], the advertisers know their bids but not the budget and valuations of other advertisers, using just the distributions F and G. They just observe their bidding probabilities without trying to derive the competitors' values from those observations. In that case, after quite heavy computations (see [193]), the average prices $\mathbb{E}[R_j|v_i]$ and winning probabilities $w_j(v_i)$ can be expressed as solutions of a system of equations that can be solved numerically. This can be done for bid-based ranking and for revenue-based ranking.

(ii) Alternatively, following [191], the advertisers can try to derive the competitors' characteristics from their observed bidding probabilities, which is done under the assumption that all advertisers have the same CTR (i.e., $q_i = q\ \forall i$). It can be shown in this context that high-valuation advertisers tend to bid less often than low-valuation ones, because they win the auction more often and reach their budget limit faster.

In the remainder of this section, we consider the first of the above scenarios. We want to determine at the highest level (the competition between search engines) what is the best solution between a revenue-based and a bid-based ranking in such a competitive context instead of a monopoly. The question is this: is there an interest for search engines to differentiate their ranking/pricing strategies? To illustrate the phenomenon and the interest of such an analysis, we consider just two SEs and two advertisers. The strategy profile of advertisers is (β_1, β_2), and we are looking for a Nash equilibrium that is a profile of proportion strategies (β_1^*, β_2^*) such that no advertiser can increase his or her revenue by unilaterally changing his or her strategy. We are going to investigate this Nash equilibrium for each combination of strategies from search engines (applying GSP, but choosing either revenue-based or bid-based ranking), to later describe the above game and ranking strategies.

As described in Section 2.3, we first compute the best response of each advertiser as a function of the strategy of his or her opponent:

$$\text{BR1}(\beta_2) = \arg\max_{\beta\in[0,1]} U_1(\beta, \beta_2) \quad \text{and} \quad \text{BR2}(\beta_1) = \arg\max_{\beta\in[0,1]} U_2(\beta_1, \beta).$$

We consider as arbitrary numerical parameters, with a uniform distribution of valuations on $[0, 20]$, $\alpha = 0.6$, $\lambda = 100$, $q_1 = 0.5$, and $q_2 = 0.6$ (hence a different CTR at the SEs), deterministic budget values $b_1 = 5$ and $b_2 = 20$, $c_1 = 0.5$, $c_2 = 0.4$, $v_1 = 10$, $v_2 = 9$, and a reserve price $p_r = 0.1$ at both SEs, corresponding to the minimum price to be paid (which is especially important when there is a single advertiser submitting a bid). Figure 6.3 presents the best-response curves when both SEs implement bid-based GSP pricing. Interestingly, there is actually an interval for the best response of advertiser 2 corresponding to each β_1, and that interval is of constant size, with bounds increasing linearly with β_1. This gives the grayed domain. For the best response of advertiser 1 in terms of β_2, we obtain the dashed curve. The only values for which the domain and the bold curve intersect are $\{0.775\} \times [0.39, 0.63]$, yielding a set of Nash equilibria.

To explain why there is an interval for the best response of advertiser 2, and not for the best response of advertiser 1, Figures 6.4 and 6.5 present the revenues of advertisers when the strategy of the opponent is $\beta_2 = 0.95$ and $\beta_1 = 0.21$, respectively. The decision yielding the maximum total revenue will give the best response of the advertiser. While for $\beta_2 = 0.95$ there is a (single) maximum at $\beta_1 \approx 0.175$, the revenue of advertiser 2 when $\beta_1 = 0.21$ is indeed maximized on an interval (see Figure 6.5). This comes from the fact that the probability of bidding is then maximal, equal to 1, and therefore independent of the budget submitted; in other words, the budget is not fully spent.

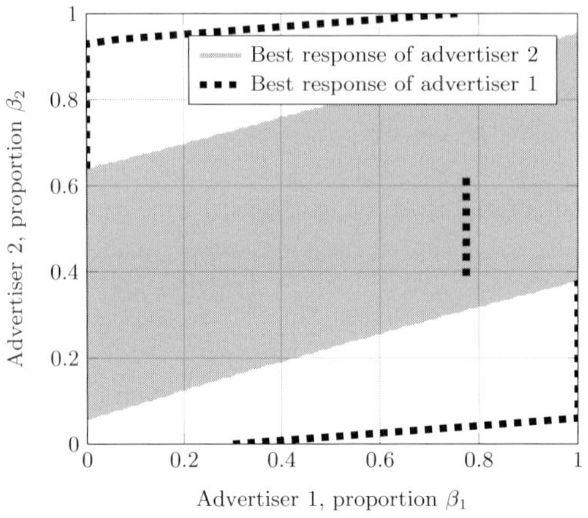

Figure 6.3 Best-response curves when bid-based pricing is implemented at both engines.

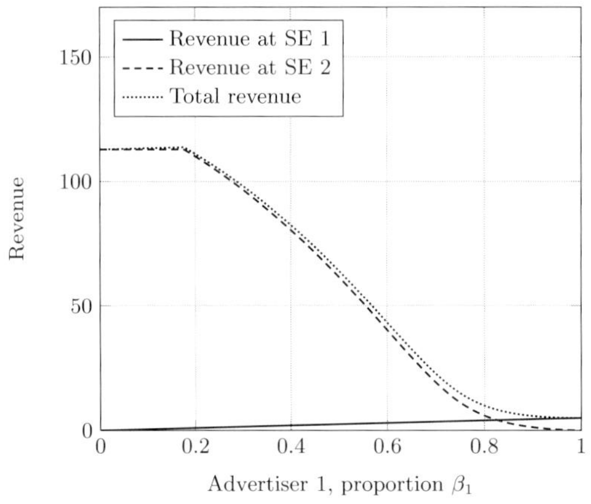

Figure 6.4 Revenues of advertiser 1 in terms of β_1, with $\beta_2 = 0.95$.

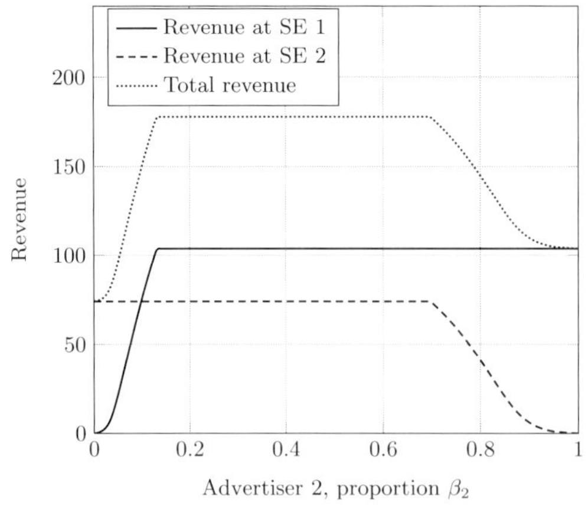

Figure 6.5 Revenues of advertiser 2 in terms of β_2, with $\beta_1 = 0.21$.

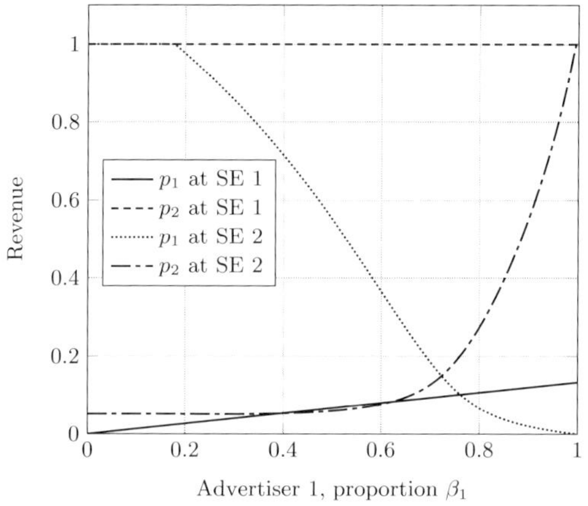

Figure 6.6 Bidding probabilities in terms of β_1 (with $\beta_2 = 0.95$).

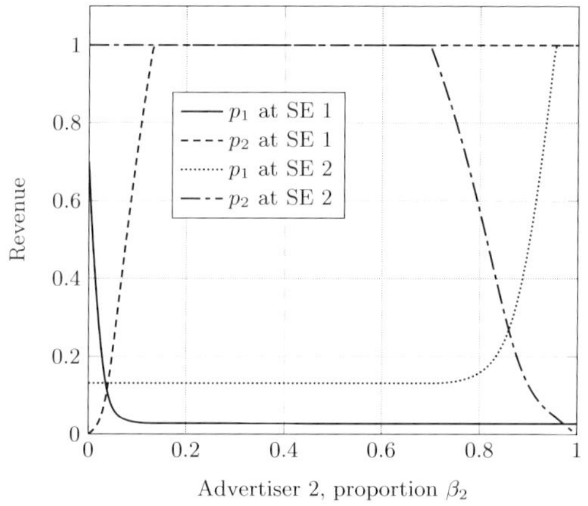

Figure 6.7 Bidding probabilities in terms of β_2 (with $\beta_1 = 0.21$).

This can be verified in Figures 6.6 and 6.7 where we draw the winning probabilities of advertisers in terms of the budget proportion allocated to SE 1, when the β_i of the competitor is fixed to the same value as in Figures 6.4 and 6.5. In Figure 6.6, there is no interval where p_1 and p_2 are constant at the same SE, resulting in no constant revenue. But in Figure 6.7 that is no longer the case.

Figures 6.8–6.10 display the best-response curves for the three other combinations of ranking/pricing at search engines: bid-based for SE 1 and revenue-based for SE 2 (called the B–R case), revenue-based for SE 1 and bid-based for SE 2 (called the R–B case), and revenue-based for both SEs (called the R–R case). The above numerical values of parameters are again used. When revenue-based ranking is implemented, the reserve revenue $r_r = 0.025$ is used, which corresponds approximately to the quantity $q_i c_i p_r$ when bid-based ranking is used in order to ensure that revenue is obtained from an ad.

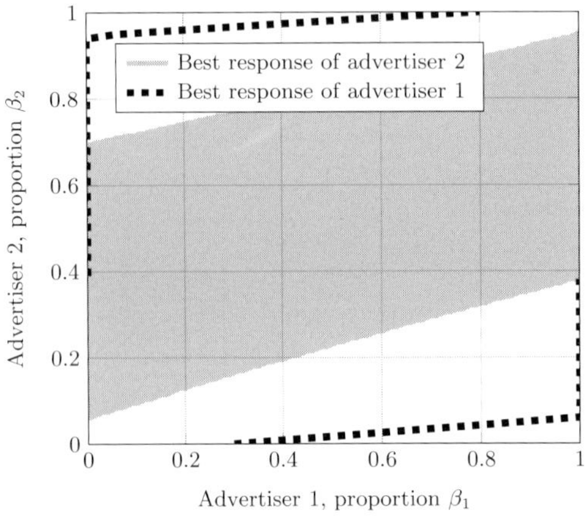

Figure 6.8 Best-response curves when bid-based pricing is implemented at SE 1 and revenue-based at SE 2.

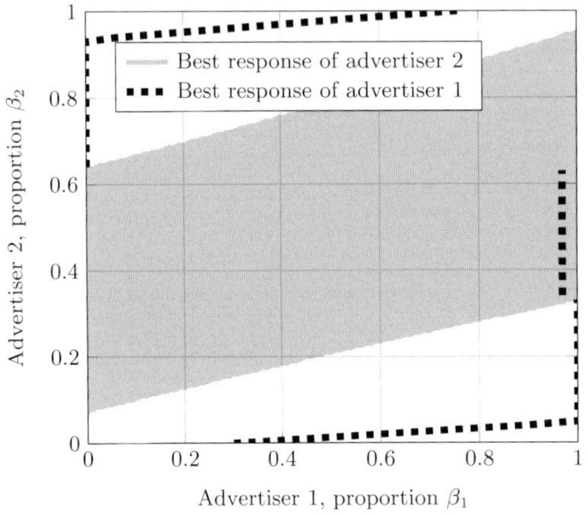

Figure 6.9 Best-response curves when bid-based pricing is implemented at SE 2 and revenue-based pricing at SE 1.

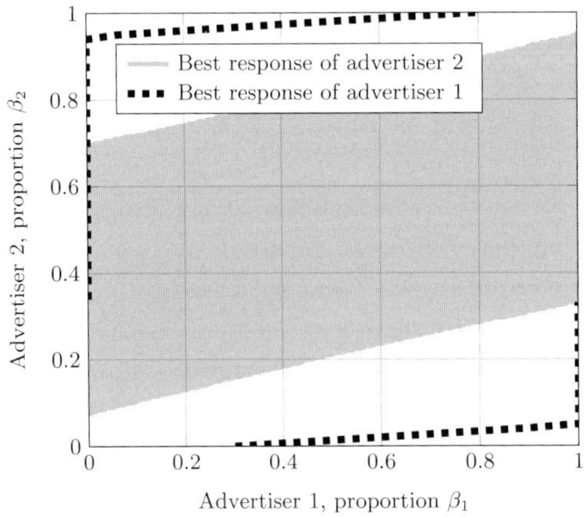

Figure 6.10 Best-response curves when revenue-based pricing is implemented at both engines.

Table 6.1 The game on ranking strategies among SEs. Each element gives revenues (Rev_1, Rev_2) in terms of the rules used by SE 1 (line) and SE 2 (column).

	B	R
B	(10.15, 3.62)	(1.20, 11.06)
R	(11.32, 1.32)	(1.50, 11.06)

From the figures, the sets of Nash equilibria can be summarized as follows:

- for the B–B case, all the profiles $(\beta_1, \beta_2) \in \{0.775\} \times [0.39, 0.63]$;
- for the B–R case, it is $\{0\} \times [0.39, 0.695]$ (the whole budget of advertiser 1 is placed on the revenue-based engine);
- for the R–B case, $\{0.97\} \times [0.34, 0.63]$; and
- for the R–R case, $\{0\} \times [0.34, 0.695]$ (the whole budget of advertiser 1 is placed on the revenue-based engine).

We can notice that, even if there are multiple equilibria, the revenues of advertisers and of search engines are unique because those equilibria correspond to the same bidding strategies, where advertiser 2 bids with probability 1 on both SEs.

Thanks to those results, the search engines can play a game on their ranking strategies where their goal is to choose between bid-based and revenue-based ranking in order to maximize their revenues. Those revenues are provided in Table 6.1.

One can note that the best strategy for SE 2 is to always play the revenue-based strategy. Actually, it can be readily checked that the combination of strategies R–R is a Nash equilibrium. But we remark, on the other hand, that we have found other sets of parameters such that B–R is an equilibrium. This kind of game has therefore to be closely looked at by search engines.

6.4 The economics of network security

Network connectivity has become a *sine qua non* condition for almost any commercial and economic activity. As a side effect, the importance of security and its economic impacts has kept increasing in the last decade.

Cybercriminality affects businesses (cyber risk is considered the most critical risk by enterprises [284]) as well as private individuals. A recent illustration is the security breach that was found in the PlayStation Network in April 2011, whose cost for Sony has been estimated to be tens of millions of dollars. That breach exposed the personal information, and possibly the credit card data, of 77 million customer accounts.

Therefore, network service providers face new economic issues linked to the provision of communications that are protected against all potential types of attack. That task is unfortunately impossible: with the appearance of new applications and services, the number of breaches to cover increases exponentially, so that many of them get discovered only when they are exploited by an attacker. The amounts of money involved in cybercriminality are enormous, be it in terms of

- damage costs (imagine the financial loss associated with a one-day rupture of service for a company like Amazon),
- investment costs (security represents about 10% of companies' IT expenses), or
- even the underlying economy of attackers; i.e., attackers selling their "services."

That last point is particularly striking: since it may become economically advantageous for a company to harm a competitor's IT system, some firms may be willing to pay to do so, which is now incredibly easy. Indeed, there now exists a market, which is almost in the open, for renting zombie computers; i.e., hackers who have managed to take control of a large number of machines rent them for a period of time, at a given price, so as to saturate the competitor's servers, i.e., run a distributed denial of service (DDoS) attack. In the same vein, but dealing more directly with people's wealth, a highly competitive black market of stolen credit card numbers has arisen in recent years [160, 308].

Those examples illustrate the fact that cybercrime is now highly organized and competitive. Building businesses such as credit card data selling or renting zombie computers takes some considerable effort, which has to be rationalized to maximize revenue. This is the reason why the interactions among all actors in the context of cybersecurity should be modeled and studied within the framework of game theory, which precisely considers the potential outcomes of situations where several self-interested agents are involved.

6.4.1 Economic models for security analysis

We first provide an overview of the game-theoretic models introduced in the literature regarding the interactions among network actors, in terms of security problems. We mainly describe here the basic mechanisms and principles; a more detailed survey can be found in [182]. Some quite simple models of network security represent the interaction between an attacker and a defender as a two-player game, where each player has two strategies (launch an attack or not for the attacker, implement a protection procedure or not for the defender) [10]. The basic assumptions are the following.

- Protecting one's system is costly for the defender – financially and/or in terms of performance – but those costs are less than the perceived cost of having one's data stolen or corrupted.
- Launching an attack has a cost, which increases if the defender implements some security procedure because the attack may become less efficient, and the attacker might be detected and sanctioned. On the other hand, a successful attack yields revenue to the attacker.

Under those reasonable assumptions, the two-player game can be represented in normal form through a payoff matrix similar to that of Table 6.2, where some example values reflecting the actors' preferences are arbitrarily chosen. It is quite straightforward to remark that, under the general assumptions above, such games have no pure Nash

Table 6.2 The payoff matrix for a simple attacker–defender game, with player 1 (the attacker) selecting a row and player 2 (the defender) selecting a column

	Protection	No protection
Attack	$(-1, -1)$	$(5, -10)$
No attack	$(0, -1)$	$(0, 0)$

equilibrium; i.e., there is no strategy profile from which no player has an interest to switch strategies: if the attacker attacks then the defender should implement the protection, in which case the attacker prefers not to attack; similarly, if the attacker does not attack then the defender should not spend money on protection, in which case the attacker is better off attacking. As we saw in Section 2.3, the outcomes that can be expected from such situations are mixed-strategy Nash equilibria, meaning that each user selects his or her strategy according to some probability distribution from among his or her available choices. In this situation (finite game) a mixed-strategy Nash equilibrium always exists. At equilibrium, the probability with which each player selects his action should be such that the opponent is indifferent between his own actions, as the unique intersection point of the best-response correspondences. For the example of Table 6.2, the attacker should attack with probability p such that $(-1)p + (-1)(1 - p)$ (the expected cost of the defender with protection) equals $(-10)p + 0(1 - p)$ (the expected cost of the defender without protection), i.e., $p = 1/10$; similarly the defender should choose the protection with probability q so that $(-1)q + 5(1 - q) = 0$, i.e., $q = 5/6$.

Some extensions of that type of model, considering more than two strategies per player, are presented in [39, 145]. Also, some Bayesian game extensions have been studied [247], where the defender has only *a priori* beliefs about the type of attacker he is facing. Here again, no pure Nash equilibrium exists, but mixed equilibria can be computed. Bayesian reasoning is also applied in [172], still on the defender's side, but in a repeated-game context: the attacker intends to perform DDoS attacks, dissimulating his traffic among the regular traffic. To detect attacks, the defending system implements a classification of user traffic, as good (non-malicious), bad (malicious), or poor (non-malicious but with properties similar to those of malicious traffic). Here the defender is a whole network system that can adopt a variety of defense postures for each router. A large number of Nash equilibria can then exist, and can be used to estimate performance metrics such as the attacking capacity of the attacker (defined as the worst-case damage caused) and the defending capacity of the system (i.e., the resilience against DDoS attacks). Network security games can also involve routing decisions, either on the attacker's or on the defender's side. The game is played on the links or interfaces of a network, and its outcome depends on whether the defender's and attacker's decisions intersect.

- In [157], the attacker selects a path to his target, trying to avoid being detected by the defender. The defending strategies consist in selecting the links to scan, and the scanning intensity (that is limited over the whole network). If the defender scans a link on the path chosen by the attacker, then the latter is detected with a given probability depending on the background traffic and the scanning intensity.

- The game is played in the opposite sense in [41]: the "defender" is a user willing to send some traffic to another node in the network, and has to select a path. The attacker can try to intercept that traffic by targeting some links where he or she scans packets. The total scanning effort is upper-bounded, and thus has to be spread over the network.

In both cases, the game is studied as a two-player zero-sum game (i.e., what one player wins is what the other loses). That game has a unique Nash equilibrium in mixed strategies, namely the saddle point of the players' payoff function, which can be computed from the solution of a flow-maximization problem in a network where link capacities would be limited.

The study of worm propagation can also benefit from a game-theoretic approach. Worms can propagate very fast: in the case of scanning worms [108], the behavior of the attacker (worm) consists in trying out many (randomly chosen) IP addresses in order to infect the corresponding host (if no patch has been installed yet on that target). We therefore have a race between the worm trying to spread as fast as possible, and the defenders needing to create and distribute the countermeasure (patch). Those situations actually correspond to Stackelberg (or leader–follower) games, where the worm selects its strategy first (i.e., it is acting as a leader), and the detection and containment techniques are then implemented (the follower's role). In [108], that containment technique consists in quarantining measures: after a given time a node is suspected of being infected, so its host connection attempts are slowed down or blocked, thus reducing the propagation speed while keeping some connectivity. The worm designer selects the rate at which infected nodes try to target other hosts (hence there is a trade-off between the propagation speed and the detection time), and the detector sets the detection threshold to classify a node as suspicious or infected. The game is again a two-player zero-sum game, where the attacker (defender) tries to maximize (minimize) the overall worm spreading speed.

Finally, the case of multi-hop networks raises interesting security issues. As we saw in Section 5.5, the nodes of those networks need their traffic to be forwarded by their peers to reach their destination (as in inter-domain or ad-hoc networks). While there has been a quite abundant literature on the design of incentives to elicit intermediate nodes to relay traffic, only a few studies involving security have been done. However, as suggested in [52], implementing the appropriate security/confidentiality levels is costly for relay nodes (in terms of computation power, battery use), thus incentives may be needed there as well. Similarly to some schemes aimed at incentivizing nodes to relay traffic, the suggested solutions here are based on sanctions for future communications, and can yield an appropriate behavior equilibrium if the game is analyzed as a repeated game. The models to do so need to encompass the value of the security level experienced in the utility functions of the sending/receiving nodes, and the cost of the security levels implemented in the utility function of the relay nodes.

6.4.2 Competition among security providers

As highlighted in the previous subsections, the provision of appropriate network security is considered to be one of the most difficult and most challenging tasks in the current evolution towards the future Internet. In this subsection, we focus on the coexistence of several security solutions provided by competing companies. Since the present chapter concentrates on interactions among service providers, we will develop in more detail a competition model that was initially introduced in [183].

The objective is to represent the fact that security architectures and systems may differ widely in terms of their attractiveness for potential attackers. On the other hand, this attractiveness affects also the security option the users choose: if, for instance, a specific solution is considered extremely safe, it will first of all attract users with high-value data to protect. But at the same time this makes this solution an interesting target for an attacker, even if its market share is relatively small, since a single successful attack yields large revenues. Hence it is natural to look for stable outcomes of such interactions through the use of tools from game theory. The model encompasses these opposite effects, by considering users with heterogeneous data values making a choice among several security possibilities. Users select among a finite set $\mathcal{I}$ of security providers to protect their data. Since not all users value data protection – i.e., evaluate the cost of having their data lost or corrupted – identically, we assume user valuations distributed among the population according to the (continuous) cumulative distribution function F; i.e., $F(v)$ is the proportion of users with data valuation v or less. Each user then makes his or her choice of a provider, considering

- the level of protection offered by each provider (measured by the probability π_i of having one's data stolen or damaged when with provider i and targeted by an attacker) and
- the price set by each security provider (we denote the price of provider i by p_i).

This gives rise to an interesting negative externality effect due to the behavior of the attacker. Indeed, an attacker will be sensitive to the expected gain when targeting a specific security system, which will depend not only on the market share of the system but also on the data value of users who selected that system (or provider). We actually consider attacks that aim at stealing or compromising data stored on randomly selected hosts; i.e., the attacker tries out some random IP addresses, so that nodes are targeted independently of their choices of provider (the attacker is not aware of the security solution chosen by his targets). We suppose that a successful attack yields the attacker a gain that is proportional to the value which the data had according to the user; hence there is an interest in attacking users with high valuations. Assuming that attacks are designed to exploit a breach in a specific security solution, their expected profit is proportional to the overall data value of users selecting that solution. Indeed, consider a type-i attack (i.e., targeting system i) reaching a user's machine: the expected revenue accruing to the attacker is

$$\mathbb{E}[v\,\mathbb{1}_{\{\text{user selects } i\}}\,\mathbb{1}_{\{\text{attack succeeds}\}}] = \pi_i \underbrace{\int_{\text{users with provider } i} v\,\mathrm{d}F(v)}_{:=V_i}.$$

Thus it is reasonable to assume that attackers will focus more on security solutions i with a high aggregated value times success probability $\pi_i V_i$. Hence we consider that the number of attacks per time unit on solution i is an increasing continuous function of $\pi_i V_i$. Said differently, the commercial success of a security provider (especially regarding users with high-value data) increases its probability of being targeted.

From the point of view of a user, the likeliness of being hit by a type-i attack is therefore an increasing function r_i of $\pi_i V_i$. Upon including in the model the performance π_i and the price p_i of the security provider, a user with valuation v can anticipate an expected overall cost of

$$v\pi_i r_i(\pi_i V_i) + p_i \tag{6.12}$$

if selecting provider i.

We consider that the individual impact of any user on V_i is negligible, so that the interactions among users' choices form a non-atomic game. The externality in that game comes from the functions $(r_i)_{i\in\mathcal{I}}$, themselves stemming from the rational behavior of attackers as a response to users' choices. In terms of game theory, the situation seen from the point of view of users is a congestion game, i.e., a game where players have a finite set of alternatives and the benefit of each alternative decreases with the number of players selecting it. Nevertheless, our model has two specificities.

- It is *weighted* [214], in the sense that not all users have the same impact on the others: high-value users have a larger marginal attractiveness to attackers.
- It has *user-specific payoff functions* [213]: each user's expected cost (6.12) depends on his or her valuation v. Therefore users with different data values can select different security solutions, and Wardrop's principle does not apply here at an equilibrium of the game among users.

Let us have a look at that specific game. First we can assume that all providers set different prices, and remark that, if a user equilibrium exists, it is completely characterized (unless for a zero-measure set of users) by the total values $(V_i)_{i\in\mathcal{I}}$ of protected data for each provider $i \in \mathcal{I}$. Indeed, given values of $(V_i)_{i\in\mathcal{I}}$ determine the values of $(r_i(V_i))_{i\in\mathcal{I}}$, and thus the cost of each user with each provider from (6.12).

It is then proved in [183] that, when the total data value over the whole population is bounded (i.e., $\int v \,\mathrm{d}F(v) < \infty$) and the functions (r_i) are continuous functions, then, for any price vector $\boldsymbol{p} = (p_1, \ldots, p_{|\mathcal{I}|})$ set by the providers, a spreading $(V_i)_{i\in\mathcal{I}}$ of the total data value among the providers is the unique one minimizing a strictly convex function, which guarantees that a user equilibrium exists, is unique if all providers fix different prices, and is tractable (using the convex optimization tools seen in Section 2.2).

Those results can then be used to study a higher-level game played by providers, consisting in fixing prices so that the resulting user equilibrium maximizes revenue. Again providers are leaders in a multi-level game, and are assumed to anticipate the users' behavior. We remark that providers do not directly care about the *value* of the data they protect, but rather about their *market share*: denoting by $n_i = \int_{\text{users with provider } i} \mathrm{d}F(v)$ the proportion of users subscribing to provider i, the revenue of that provider is indeed $p_i n_i$.

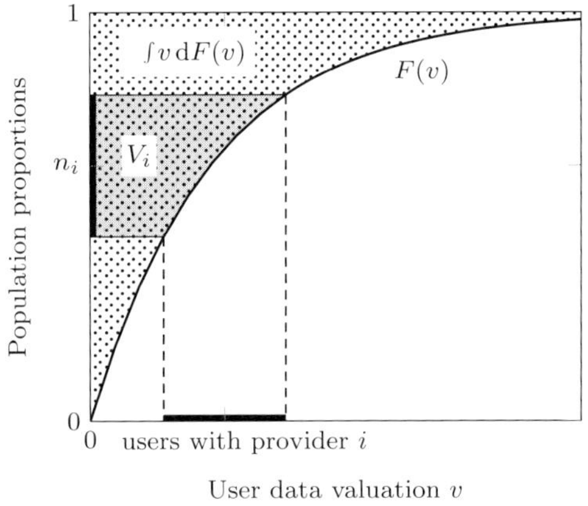

Figure 6.11 Values of interest in the security game model.

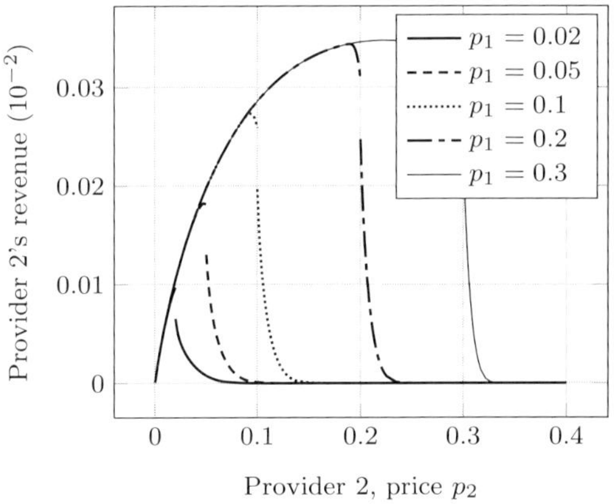

Figure 6.12 The revenue of provider 2 ($\pi_2 = 0.005$) when facing provider 1 ($\pi_1 = 0.01$) and free provider 0 ($\pi_0 = 0.05$), with $r_i(x) = 1 - e^{-x}$.

The variables of interest in the model are drawn in Figure 6.11, to summarize the interactions among the three types of actors in the model through those variables: for a particular provider $i \in \mathcal{I}$

- *users* care about the risk $(\pi_i r_i(\pi_i V_i))_{i\in\mathcal{I}}$ and the price $(p_i)_{i\in\mathcal{I}}$ when selecting a provider j with minimal total cost $v\pi_j r_j(\pi_j V_j) + p_j$;
- *attackers* focus on the target values (V_i) (balanced with the protection efficiencies (π_i)); and
- each *provider* i is interested in its market share n_i, through the product $p_i n_i$.

It is established in [183] that, if a bounded-price alternative exists (e.g., a free security solution), then providers will not raise their prices to extremely high values since their revenue will decrease to 0. This is illustrated in Figure 6.12, where we consider two security providers competing on prices, but with a free alternative (denoted by provider 0). It is actually quite likely that a price-war situation arises, i.e., that successive price adaptations to the competitors' behavior lead to outcomes where providers make no revenue. However, when studying the interaction among security providers as a repeated game, i.e., a game played repeatedly over time, equilibrium prices yielding positive revenues can be reached (due to the folk theorem, see Section 2.3).

6.4.3 Collaboration/competition issues

In the case of competing security service providers, the question of cooperation is probably more relevant than in many other fields. Indeed, due to the interactions among users, low security provided by a competitor induces a risk for one's own customers, and therefore a lower security level. This positive externality of security holds for epidemic attacks such as those based on worms or massive e-mailing.

Coalition formation can thus become efficient for providers, in terms of reputation and revenue, since joint efforts can yield a higher overall protection than separate ones. That point is raised in [212]: "businesses have a strong incentive to seek profit from users (consumers) while cooperating – and competing – in the provision of privacy and security." It is therefore very interesting to model and investigate the incentives for forming such coalitions, and whether or not full cooperation is the best solution for all providers. Answers to such questions, especially regarding the sustainability of coalitions and the effect of revenue distribution, can be obtained using the results on coalitional games.

Those collaboration issues also apply at the user level, because of the positive externality among their security decisions: if all the other nodes/neighbors are protected, they are less likely to get infected and to transmit the infection to a given host; i.e., that host benefits from the investments of its peers. This positive externality leads to suboptimal global investments in security (why invest while I can benefit from the others' efforts?), and some schemes can be developed to incentivize individual users to invest [120, 142].

7 Relations between content/application providers and access service providers

7.1 The evolution of economic relations between content/application and network providers

The previous chapters discussed relations of network access, service, or content providers with customers, and the competition for those customers. However, the different types of providers also need to interact in order to reach and attract users.

- Content providers (CPs) and service providers cannot reach users without sending traffic through network providers; they need agreements (or regulation) to ensure that traffic transfer with a "sufficient" QoS occurs.
- Network providers need the network to be attractive for users in order to increase demand and (as a consequence) revenue; this happens typically through applications and services those users want to consume. Thus rejecting traffic from some CPs might be done at the eventual expense of network providers.

Therefore, even if sometimes those different commercial entities compete or disagree (see the network neutrality debate, for example), network and content/service providers have to cooperate to run a successful business.

Up to now, CPs have been (or were) connected to the Internet similarly to users, paying a flat-rate fee for that service, mostly independently of the externality (positive through the added value to the network, and negative through the increased congestion) they create. Unfortunately, information about how much CPs pay for access is rare, if not inaccessible, probably because they are reluctant to reveal it to competitors. Those flat-rate prices being charged on both sides (content and users), while network providers usually charge between themselves according to volume, make their economic relationships more risky because the volume parameter is not taken into account and can lead to losses if not properly anticipated. This is one of the reasons why network providers want to change the pricing schemes to better encompass externalities due to traffic volumes.

Note that, in addition to the access price they pay to network providers, many content or service providers make important investments in the network infrastructure by using caches: small companies use services offered by CDNs, while bigger ones develop their own CDNs internally [270]. As illustrations of this last point, Facebook is now hosting its content on its own data centers, not using other cloud services; and Google is acting in the same way, investing in CDN activity but also in a global network in order to be able to peer with ISPs without having to pay for transit (in other words, Google is becoming an ISP). CDNs, by virtue of their being placed at the edges of the network, can provide a better QoS to users and avoid paying transit charges to content providers; this of course requires ISPs to accept having CDN servers within their networks. Because doing so may reduce their revenues, this acceptance or rejection deserves to be closely studied. This emphasizes the role of regulators ensuring that fair behavior of actors, which is beneficial to society, occurs.

The foregoing highlights what we believe to be the most important issues related to interactions between content/service and network providers.

- The trend for vertical integration in order to internalize the costs for big actors. This can be of interest for big content/service providers, but also for ISPs, which also could gain from offering their own content, their own services (cloud, CDN, etc.).
- The "fight" for new economic transaction rules between those main actors, which has brought about the network neutrality debate.

Those two items are the topics of the two next sections of this chapter. We also add a section on search neutrality, a topic related to network neutrality but where the "non-neutral" behavior could come from a search engine ranking some content higher in the organic (supposed to be objective) results of a search because it can earn more money by so doing. This type of action may also be unfair for some content providers.

Before proceeding with those aspects, we can note several recent propositions that could change the relations between content and access providers. First, the Federal Communications Commission (FCC), the telecommunications regulator in the USA, has proposed free WiFi Internet access[1] in order to ensure that Internet access is available to the whole population (engendering lively debate). This was of course supported by the main service/content providers such as Google and Microsoft (because it reduces the dominant position of ISPs), and objected to by access providers. The arguments in favor were to "make possible a nation of smart networked devices that will compose the coming Internet of things" and to fuel content innovation. Secondly, in the opposite direction and despite claiming to reject this type of non-neutral behavior, it has been noted that some big content/service providers are already starting to pay network providers to deliver their traffic to users: this is what happens with Google paying France Telecom-Orange[2] (Google being claimed to generate half of Orange's traffic) despite its position as a defender of network neutrality. A strategic goal for Google might be to create barriers to entry for newcomers not able to afford such a cost.

The relationship issue is also related to the question "Is content king?", as discussed in [237], where it is argued that connectivity is actually what matters the most, being more important than content itself. At the other extreme, some believe that Internet providers are like the paperboy delivering newspapers, and not the primal concern for users.

[1] see http://www.popsci.com/technology/article/2013-02/fcc-proposes-national-free-super-wifi-network.
[2] See http://gigaom.com/2013/01/18/google-should-be-ashamed-for-paying-carriers-to-handle-its-traffic/.

7.2 Value chain, vertical integration

7.2.1 Value chain and multi-sided markets

The world of telecommunication networks can be seen as a **value chain**. A value chain is usually defined as the chain of activities in a specific industry that is required in order to deliver a product or a service to customers (the reader can have a look at [251][3]). In the context of telecommunications, the value chain is sketched in Figure 1.5 in Chapter 1 (though it is sometimes represented in a more intricate and complicated way than as a strict chain). The chain of actors and activities includes

- network architects and device constructors, providing the network infrastructures and devices to all the other actors;
- access network providers, offering network connectivity to users;
- transit providers managing the network core, which are usually not directly connected to users but transfer traffic between access providers and/or other transit providers (those providers are usually organized in a hierarchy, see Chapter 1);
- service providers such as search engines facilitating access to any content or application whatsoever, CDNs, P2P networks, clouds, social networks, security providers, etc., all simplifying access, quality, diffusion, and connectivity; and
- CPs proposing (or creating) information and entertainment.

This can be connected to the **supply chain**, consisting in the activities and resources involved in moving a product or service from supplier to customer; or **industrial organization**, where firms are the basic units for the analysis. Analyzing the value chain as a whole is of more relevance than looking at just relations between a seller (or competing sellers) and customers. This difference can be summarized as a **multi-sided market**, where an actor has to "negotiate" with multiple levels of the value/supply chain, with respect to a single-sided market, where one finds just one type of sellers and one type of buyers. Our goal in this section is not to survey the existing works on value chains and supply chains, but rather to highlight the relevance of this type of modeling in the telecommunications area (the interested reader is encouraged to look at [228]).

The impact of interactions between different levels of the supply chain on *value creation* can be analyzed like in [127], highlighting the interest of information technology (IT) service platforms that enable the development of application services. The different stakeholders in this context are (i) the application service users, (ii) a platform provider offering the environment where the services/applications can be developed, and (iii) the service developers creating the applications. There are parameters impacting the value perceived by the actors of the chain. In this case, they are the QoS, range of offers, number of users (as a positive externality), and cost. Users get a positive value from the QoS provided by the platform, from the number of services, and from the user base (a "social network" effect), but of course have to pay a cost for using the service. Developers also benefit from the QoS provided by the platform, from the base of customers (through

[3] See also, for instance, https://en.wikipedia.org/wiki/Value_chain.

sales), but there is a negative externality due to the presence of other developers (acting as competitors); they experience a development cost as well. The service platform takes a proportion of developers' revenues (sales) as a tax, but incurs costs to support users and maintain the QoS. The multi-sided interactions/game need to be analyzed. This is actually the type of thing that was used for application stores in Section 4.3.

To illustrate the effect of two-sided markets, we summarize a simple model from [76] considering an ISP, with CPs and users being charged to access the network. The price paid by CPs is p_c, while that paid by users is denoted by p_u (each time, that network access price is paid to the ISP). There is a so-called "native" demand for access on both markets (users and CPs) depending on prices, $D_c(p_c)$ and $D_u(p_u)$ for CPs and users, respectively. But the total demand q_i at market $i \in \{u, c\}$ depends also on the demand from the other market $j \in \{u, c\}$, $j \neq i$, as a positive externality (the more users, the more CPs are expected, and vice versa). Let us consider the specific form

$$q_i = D_i(p_i) + e_{ji} D_j(p_j),$$

where e_{ji} represents this externality effect of market j on market i. The goal of the ISP is then to maximize its revenue, i.e., to solve the problem

$$\max_{p_c, p_u \geq 0} p_c q_c + p_u q_u.$$

But, for a fixed price p_i on the market i, the optimal price on the other market j may be different, being given by the solution

$$\begin{aligned} p_j^*(p_i) &= \arg\max_{p_j \geq 0} p_j q_j + p_i q_i \\ &= \arg\max_{p_j \geq 0} p_j D_j(p_j) + e_{ij} p_j D_i(p_i) + p_i D_i(p_i) + p_i e_{ji} D_j(p_j). \end{aligned}$$

Example 48 To better understand the mutual impact among markets, let us assume linear demands $D_c(p_c) = \max(0, d_{c,0} - \alpha_c p_c)$ and $D_u(p_u) = \max(0, d_{u,0} - \alpha_u p_u)$, with α_c, α_u, $d_{c,0}$, and $d_{u,0}$ positive constants. Just to illustrate, let us forget the constraints associated with the max. Taking first-order conditions, by differentiating with respect to p_c and p_u, gives the system of equations

$$\begin{aligned} 2\alpha_c p_c + (e_{uc}\alpha_u + e_{cu}\alpha_c) p_u &= d_{c,0} + e_{uc} d_{u,0}, \\ 2\alpha_u p_u + (e_{uc}\alpha_u + e_{cu}\alpha_c) p_c &= d_{u,0} + e_{cu} d_{c,0}, \end{aligned}$$

giving as a solution

$$\begin{aligned} p_c &= \frac{-e_{uc}\alpha_u d_{u,0} + e_{cu}\alpha_c d_{u,0} + e_{cu}^2 \alpha_c d_{c,0} - 2d_{c,0}\alpha_u + e_{uc}\alpha_u e_{cu} d_{c,0}}{-4\,\alpha_c\alpha_u + e_{uc}^2\alpha_u^2 + 2e_{uc}\alpha_u e_{cu}\alpha_c + e_{cu}^2\alpha_c^2}, \\ p_u &= \frac{-e_{cu}\alpha_c d_{c,0} + e_{uc}\alpha_u d_{c,0} + e_{uc}^2 \alpha_u d_{u,0} - 2d_{u,0}\alpha_c + e_{cu}\alpha_c e_{uc} d_{u,0}}{-4\,\alpha_c\alpha_u + e_{uc}^2\alpha_u^2 + 2e_{uc}\alpha_u e_{cu}\alpha_c + e_{cu}^2\alpha_c^2}. \end{aligned}$$

For arbitrary values $d_{u,0} = 50$, $d_{c,0} = 10$, $e_{cu} = e_{uc} = 0.5$, $\alpha_u = 0.5$, and $\alpha_c = 0.6$, we get $p_c \approx 5.2925$ and $p_u \approx 52.09$. The case when $e_{uc} = e_{cu} = 0$ gives, on the other hand, $p_c \approx 8.33$ and $p_u = 50$, leading to some differences in prices that should not be ignored. The "optimal" revenue when a two-sided market is considered is 1525.07, while the revenue is 1291.67 when it is ignored.

7.2.2 Vertical integration

Within the supply/value chain, there might be interest in vertically integrating services. The term **vertical integration** means that a company owns several steps of a supply chain. Thanks to vertical integration, a company can lower the transaction costs (by controlling them), eliminate **double marginalization** (defined as the "exercise of market power at successive vertical layers in a supply chain," which occurs when companies at successive levels have monopoly power and each company reduces the output from the competitive level to the monopoly level, hence generating overall losses), and better synchronize supply and demand. In other words, it increases efficiency and gives more opportunities for making profit. It is often believed that vertical integration will make innovation more relevant and quicker.

During the twentieth century vertical integration used to be the common practice in telecommunications (i.e., in telephony), with operators building their own telephones, installing and running their own lines and necessary equipment, and offering telephony/fax/telex services. This norm has been counteracted by regulation, increased competition (which is known to diminish the vertical effects), and diversification of the range of means of telecommunication. For example, in the USA, the regional operating companies were separated from the big operator AT&T under the antitrust rule in 1984. During the same year, British Telecom was privatized, and competition increased.

But vertical integration is still a goal that is being sought in telecommunications. The most typical form is bundling content and access into a single service. Below are some notable examples of integration.

- ISPs propose portals with news and other applications, incentivizing users to go through their own services or to click on the ads they display (increasing their revenue *à la* Google).
- Mentioning the notion of convergence of networks is also relevant in this area. ISPs also provide TV channels (e.g., Orange in France). In the UK, Vodafone acquired the Cable & Wireless Worldwide business for more than one billion pounds. Vodafone will then use the fixed network to improve its backhaul activity and service capabilities, generating an advantage with respect to competition, to the (expected) benefit of users.
- Companies at different levels of the supply chain merge to acquire a stronger position. A striking illustration is the merging of AOL (an ISP) and Time Warner (a media and entertainment leader) in 2000, which was the largest merger in American business history, worth $350 billion, but eventually a failure. More recently, the FCC and the US Department of Justice approved in January 2011 a merger combining Comcast, a major ISP and cable TV operator, and NBCUniversal, a TV program maker.
- Google is one of the most well-known examples of vertical integration, as a company that initially focused on a single activity but later expanded its range of services. We already highlighted this broad and increasing range of activities by the company at the end of Section 4.1. Briefly, Google started in 1998 with its search-engine service, making money through advertising. Google extended the range of its (free) services, offering Gmail e-mail services, Google Calendar, YouTube, Google News, Google Earth and Google Maps, the browser Chrome, and the operating system Android. The

general principle is to incorporate Google search tools into those services, generating as a consequence more advertising revenue. It also induces more competition at those service levels, thus providing a cost reduction for users, whose demand for services will be increased, potentially creating even more advertising revenue. In addition, as of 2011, Google owns more than 900,000 servers, which are used for performing searches but also for cloud computing services. Moreover, in order to reduce transit costs and again to attract users to telecommunication services and thus to Google ads, the company is starting to run its own ultra-high-speed fiber network[4] and is launching a free in-home wireless broadband service.[5] Google also acquired the mobile device maker Motorola in order to build its own smartphones and television set-top boxes.

- Microsoft, which initially focused on providing operating systems, has also moved towards vertical integration. A notorious issue was the incorporation of its web browser into its operating system, the topic of a major debate that almost ended in the company being dismantled. It also commercializes major and almost inescapable software included in the Windows Office suite, but also developed its own search engine Bing to compete with Google and get advertisement revenue, the Xbox game console, the Surface Tablet, and other hardware devices.
- Apple is also a major example of vertical integration, combining hardware (computers, iPod, iPhone, and iPad) and software services (its own operating systems, iTunes, and Apple Store taxing revenues from application developers).

The bundling through triple-play or quadruple-play offers from operators can be seen as a form of vertical integration, but is rather a mixture of vertical and horizontal (at the same level) integration, which is sometimes called **verizontal integration**.

We can remark that vertical integration has experienced a noticeable amount of success in many closely related areas, such as cable television in the 1980s in the USA, with cable operators investing a lot in the network in order to be able to respond to demand for content, and in the media business.

Vertical integration creates concerns, though; hence there have been actions by regulatory bodies to prevent "excessive" integration. One notable example mentioned above was the action in 2001 against Microsoft, which did not provide customers with a choice of Internet browser. This was a main concern because Microsoft had a quasi-monopoly on operating systems (thus leaving users with hardly any choice in terms of browser); hence the company came close to being dismantled under the USA's antitrust law. This issue of choice of web browser regularly comes back, such as recently with the introduction of Windows 7 and Windows 8 operating systems, with Microsoft offering Internet Explorer each time as the default web browser. As another example, the Canadian regulator, the CRTC, also prevented excessive vertical integration by denying Bell's attempt to acquire Astral Media (a pay-TV network) in 2012. An "excessive" vertical integration is a merger that can abusively reduce competition: this is what worried the CTRC because Bell was already the largest telephone company and the largest ISP, owning big

4 See https://fiber.google.com/about/ or http://www.thehindubusinessline.com/industry-and-economy/info-tech/.google-unveils-worlds-fastest-internet-connection/article3696275.ece.

5 See http://www.google.com/tisp/.

TV programming services such as CTV and TSN. If the level of integration is too high, without strong competition at every given step of the supply chain, the company is likely to favor its own services at other steps and weaken competition up to extinction. It may create barriers to entry for new competitors.

A parallel with (and return on experience from) the vertical integration in cable TV in the USA can be made [298, 299, 306]: while financing of programs can benefit from vertical integration, the main concern is probably the lack of horizontal competition. This was why the FTC made the merger of Time Warner and Turner Broadcasting in 1996 conditional on the obligation for Time Warner to include in its offer a news channel in addition to Turner's CNN. In cable TV, it is clear that favoring one's own content was common practice, but it is also rightly remarked that the Internet programming market has still not reached the level of maturity of cable TV, making it dangerous to say that the same type of conclusions can be drawn. A difference is the almost-zero Internet capacity cost to store content and make it available to users. Moreover, the presence of aggregators for content (P2P, etc.) will make the control of the whole value chain more difficult for vertically integrated ISPs.

There has been a flood of works about when vertical integration is beneficial or harmful, following the Chicago school of antitrust analysis [168, 306]. According to those works, there are preconditions to be satisfied for vertical integration to be harmful. The preconditions are as follows.

(i) The integrated actor has a powerful position in a concentrated market.
(ii) The "secondary" market in which the provider will be integrated must be concentrated and protected by barriers to entry.

How can we measure whether such conditions are satisfied, especially in the telecommunications world? We need to measure the degree of market concentration within a market. This can be done by using the **Hirschman–Herfindahl index (HHI)**. The HHI measures the concentration of a market among a number of competitors (i.e., the level of competition), and is one of the most commonly used indices, being actually the standard concentration measure for antitrust enforcement purposes in the USA. To define it formally, consider n (horizontal) competitors with market shares $\{\alpha_1, \alpha_2, \ldots, \alpha_n\}$, which satisfy the constraint $\sum_{1\leq i\leq n} \alpha_i = 1$ and $\alpha_i \geq 0\ \forall i$. The HHI is then defined as

$$\mathrm{HHI} = \sum_{1\leq i\leq n} \alpha_i^2, \tag{7.1}$$

and lies in the interval [0, 1]. Higher values of the HHI indicate a larger degree of concentration (and hence a lower level of competition). It is commonly considered (somewhat arbitrarily, but in order to give a good idea) that the market is concentrated if the HHI is above 0.18. The antitrust authorities are likely to challenge a vertical merger only if this is the case in the primary market, i.e., the market in which the company's position is already established. The same type of result must then be checked for the secondary market, defined as the market into which the company seeks to carry out vertical integration. This must also be looked at in relation to the geographic parameter

Table 7.1 National market shares for broadband access in the USA (data for 2004). Note that, due to rounding, the sum of market shares is not exactly 1.

Provider i	Market share α_i (approached value)
Comcast Cable Communications	0.1959
SBC Communications	0.1430
Time Warner Cable	0.1096
Verizon Communications	0.1008
Cox Communications	0.07256
BellSouth	0.05872
Charter Communications	0.05278
Earthlink	0.03821
Adelphia Communications	0.03810
Cablevision Systems	0.03687
Qwest Communications	0.02801
Bright House Networks	0.02031
Covad Communications	0.01493
Sprint	0.01378
Mediacom Communications	0.01028
Insight Communications	0.009273
Alltel	0.006807
RCN	0.006219
Hughes Direcway	0.006163
Citizens Communications	0.005939
Cable One	0.004986
Century Tel	0.004006
Cincinnati Bell	0.003670
Others	0.01961

(how regionally located providers, with strong positions or not, influence the whole market) though, with the regional factor complicating the analysis.

To illustrate the use and values of HHI, we can for example look at the data about national broadband access market shares in the USA summarized in Table 7.1 (values taken from [168]). We can see that the HHI is considered low enough to ensure that any vertical integration will not harm economic efficiency. Indeed, the HHI computed as the sum of squares of market shares is then $\sum_i \alpha_i^2 = 0.09917$, far below the 0.18 threshold given above.

Looking at the Internet broadband market shares in France (figures as of June 2012, taken from Wikipedia) given in Table 7.2, we can easily see that the HHI is 0.3342, thus vertical integration in that type of market needs to be placed under scrutiny.

The global web-browser market shares are described in Table 7.3 (data from http://www.netmarketshare.com/). The HHI is here 0.3826, which is the reason why the already existing level of vertical integration including Internet Explorer is so often scrutinized.

Finally, looking at search engines in use worldwide, the global market shares are described in Table 7.4 (data from http://www.netmarketshare.com/). The HHI is here

Table 7.2 French broadband market shares in June 2012

Provider i	Market share α_i
Orange	0.4170
Free	0.2301
SFR	0.2295
Bouygues Telecom	0.0577
Others	0.0657

Table 7.3 Global browser market shares in April 2013

Browser i	Market share α_i
Internet Explorer	0.5581
Firefox	0.2030
Chrome	0.1635
Safari	0.0538
Opera	0.0173
Others	0.0043

Table 7.4 Global market shares for search engines in April 2013

Search engine i	Market share α_i
Google	0.8322
Yahoo!	0.0801
Bing	0.0540
Baidu	0.0165
AOL	0.0038
Ask	0.0031
Excite	0.0001
Lycos	0.0000

0.7022; vertical integration by Google is thus a concern with such a dominant position in its primary market.

According to the Chicago school of antitrust analysis, vertical integration, or vertical arrangements between companies, are often economically efficient even if there is a monopoly of the vertically integrated entity at a given level of the supply chain (hence the principle of looking at the primary *and* secondary markets). Indeed, even the provider with a monopoly position has an incentive to ensure that consumers will have the largest valuation for the whole value chain.

The Chicago school of antitrust analysis has had an important impact in the USA, with the official Vertical Merger Guidelines promulgated in 1984 following the guidelines of the school on whether or not a vertical merger will have an anticompetitive effect.

7.3 The network neutrality issue

7.3.1 Introduction and historical facts

We have already sketched some concerns about the relations between ISPs and content/service providers in the first chapter and in Section 7.1.

- Especially due to vertical integration, some ISPs may favor their own content (or content they are being paid to favor) over external content, e.g., by providing a better QoS.
- With the constantly decreasing transit prices, ISPs are worried that revenues will become more hazardous and that investment in the network infrastructure will become more difficult. They are thus asking that CPs should pay transit fees (i.e., pay for traffic transferred by distant ISPs). Indeed, distant providers, particularly transit ones, have to transfer traffic for content providers while getting decreasing revenues from those transfers, which, on the other hand, require more and more resources. Moreover, users and content/service providers pay flat-rate fees while transit costs between ISPs are volume-based; ISPs thus have to estimate future traffic volumes to ensure sufficient revenue, which makes their revenue management more risky. Table 7.5 lists the values of Internet transit prices, illustrating their constant decrease.

Those suggestions to discriminate against traffic or to charge content providers seem in contradiction with the traditional vision of the Internet providing a universal connectivity and serving all packets in the same way. This has engendered the so-called *network neutrality debate*.

But discussing network neutrality first requires one to define what a neutral (or non-neutral) behavior is. Surprisingly, there is no well-defined notion of such neutrality. Authors sometimes make a distinction between *strong* and *weak* neutrality, where a strongly neutral network is one that does not allow one to manage packets differently in any way whatsoever, while a weakly neutral network just prohibits user discrimination but allows discrimination between application types. In the former case, the network is understood to be as "dumb" as possible, just carrying packets, and the "intelligence," if any, is rather placed at the source and destination. The idea in the latter case, on the other hand, is that some applications may have more stringent QoS requirements (typically, video needing limited delay and latencies in comparison with e-mail services), hence there is no harm but only benefits for users when there is a need to discriminate in the case of congestion. A quite generally used definition of neutrality was introduced by Professors Timothy Wu and Lawrence Lessig, namely that "Network neutrality is best defined as a network design principle. The idea is that a maximally useful public information network aspires to treat all content, sites, and platforms equally." A kind of "official" definition has been summarized by the four following items, and was provided by the FCC in the USA in 2005 [97]:

(i) no content access can be denied to users;
(ii) users are free to use the applications of their choice;

Table 7.5 Internet transit prices per Mbps (the last values are *projected* prices). Data from drpeering.net.[6]

Year	Price per Mbps (in US$)
1999	800
2000	675
2001	400
2002	200
2003	120
2004	90
2005	75
2006	50
2007	25
2008	12
2009	9
2010	5
2011	3.25
2012	2.34
2013	1.57
2014	0.94
2015	0.63

(iii) they can also use any type of terminal, provided that it does not harm the network;
(iv) they can competitively select the access and content providers of their choice.

To give more details on the issues at hand, let us first describe the historical developments and arguments of the various actors. Non-neutral initiatives came from the increasing traffic asymmetry between ISPs, mainly due to some prominent and resource-consuming CPs that are usually connected to a single ISP. A common example is YouTube traffic accounting for a non-negligible proportion of global Internet traffic accessed by all users while being hosted by a single Tier-1 ISP.[7] ISPs started complaining about having to transfer large amounts of traffic from some CPs that were not directly connected to them (and thus not paying anything to them) to their customers or to other network providers and not participating in the necessary network maintenance and investments. The first notable complaint came at the end of 2005, from Ed Whitacre (CEO of AT&T) [240]. He claimed, deploying an argument that we have already highlighted above, that distant CPs were using the AT&T network for free in order to reach end-users in an unfair way because, on the other hand, transit prices were decreasing. Those decreasing prices were also put into perspective with the increasing revenues that

6 See http://drpeering.net/white-papers/Internet-Transit-Pricing-Historical-And-Projected.php.

7 See also http://techcrunch.com/2011/05/17/netflix-largest-internet-traffic/, for instance, where the distribution of traffic in the USA and Canada is described, showing also that Netflix is actually starting to be the largest traffic consumer.

content or service providers were making, thanks notably to an estimated spending of about US$34.5 billion on online advertising in the USA alone (see Chapter 4 for a more detailed description of the increasing advertising revenues). The conclusion was that CPs should participate in the infrastructure investments. The threat was to lower the quality of service of CPs that did not pay any fee to the ISPs, or even to block their traffic. The debate became more heated in 2007 when US provider Comcast blocked (resource-consuming) P2P applications such as BitTorrent, arguing that P2P content is mostly illegal, as it had earlier in 2005 when Madison River Communications blocked ports used by the VoIP service of Vonage.

These possibilities of degrading or blocking some traffic have led to protests from CPs and user associations, and many consultations have been conducted worldwide to determine whether or not the network should be neutral [20, 303], bringing forth at the same time a lot of questions on the economical, legal, technical, and political levels.

7.3.2 Arguments of proponents and opponents of neutrality

It is interesting at this point, before illustrating how modeling can be used, to enumerate the arguments from both sides (which are listed here each time in an arbitrary order). We have tried to be as objective as possible, so that the reader can form his or her own opinion. This part is an extension of the brief description that can be found in Subsection 1.2.4.

Arguments of the proponents of neutrality: an idealistic and humanist view

(i) The *layer-model argument*. One of the known (or at least believed) reasons for the success of the Internet is its organization in *layers*, as defined by the OSI model. The network is then abstracted as described in Table 1.2 into seven layers (physical, data-link, network, transport, session, presentation, application) such that each entity at a given level/layer interacts only with the layer beneath it, and provides "capacities" to the layer above it. The advantage is that the relations between layers are mediated through standardized protocols, meaning that an entity can manage its operations on its own, provided that the protocols employed to interact with neighboring layers are respected. This organization allowed a high degree of interoperability and the development of new ideas at all levels. But, if neutrality were no longer to be enforced, it is feared that this successful and cheap model would be destroyed. Indeed, it is believed in that case that ISPs will develop exclusivity arrangements with content and application providers, using proprietary protocols between layers, destroying the model and reducing competition (as an underlined goal) because competitors will not be able to use those protocols. It is claimed that ISPs will then increase their market power and easily be able to discriminate between content or service providers by "naturally" favoring the content or service providers affiliated to them (i.e., using their proprietary protocols). The use of such proprietary protocols is also thought to be a drag on the development of services (i.e., innovation) by creating barriers to entry.

(ii) The *freedom-of-speech argument*. This often-invoked important argument is closely linked to the first item: thanks to the layered architecture, the network can hardly discriminate according to the identity of an entity. The technical aspects of the layered architecture bring (i) relative anonymity, (ii) a decentralized implementation, (iii) the possibility of reaching multiple access points, (iv) the lack of a simple system to identify content, and (v) tools for encryption. A network being open and hence without proprietary (i.e., restricted) rules complicates its control by governments or other types of organizations, which could otherwise determine the identity of users much more easily. The open network, for instance, has been shown to be very helpful, if not a key component, in the Arabic revolutions in the early 2010s, which is the reason why oppressive countries are trying to make the network non-neutral in order to control or exclude data from specific applications (e.g., Twitter, Facebook) or from some origins (news websites). The argument is thus that changing the current mode of functioning of the network could harm the development of democratic values worldwide.

(iii) The *universality argument*. The universality principle driving the Internet is usually defined as "all consumers are entitled to reach meaningful content, whatever the technical limitations of their service; they are also entitled to attach any type of device to the network; and to run any application." A goal is to keep the whole population informed (by allowing unlimited access to information and culture) and able to communicate. An idealistic view is that this could decrease the gap between rich and poor people (even though, in practice, open access has often been more profitable to already educated people). We can remark here that the universality principle is not in contradiction with *weak neutrality*, which corresponds to "reasonable network management." Service differentiation, based on price or, for some applications, depending on their QoS requirements, could be offered as soon as a to-be-defined "minimal" network service of "sufficient" quality is offered. There is a fear that, if no regulation on neutrality is imposed, the Internet will become like the cable TV network in the USA, with some channels preselected to be broadcasted by the cable operator under the conditions imposed by this cable operator (hence the position of this operator would be too powerful), and subscribers to any particular operator not having access to the full range of channels/services [298, 299].

(iv) *Economic arguments*. Even from an economic point of view, proponents of neutrality believe that non-neutral behavior could be harmful.

1. The first argument developed is that, again, without neutrality last-mile ISPs have too much market power; they will discriminate between content providers, favoring the ones that are connected to them. But that will have an economic impact since end-users will experience less choice, and then will potentially be less interested in the network, reducing demand and hence revenues both for content providers and for ISPs.
2. The potential entry fees or restrictions on the application/service side will (or at least might) also reduce innovation by application developers, especially new ones, hence again decreasing demand from end-users, who become less interested in joining the network, and as a consequence potentially decreasing

the revenues of ISPs. It is said as an example that Google-like businesses starting in a garage will become less likely to happen.

3. It is also often argued that, if revenues decrease and infrastructure investments are no longer possible, why are ISPs' costs so difficult to get? With transparency, data could be verified, and ISPs could be trusted. While transit prices keep decreasing, it has actually been shown that the broadband revenue is on average increasing (with an increasing number of subscribers), while the **average revenue per user (ARPU)** per unit time is relatively stable [270]. Those data illustrate that ISPs' revenues are approximately proportional to the number of subscribers. Proponents of neutrality complain that ISPs actually want to earn money from both ends, end-users and CPs (instead of just from end-users) in order to increase their revenue.
4. Enforcing the universality principle is also said to have been implemented in many other economic areas, such as roads, canals, mail, etc., with potentially paid highways coexisting with free small roads, or fast but expensive mail service with cheap regular mail services. There has always been regulation imposed when dealing with some kind of "necessary" goods/services, even if resources are owned by private companies, and up to now this has been implemented without major concern.

Arguments of the opponents of neutrality: an economic view

Regarding the arguments from opponents of neutrality, the interested reader can also have a look at [168], whence most of the explanations from economists have been taken. The main arguments are as follows.

(i) The *(initial) investment argument*. As has already been implied, neutrality has been argued by ISPs to constitute a disincentive to invest in the network infrastructure because the transit prices keep decreasing. ISPs complain that proponents of neutrality are focusing on the network content and services innovation, whereas getting a sufficient QoS for the connection is of importance too; thus ISPs should be taken care of too.

(ii) *Shift of powerful position*. Imposing a regulation on neutrality adds or shifts power (even) closer to the CPs and service providers, especially if there is limited competition. It is argued that competition occurs rather at the ISP level, while competition is, on the other hand, limited for most types of service/content. As an interesting remark, even governments see the position of CPs as an issue: countries increasingly complain about the small amount of taxes paid by service providers, which, thanks to globalization and usually not needing to be located nearby their users, can locate their headquarters in tax havens. There is a related worry about the online advertising revenue not being taxed as is usually done.

(iii) *Vertical integration*. We have already discussed vertical integration in the previous subsection, providing examples, explaining its interest, and evoking the potential impact on social welfare. None of the preconditions claimed as necessary to preclude vertical integration by the Chicago school of antitrust analysis can be satisfied in the current telecommunications industry, as shown in Table 7.1 for the USA, and therefore vertical integration is said to be more efficient than keeping the current neutral model. It is also often said that the Internet is more like a media business, and should rather be treated that way instead of as a telephony/telecommunication economic business. In other words, this means that ISPs are there only for content distribution; i.e., broadband is not the primary goal for end-users, but rather content/services is/are what is sought. In the media business, again, vertical integration has been implemented with evident success. Opponents of neutrality are thus strong defenders of vertical integration. It is even argued to be not necessarily against competition, because competition can be fostered at the network (ISP) level, as was done in the cable/satellite TV business in the USA. Indeed, cable operators were required to improve their infrastructure in order to provide a better QoS when facing competitors (satellite TV operators); getting money from content does not mean that the network did not have to be upgraded. In addition, vertical integration fosters content innovation by competitors in order to attract more customers. All this means that it is competition that best spurs innovation (see [306] for a theoretical and empirical discussion about the relation between market concentration and innovation). Moreover, it is believed that there is no reasonable threat to cut off access to major CPs or service providers such as Amazon, Facebook, etc., otherwise the pressure from unhappy subscribers to change providers would be too strong.

(iv) *Service differentiation/segmentation.* From an economic point of view, service differentiation is the most efficient way to proceed: it allows one to increase revenues, but also social welfare, by adapting prices and services to user preferences. Service differentiation has shown its efficiency in many other areas, such as the airline industry, hospitality, etc., where its success is undisputed [285]. Service differentiation would mean treating packets differently according to their type or origin, and hence being non-neutral. If it were not implemented, it is argued that the declining costs for users (due to Bertrand competition) would actually result in a limited amount of surviving companies and natural monopolies, an undesired outcome. It is additionally argued that the Internet already implements some differentiation anyway, even though that is not desired, with protocols such as TCP giving a larger throughput to users with a smaller round-trip time to the server.

(v) Regarding the *layered architecture*, it is fully admitted that the (past) success of the Internet is related to the layered architecture. However, the current model is considered as not necessarily being written in stone: it could evolve with economic changes and needs. The Internet in this sense is not a historic monument that could not move from neutrality, but rather an evolving infrastructure. Using proprietary protocols could simplify the advent of some applications and foster innovation because there would be a larger range of networking tools. Actually, Internet users are not particularly skilled in general, and do not care about the layered architecture as long as quality is present. The "optimal" architecture may be different from the layered one in the near future, and there is no reason necessarily to keep the existing one in place.

(vi) *The universality principle may harm investment.* If universal access is granted, the question then becomes "why should ISPs invest in capacity and technologies, since they will have a very limited lever?" Moreover, granting access to users means a very small access price at least for the poor (something that the French government tried to impose in 2011[8] even though we doubt that poor people will all jump on Internet access at 20 euros per month for broadband plus wired telephony). It is conceded that non-demanding applications such as web browsing and e-mail could be granted for free or at a small fee (since their use does not create congestion and does not require further investment), but not bandwidth-consuming applications. The universality principle lauded by governments is, on the other hand, scoffed at by the same institutions. France, for instance, passed a law in 2009 disconnecting users from the Internet if they illegally download copyrighted material (the so-called HADOPI law). This raised a debate at the legal level (the French Constitutional Council asked the government to modify its implementation because it was in contradiction with the Declaration of the Rights of Man and of the Citizen coming from the French revolution: the ISPs will not decide by themselves to filter traffic, but will implement the decision if it comes from a judicial authority). This differentiates the treatment of legal and illegal traffic, but, with the notion of legality being considered differently in other countries, this could work against freedom of speech.

(vii) The *security aspect*. It is advocated that proprietary protocols (and vertically integrated services) are usually more secure than protocols from an open network, because they are less based on trust and cooperation between nodes. This is an attractive argument for regulators.

The neutrality principle is then probably more a question of transparency and efficiency rather than a question of non-discrimination.

Which arguments are the most relevant?

The previous lists of arguments show (in our opinion) that we actually have two different worlds with two different goals:

- an idealistic (neutral or weakly neutral) network as imagined initially by scientists, with an organization in layers, available at a low cost, and for which end-to-end connectivity and universality are the key issues; and
- a purely economic (non-neutral) view of the network, looking at an efficient economic management.

A strict comparison is thus difficult since it depends on one's view of what the network should be. Network neutrality is thus rather a political question about whether we want a commercial or a non-commercial network.

We can also remark here that, whatever the goal, the need for regulation may differ depending on the country considered. Indeed, there are differences in the competition

[8] See http://www.choisir.com/box/fai/bientot-un-tarif-social-internet-pour-les-pauvres.html.

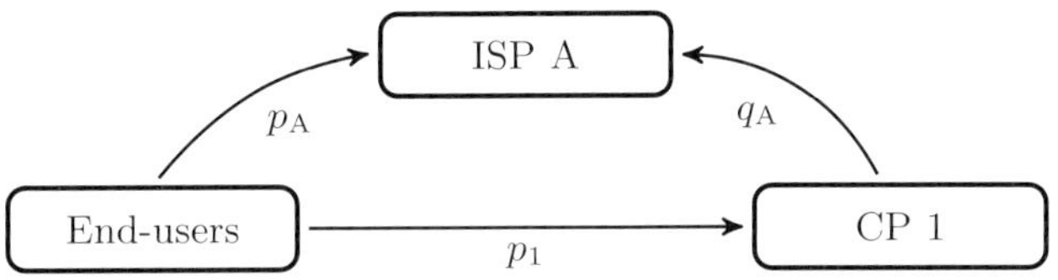

Figure 7.1 Charging for interactions between stakeholders. Prices p_1 and p_A are positive flat rates, whereas q_A is a positive per-volume unit price.

between ISPs that could lead to a different game output if a careful analysis is not performed: in the USA, for instance, competition is not as fierce as in Europe for broadband Internet access. Moreover, the fact that the main content/service providers come from the USA may have an impact on the political decisions.

7.3.3 Modeling content and network providers' interactions and analyzing neutrality issues

There exist many works trying to model network-neutrality-related issues that perform analyses drawing on game theory (see, among others, [14, 22, 43, 50, 103, 128, 174, 175, 222, 233] and the references therein). We illustrate in this section how modeling can help one to draw conclusions.

One content provider and one access provider

The model we present is based on the one in [65], with one CP and two ISPs, which we first simplify by considering one CP, whose parameters will be indexed by 1, and a single ISP, named A. The *flat-rate* subscription fees charged to users to access the CP and ISP A are respectively denoted by p_1 and p_A. We are going to study and compare neutral and non-neutral outcomes; therefore, we introduce a **side payment** $q_A > 0$ that the CP has to pay *per unit of volume* to ISP A. In the neutral case, q_A is just fixed to zero. The charges imposed by actors on other players are summarized in Figure 7.1.

We assume a continuum of end-users, of mass 1 without loss of generality. We assume that users first choose whether to subscribe to the ISP (depending on price), and then, in the affirmative case, whether to subscribe to the CP. We separate this choice from broadband access, because users also want to access the network for other reasons, such as e-mail, web browsing, etc. We consider the discrete-choice/stickiness model of Section 5.3 (which has been considered in many of our examples) for the choice of subscribing or not to ISP A, with a "cost" p_0 of not subscribing to the Internet, leading to a mass (or proportion) of users subscribing to Internet access (through A) of

$$\sigma_A = \begin{cases} \dfrac{p_A^{-\beta}}{p_A^{-\beta} + p_0^{-\beta}} & \text{if } p_A > 0 \\ 1 & \text{if } p_A = 0. \end{cases} \tag{7.2}$$

Now, the willingness to pay a subscription to the CP is assumed to follow an exponential distribution with mean value $1/\alpha > 0$ over the population, independently of the choice of ISP, leading to a proportion $e^{-\alpha p_1}$ of the ISP subscribers deciding to subscribe also to the CP, and hence a total mass $\sigma_A e^{-\alpha p_1}$ subscribing to both. We denote by D_0 the average volume the CP subscribers download from the CP, giving a data volume of

$$D_A = D_0 \sigma_A e^{-\alpha p_1}, \tag{7.3}$$

which will be needed in order to compute the *volume-based* transit costs for transmission from the CP to the ISP.

The ISP's revenue is

$$U_A = p_A \sigma_A + q_A D_A$$

and the CP's revenue equals

$$U_1 = (p_1 \sigma_A e^{-\alpha p_1} - q_A D_A) = (p_1/D_0 - q_A) D_A,$$

where we include the subscription gains and volume-based side payments.

The user welfare *associated with the existence of the CP* is then (looking at the "gain" $x - p_1$ for a subscriber willing to pay x)

$$\begin{aligned} \mathrm{UW}_{\mathrm{CP}} &= \sigma_A \int_{p_1}^{\infty} \alpha e^{-\alpha x}(x - p_1)\mathrm{d}x \\ &= \sigma_A \frac{e^{-\alpha p_1}}{\alpha} \\ &= \frac{D_A}{\alpha D_0}. \end{aligned}$$

User welfare can be decomposed into two components: the user welfare due to the existence of the CP (computed above), and the user welfare due to the presence of the ISP. For the latter part, we take a form derived from the one computed in Chapter 5, where we expressed in (5.21) the relative gain that users get from the ISP(s) being present. Since the expression was not in a monetary-equivalent unit (but rather some perceived value without any specified unit), we convert it by applying the transformation $p_0 \times (\exp(\cdot) - 1)$, which yields

$$\mathrm{UW}_{\mathrm{ISP}} = p_0 \left(\frac{p_0}{p_A}\right)^{\beta}.$$

(We remark that, since user utility functions are not quasi-linear – i.e., expressed in a monetary-equivalent form – it is difficult to define $\mathrm{UW}_{\mathrm{ISP}}$, and other choices of definition can be made.) The global user welfare generated by the system (ISPs and CP) is therefore

$$\mathrm{UW} = \mathrm{UW}_{\mathrm{CP}} + \mathrm{UW}_{\mathrm{ISP}}. \tag{7.4}$$

Whatever the value of q_A assumed fixed first, CP 1 and ISP A choose respectively their prices p_1 and p_A maximizing their revenue, but ISP A does it first on a larger time scale, anticipating the decision of CP 1. The ISP is therefore the leader in a Stackelberg game; hence we perform an analysis of the interactions among providers using backward

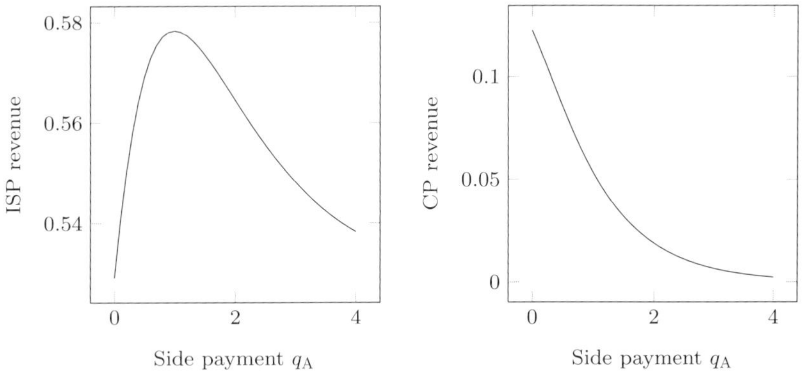

Figure 7.2 ISP and CP revenues, when the side payment q_A varies.

induction. The first-order condition then gives for CP 1 (with p_A and q_A fixed)

$$\frac{\partial U_1}{\partial p_1} = \frac{D_A}{D_0} - \alpha(p_1/D_0 - q_A)D_A = 0;$$

i.e.,

$$p_1 = 1/\alpha + q_A D_0.$$

We can remark here from the value of p_1 that inserting side payments induces a larger subscription fee for the CP, such that its revenue is $U_1 = D_A/(\alpha D_0)$ whatever the side payment, which is exactly the CP-related user welfare UW_{CP}. In other words, the interest of users connected to the Internet and that of the CP coincide.

To compute the optimal price p_A for ISP A, we also compute the derivative

$$\frac{\partial U_A}{\partial p_A} = \sigma_A + p_A \frac{\partial \sigma_A}{\partial p_A} + q_A D_0 e^{-\alpha(1/\alpha + q_A D_0)} \frac{\partial \sigma_A}{\partial p_A}.$$

In the neutral case ($q_A = 0$), p_A does not depend on p_1, and we get $\partial U_A/\partial p_A = \sigma_A + p_A \partial\sigma_A/\partial p_A = p_A^{-\beta}(p_A^{-\beta} + (1-\beta)p_0^{-\beta})/(p_A^{-\beta} + p_0^{-\beta})$.

- If $\beta \leq 1$, this derivative is always positive, hence setting an infinite price is the "best solution." In other words, the sensitivity to prices is not large enough to deter the ISP from increasing its price.
- If $\beta > 1$, the first-order condition gives

$$p_A = p_0(\beta - 1)^{-1/\beta}.$$

The non-neutral case is not tractable and we need to resort to a numerical evaluation. Figures 7.2 and 7.3 display the ISP revenue, CP revenue, user welfare, and social welfare (the sum of the user welfare and the provider revenues) in terms of q_A for the optimally chosen p_1 and p_A (with the latter being determined numerically), when $\alpha = 1$, $p_0 = 1$, $D_0 = 1$, and $\beta = 1.5$. To compare the outcome with the neutral case, we just need to compare with the point at the origin ($q_A = 0$). The introduction of a side payment clearly increases the ISP revenue, user welfare, and social welfare here, up to an optimal value above which demand decreases too significantly. We remark that the side payment

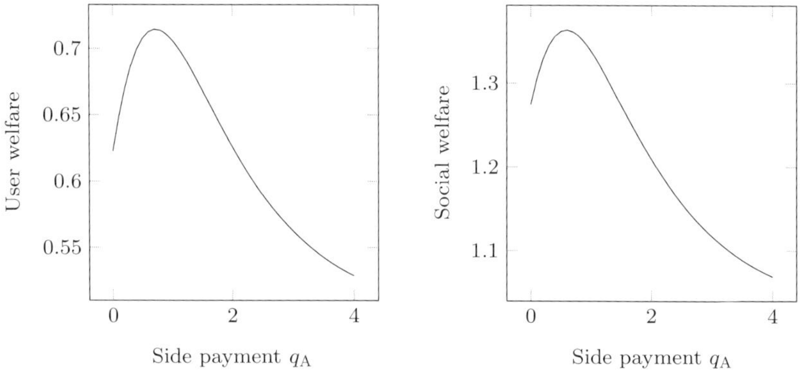

Figure 7.3 User and social welfare, when the side payment q_A varies.

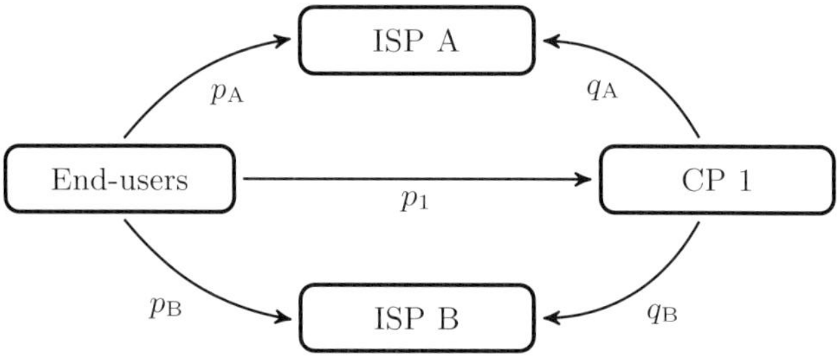

Figure 7.4 Charging interactions with two ISPs. Prices p_1, p_A, and p_B are positive flat rates, whereas q_A and q_B are positive per-unit-volume prices.

optimizing the ISP revenue is slightly larger than the one optimizing the user welfare. On the other hand, the impact of the side payment on the CP revenue is negative.

One content provider and two access providers

We now consider two ISPs in competition instead of just one. They are named A and B, all parameters for B being defined as for A above. Why consider competition between ISPs instead of between CPs? Actually, ISPs complain that they endure competition (which is particularly true in Europe) at the network access level, while for most types of services there is often a dominant actor (Netflix, Google, etc.) and less competition, a reason why it has been argued that side payments will become relevant. Because of that competition, ISPs say that they are forced to decrease their access prices and thus forced to get money from CPs. Our model can help to study the relevance of this argument. A new sketch of charges imposed among players for that setting is shown in Figure 7.4. We assume again that users first choose their ISP, and then subscribe to the CP (or don't). We still focus here on a discrete-choice/stickiness model for the ISP selection, but note that a model based on Wardrop principles has also been considered in [43], such that users simply select the cheapest ISP (leading to a Bertrand competition). With the stickiness model, the proportion of users subscribing to ISP $i \in \{A, B\}$ is

$$\sigma_i = \begin{cases} \dfrac{p_i^{-\beta}}{p_A^{-\beta} + p_B^{-\beta} + p_0^{-\beta}} & \text{if } p_A > 0 \text{ and } p_B > 0 \\ 1 & \text{if } p_i = 0 \text{ and } p_j > 0 \\ 1/2 & \text{if } p_A = 0 \text{ and } p_B = 0 \\ 0 & \text{if } p_i > 0 \text{ and } p_j = 0. \end{cases} \tag{7.5}$$

Using exactly the same arguments as for a single ISP, the user welfare associated with the existence of the CP is

$$\mathrm{UW_{CP}} = \frac{D_A + D_B}{\alpha D_0},$$

and the providers' revenues are, for the ISPs ($i \in \{A, B\}$),

$$U_i = p_i \sigma_i + q_i D_i,$$

and, for the CP,

$$\begin{aligned} U_1 &= (p_1 \sigma_A e^{-\alpha p_1} - q_A D_A) + (p_1 \sigma_B e^{-\alpha p_1} - q_B D_B) \\ &= (p_1/D_0 - q_A) D_A + (p_1/D_0 - q_B) D_B. \end{aligned}$$

The decisions on prices are still analyzed by backward induction, the decision at a given time scale being made anticipating the output at the later time scales (as before, ISPs play first – hence there is a game among them – and the CP adapts its price to the ISPs' prices).

We again look first at the smallest time scale (decision on p_1) for other fixed values. For convenience, we define $P_i := p_i^{\beta}$. A solution of the first-order condition gives (see [65] if details are required)

$$p_1^* = \begin{cases} \dfrac{P_A}{P_A + P_B}\left(D_0 q_B + \dfrac{1}{\alpha}\right) + \dfrac{P_B}{P_A + P_B}\left(D_0 q_A + \dfrac{1}{\alpha}\right) & \\ & \text{if } p_A > 0 \text{ or } p_B > 0 \\ D_0 \dfrac{q_A + q_B}{2} + \dfrac{1}{\alpha} & \text{if } p_A = 0 \text{ and } p_B = 0. \end{cases} \tag{7.6}$$

Here again, the CP's revenue when using this optimal price corresponds to the CP-related user welfare $(D_A + D_B)/(D_0 \alpha)$.

Knowing this reaction of the CP to ISPs' prices, these ISPs play a game on their choice of p_A and p_B. In the neutral case (i.e., $q_A = q_B = 0$), it can be shown, using the following formulation for the revenue of ISP A on plugging in the expression of the optimal price p_1 (with a symmetric formulation being obtained for B), that

$$U_A = \begin{cases} \dfrac{P_0 P_B p_A}{P_0 P_A + P_0 P_B + P_A P_B} & \text{if } p_A > 0 \text{ and } p_B > 0 \\ 0 & \text{if } p_A = 0 \text{ or } p_B = 0, \end{cases}$$

so that $(p_A, p_B) = (0, 0)$ is a Nash equilibrium since no player can strictly increase his or her revenue by unilaterally changing his or her action. But such a player's strategy is strictly dominated by any other as soon as the adversary's price is not zero. So it is not

Table 7.6 Outcomes to expect from the ISP price competition game on p_A and p_B in the neutral case

$\beta \le 1$	$1 < \beta \le 2$	$\beta > 2$
No equilibrium (prices tend to infinity)	Nash equilibrium $p_A = p_B = p_0((2-\beta)/(\beta-1))^{1/\beta}$ $U_A = U_B := U^{\text{neutral}} = p_0/[(2-\beta)^{1/\beta}(\beta-1)^{1-1/\beta}\beta]$	Nash equilibrium $p_A = p_A = 0$ (price war) $U_A = U_B = 0$

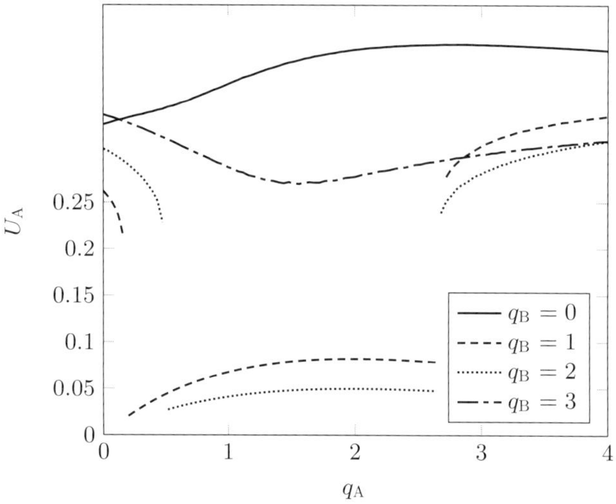

Figure 7.5 ISP A's revenue at equilibrium as a function of the side payment q_A, with $q_B \in \{0, 1, 2, 3\}$.

likely to be chosen by ISPs if another equilibrium exists. Actually, in this neutral case it can be shown (see [65] for a proof) that the (other) Nash equilibria can be described as presented in Table 7.6, in a way similar to the results in Table 5.5. In the case of (positive) side payments, we are here too not able to get analytical results. But equilibria can be determined numerically and the resulting utilities compared with the neutral case. We present a part of the results in [65], still with $\alpha = 1$, $p_0 = 1$, $D_0 = 1$, and $\beta = 1.5$ (other values give similar outcomes). Numerical computations show that the revenue of the CP and the user welfare it creates are always equal at equilibrium (which is easy to prove when $q_A = q_B$, but not in the general case). We display the revenues of providers in Figures 7.5 and 7.6. Discontinuities can be observed, corresponding to situations when there is a price war: the equilibrium subscription prices of both ISPs fall to 0 for some side payments (this is, for example, the case when $q_A = q_B = 1$, but never when $q_B = 0$ or $q_B = 3$). We can briefly remark that the revenue of ISPs is not monotonic with the side payment, that the maximal revenue of an ISP, say A, may be

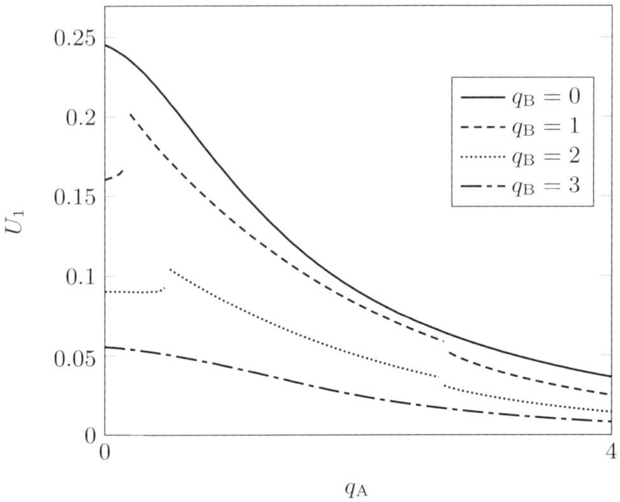

Figure 7.6 The CP's revenue at equilibrium as a function of the side payment q_A, with $q_B \in \{0, 1, 2, 3\}$.

obtained for a null or positive side payment, and that the CP's revenue has a tendency to decrease with side payments.

For this model with competing ISPs, we have not yet evoked the decision level corresponding to setting the side payment. Side payments can be determined by the CP, or by ISPs, through a game. The conclusions from numerical investigations in [65] for these three cases are as follows.

(i) If the side payments are decided by the CP, it is interesting to note that strictly positive side payments can be optimal for the CP, something that is counter-intuitive at first sight, especially since it was not the case when we had a single ISP. Actually, side payments exacerbate the competition between ISPs on access prices, which are reduced at equilibrium with respect to those in the neutral case; this is beneficial to end-users, and finally to the CP, which can reach more customers.
(ii) If side payments are decided by the ISPs, through a game, $(0, 2.80)$ and the symmetric point $(2.80, 0)$ are Nash equilibria. In comparison with the neutral case, the ISPs' total revenue increases by about 15% (in agreement with their request), while the CP's revenue decreases by 75% of its value.

7.4 Search neutrality

7.4.1 The debate

Search engines play a pivotal role in the Internet economy, as the entry points to websites for most Internet users. As an illustration, in the USA alone about 20 billion requests from home and work computers are treated by search engines each month.[9] Search engines return a ranked list of links (the so-called organic results) to documents available on the World Wide Web given any keyword. The list is obtained from a link-analysis algorithm, which assigns a weight to documents [29, 30], the goal being to provide the most relevant results. In this section we will focus exclusively on those organic results, ignoring the advertisement links aimed at yielding revenue to search engines, which we treated in Section 4.2.

But the ranking of organic links by search engines is now being questioned by various actors of the Internet and regulators, claiming that relevance is not its only factor, and that some revenue-making components are being taken into consideration [74]. This question has become a lively debate worldwide [74, 139]. The term **search neutrality** was coined in 2009 by Adam Raff, co-founder of Foundem (a price-comparison company), after an impassioned argument about Google penalizing his company in its ranking. The term is evidently inspired by *network neutrality*, because of the similarities in their stakes, namely the limitations on users' access to all relevant services on the Internet. Google actually acknowledged attaching a penalty to results such as Foundem's web site, using the argument that it is a **vertical search engine** – i.e., a search engine focusing only on a part of the Internet – and that vertical search engines are perceived by users as spam. But Google also offers price-comparison and other specialized search services, and penalizing other companies that are active in that business can be seen as hindering competition; Google finally decided to whitelist Foundem manually, but kept its penalty policy towards other vertical search engines, leaving the debate open.

In addition (still focusing on Google, because it represents more than 80% of the search market), Google offers many other services – e-mail, maps, calendar, video, shopping... – and would naturally be tempted to direct users towards them rather than towards their competitors. For example, Google favors (or is accused of favoring) YouTube content because money can be generated from those links.[10] In this section, we therefore focus on the analysis of the reality of such biases, and discuss their impact on competition and on user welfare.

The main arguments in favor of search neutrality (i.e., imposing that ranking be based solely on relevance) are in terms of user and social welfare. First, neutrality should benefit users by providing them with the most relevant results, instead of driving them towards what yields the largest revenue to the search engine. For the same reason, neutrality benefits the global economy by facilitating access to the best-performing actors/services (rather than those paying to be ranked well), so that new businesses can emerge more easily. As a side effect, it is also claimed that search neutrality elicits efforts from web sites to improve their content quality, rather than encouraging them to pay in order to be ranked better.

On the other hand, the opponents of search neutrality consider that users are interested in the differences among search engines, and select their preferred one, while such a

[9] www.comscore.com.

[10] See, for instance, http://www.guardian.co.uk/technology/blog/2011/sep/21/eric-schmidt-google-senate-hearing or, for measurements, http://www.benedelman.org/searchbias/.

differentiation would disappear if neutrality were enforced. Also, neutrality would prevent search engines from manipulating rankings, which it is claimed they do mainly to improve the results by avoiding spam: neutrality would then lead to worse results for users. Finally, imposing transparency of the ranking algorithms (as is often advocated by proponents of search neutrality) raises issues in terms of intellectual property, and facilitates the job of spammers exploiting the workings of those algorithms.

This section does not aim at resolving the debate, but rather at providing an objective view regarding the extent of non-neutrality in current search engines, and the potential economic impacts of imposing neutrality.

7.4.2 Do search engines return biased results?

In order to have an objective view of whether **search biases** occur, some experiments were conducted in 2010 [90] and 2011 [304], focusing on own-content bias, i.e., the tendency to favor one's content. The experiments consist in comparing the results provided by different search engines in response to the same requests, and measuring the extent to which the positions of the content affiliated to one of them differ. In particular, the comparison considered the respective positions of Google content and Microsoft content in the result pages of Google and Bing (the Microsoft search engine). Other search engines were considered as well, namely Yahoo!, Ask, and AOL in [90] and Blekko (a search engine not having *a priori* conflicts of interest) in [304].

Let us focus on the 2011 experiment in [304], where the comparison was performed for a set of 1,000 search queries (selected randomly from a list made public by AOL, of 20 million search queries run by AOL users). The results revealed significant discrepancies among search engines, tending towards an actual favoring of their own content. An illustration is provided in Figure 7.7, showing the proportion of searches for which own content ranked by a search engine is not as well placed in the results pages of any other search engine (recall that the three search engines here are Google, Bing, and Blekko). For example, the figure indicates that, when Bing ranks some Microsoft content among its top three results, then in 99.2% of cases neither Google nor Blekko ranks Microsoft content in their top three results. All values remain above 90%, indicating a clear tendency to favor one's content: cases in which a search engine other than the affiliated one agrees to show Google or Microsoft content are very rare.

That tendency also appears through another measure on the same experiment: Microsoft content is 26 times more likely to be displayed on the first page of Bing than on either of the two other search engines (57 times more likely if we consider only the first position). Similarly, Google content appears 17 times more often on the first page of a Google search than on the other search engines (22 if we focus on the first position). Therefore *search engines undeniably show biased results, favoring their own content*. Let us, however, remark that the symmetric bias, consisting in blocking access to competitors, is less obvious: for example, the experiment suggests that Bing references Google content more often than Blekko (which is assumed to have no interest in biasing results) does. But Google shows Microsoft content less than Blekko (the opposite of what was observed in the 2010 experiment [90]).

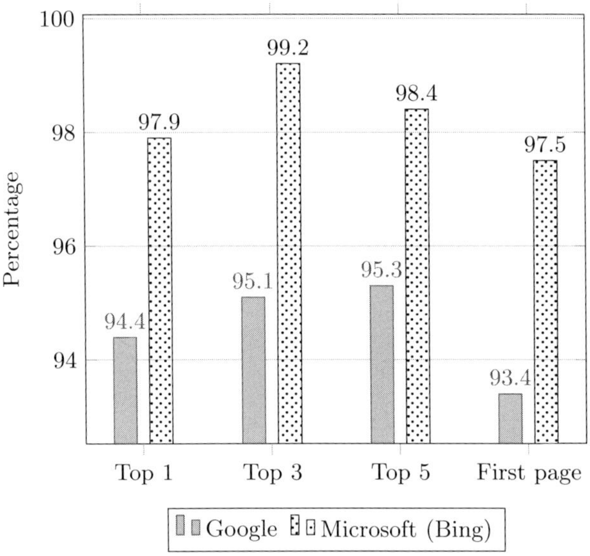

Figure 7.7 Percentages of Google and Bing search results with their own content not ranked similarly by any rival search engine [304].

7.4.3 Do we need regulatory intervention?

While the findings from [90] and [304] are quite similar in nature, their interpretations and conclusions are drastically different.

Edelman and Lockwood [90] stress the negative impact on user welfare of search biases, and suggest that search algorithms be controlled to ensure some neutrality and foster competition and innovation. That stance is naturally shared by the complainants against Google – in particular, vertical search engines such as Nextag or Foundem – whose advocates strongly defended it during a September 2011 US Senate hearing.

On the other hand, Wright [304] argues that the own-content bias might not be simply aimed at improving short-term revenues, but could also stem from the natural competition among search engines willing to satisfy users by differentiating their products. In other words, Google does not favor Google content to rule out competition, but to please its users. The reasoning is that a user choosing Google to perform a search is more expected to like Google products than a user choosing Bing, and vice versa. Hence it is logical to show Google content for a Google search. Another argument used by Wright is that antitrust regulation aims at protecting "competition, not competitors," and that competition is not currently harmed.

- Users can always choose to change search engines, encouraging search engines to provide satisfying results.
- The own-content bias is not of sufficient scale to fall within the antitrust framework, in the sense that it has not been proved to prevent competition. In his experiment, Wright

observed that the relative differences among search engine biases are significant, but the absolute levels of bias remain small: for example, Google references its own products while the others do not in only 8% of the total number of Google queries. For this reason, Wright qualifies anticompetitive exclusion arguments as unsustainable.

In summary, for Wright no regulatory intervention is needed, the search bias being the result of a product differentiation process that benefits users.

Let us remark, however, that the experiments in [90, 304] do not directly address the situation of vertical search engines, or companies offering specialized search services. The propensity of search engines to penalize those potential competitors was not studied, even though it is among the key issues in the debate, as pointed out by Adam and Shivaun Raff:

> The competitors Google is referring to when it says "competition is just a click away" are rival horizontal search engines like Yahoo and Bing, but the businesses being harmed by the anti-competitive practices [. . .] are the thousands of businesses that compete with Google's other services – in price comparison, online video, digital mapping, news aggregation, local search, travel search, financial search, job search, property search, social networking, and so on. (Adam and Shivaun Raff (co-founders of Foundem), June 2012)

The concern about non-neutral search engines is strongly shared by the European Union's regulators, who are progressing towards an antitrust settlement deal with Google, after a complaint issued in 2010 by several specialized service companies. The European consumers' organization (BEUC) has acknowledged the risks of search bias, and suggested remedies. It has been agreed that Google's search algorithm should follow some general principles in order to guarantee more fairness of its search results.

> Google must be even-handed. It must hold all services, including its own, to exactly the same standards, using exactly the same crawling, indexing, ranking, display, and penalty algorithms. (BEUC, March 2013 [242])

In particular, measures consisting in labeling among organic results those pointing to own content or sponsored content are not sufficient. This is in accordance with results from recent studies [138] showing that such labels have hardly any effect on users' clicking behavior; instead, the position of the link among the organic results appears to be the main factor.

In April 2013, the European Commission began market testing to estimate the extent to which the Google ranking algorithm respects these guidelines. Depending on the results of that testing, a settlement could be finalized or Google may face a fine that could be as large as $5 billion.

7.4.4 Neutral versus non-neutral search engines: a simple model

Let us now investigate through a mathematical model (introduced in [64]) the incentives for a search engine to deviate from a ranking that is based solely on relevance. Such a deviation will be called biased, or *non-neutral*. We remark here that the problem is similar for merchants' web sites including a search tool: the way items are ranked for a given search could be based only on relevance (including price, for example) or on other criteria of interest to the merchant. Also, the ranking of ads yields the same kind of questions, despite the fact that here advertisers are explicitly charged for the position of their ad: the trade-offs between the price charged and the relevance (estimated by the click-through rate) led to the different auction designs discussed in Section 4.2. We will mostly stick to the framework of a (classic) search engine in what follows, but the parallels with the merchant and advertisement sites should be kept in mind.

To compare rankings that are based on **relevance** with other types of rankings, we first have to assume that an objective measure of relevance exists for each web page potentially ranked by the search engine, and that the search engine is able to compute it. This assumption seems reasonable since search engines deploy complex algorithms to rank web pages, and should be able to fine-tune them for the best interests of their users (e.g., using the feedback on rankings that users make through their clicking decisions). Note that we understand here the *relevance* of a web page as the *value* that this web page has to the user performing a search: that value may vary among users, depending on some parameters that the search engine has access to, such as

- the user's machine type (smartphone, tablet, laptop),
- the operating system and the browser running,
- the user's browsing history,
- the user's geographic location, etc.

Also, the specific search engine used may be among those criteria, which may justify own-content bias as argued before ("*Google users tend to prefer Google products*"). The use of any of those data to affect the ranking is called **personalization** of search results. We do not consider here personalization as totally contradictory with search neutrality, since personalized results need not be voluntarily biased in the interest of the search engine.

Let us focus in this model on a specific search, performed by users of a given type. Assume that there is a number m of web pages to be ranked by the search engine. The relevance of page i is denoted by r_i, and its expected gain for the page owner is denoted by g_i (that gain can stem from advertising on the web page, or from the expected revenue from sales on a merchant's site). The question for the search engine is how to rank those pages, i.e., to choose a permutation $\boldsymbol{\pi} = (\pi_1, \ldots, \pi_m)$, with π_i the position allocated to page i. The search engine then has to consider two opposite effects.

- The ranking designer may prefer to rank first the pages yielding a larger benefit to him or her: for example YouTube content first for a search engine like Google. Our model would then have to indicate what part α_i of the revenue g_i goes to the search engine, a larger α_i corresponding to pages affiliated to the search engine.
- Basing the ranking on relevance (the r_i values), placing first the links that are most likely to please customers, is also likely to bring more interest in the search engine and hence lead to more revenue in the long run, notably via advertising slots. For

example, Google's search-engine algorithm was perceived (at least when it was first introduced) as very quick and efficient, leading almost to a monopoly, with about 70% of the searches (and more than 80% of users) in March 2013.[11]

In summary, ranking according to relevance alone might not produce the largest revenue, while ranking based on immediate returns can divert customers from the site because using it is then not considered efficient from the users' point of view. A trade-off between those two considerations has therefore to be investigated by the engine manager, but also its impact on user welfare and overall social welfare should be studied to determine whether regulation is needed.

We now refine the model to represent this trade-off, so that the ranking affects the clicking behavior of users, similarly to what we did in Section 4.2 to model the click-through rates on ads: displaying a link at position i in the organic results leads to a probability of that link being clicked that is denoted by θ_i, with $\theta_1 > \theta_2 > \ldots > \theta_m$. We remark that this suggests that the probability of a link being clicked is independent of the page itself, but actually the model also encompasses the case when click probabilities are separable, i.e. they are the product of the position-specific component θ_i and a page-specific component q_i. This is done by considering that r_i and g_i already include the page-specific component: $r_i = q_i \hat{r}_i$, with $\hat{r}_i$ the perceived relevance *when clicked* (and similarly for g_i).

The relevance of the ranking for users can be defined as the average relevance experienced when performing a search for that keyword, i.e.,

$$r = \sum_{i=1}^{m} \theta_{\pi_i} r_i.$$

The users' sensitivity to relevance is represented by the search engine's success: we assume that the arrival rate of requests for the search considered (and from that type of user) is an increasing function $\lambda(r)$ of the expected relevance r. Note that this implicitly covers the competition among ISPs: if the ranking relevance is bad, users will switch to alternative search engines.

Summarizing, for a given ranking defined by the permutation $\boldsymbol{\pi}$, the expected revenue per time unit of the search engine equals

$$u_{\mathrm{SE}}(\boldsymbol{\pi}) = \lambda(r)\left(\beta + \sum_i \theta_{\pi_i}\alpha_i g_i\right) = \lambda\left(\sum_i \theta_{\pi_i} r_i\right)\cdot\left(\beta + \sum_i \theta_{\pi_i}\alpha_i g_i\right), \tag{7.7}$$

with β the expected advertising gain per search performed (which is assumed to be independent of the ranking), i.e., the average amount that the search engine can make by selling advertisement slots, the average revenue per search gained from the page owners being $\sum_i \theta_{\pi_i}\alpha_i g_i$.

Let us analyze a simple specific case.

[11] Sources: www.comscore.com and netmarketshare.com.

Table 7.7 Relevance and gain values for the three web pages corresponding to the search considered

i	Relevance r_i	Gain g_i
1	3	3
2	2	8
3	1	2

Example 49 (Revenue-maximizing rankings) Consider a specific request, and assume that the arrival rate for that request (the number of requests per time unit) is proportional to the expected relevance of the results, so that we can take $\lambda(r) = r$. Suppose that three web pages have to be ranked by the search engine, with relevance and gain values as displayed in Table 7.7. As for the position-specific click-through rates, we take $\theta_i = (1/2)^i$. With those values, a ranking according to relevance – called neutral ranking – would lead to showing page i in position i, for $i = 1, 2, 3$. Then the corresponding average relevance would be

$$r_{\text{neutral}} = \frac{1}{2} \times 3 + \frac{1}{4} \times 2 + \frac{1}{8} \times 1 = \frac{17}{8} = 2.125,$$

and the average cumulated revenue of web-page owners per search (without counting the possible part α_i left to the search engine) equals

$$g_{\text{neutral}} = \frac{1}{2} \times 3 + \frac{1}{4} \times 8 + \frac{1}{8} \times 2 = \frac{15}{4} \simeq 3.75,$$

yielding an expected cumulated revenue of web-page owners per time unit of

$$v_{\text{neutral}} = \lambda(r_{\text{neutral}}) g_{\text{neutral}} = \frac{17}{8} \times \frac{15}{4} \simeq 7.97.$$

This can represent the "commercial value" of the ranking, i.e. how much can be raised from the organic results.

Now consider two possible situations with respect to what the search engine gains from the organic results.

- ***If the search engine gets a fixed share*** $\alpha > 0$ ***of the revenue made by the web-page owners***, i.e., $\alpha_i = \alpha$ for $i = 1, 2, 3$, then it is clear that page 3 will always be ranked in third position since it yields less revenue and in addition is less relevant than the two others. The search engine therefore chooses between two possible rankings, namely the neutral one and the one given by $\pi = (2; 1; 3)$. For the former ranking, the revenue of the search engine as given in (7.7) equals $r_{\text{neutral}}(\beta + \alpha g_{\text{neutral}}) = (17/8)\beta(1 + (\alpha/\beta)(15/4))$. On the other hand, the latter ranking leads to an expected relevance per search of $15/8$ and an average cumulated revenue of 5, and hence a revenue to the search engine of $(15/8)\beta(1 + 5\alpha/\beta)$.

Table 7.8 A comparison of the candidate rankings when the search engine owns page 3, for the relevance and gain values of Table 7.7

Ranking	Relevance (r)	Revenue per search from organic results (g)	"Commercial value" of organic results ($\lambda(r)g$)	Engine revenue per time unit (u_{SE})
1; 2; 3	2.125	3.75	7.97	$2.125(\beta + 0.25)$
1; 3; 2	2	3	6	$2(\beta + 0.5)$
3; 1; 2	1.5	2.75	4.125	$1.5(\beta + 1)$

Therefore the search engine should prefer non-neutral ranking if and only if

$$\frac{15}{8}\beta\left(1 + 5\frac{\alpha}{\beta}\right) > \frac{17}{8}\beta\left(1 + \frac{15}{4}\frac{\alpha}{\beta}\right),$$

i.e., if and only if $\alpha/\beta > 8/45 \simeq 0.178$. In other terms, if the search engine gets a sufficiently large part of the revenues from the web pages it ranks, then non-neutral ranking may be favored. It is the case for this specific example: the search engine then obtains a larger revenue and the "commercial value" of the ranking also increases (from $\lambda(r) \cdot g = 7.97$ to $\lambda(r) \cdot g = 9.37$), but this is at the expense of the users' perceived relevance decreasing from 17/8 to 15/8).

- ***If the search engine owns page* 3 *and does not charge the others***. This can be seen as a case of vertical integration, and for our model means that we have $\alpha_1 = \alpha_2 = 0$ and $\alpha_3 = 1$. In that case, since page 1 is more relevant than page 2 and the search engine is not interested in the revenues from those pages, page 1 should always be ranked before page 2, which leaves three possible ranking choices, which are compared in Table 7.8. The selected rankings depending on the value of β can then be found from the last column of the table. The ranking $(3; 1; 2)$ $((1; 3; 2))$ maximizes the search engine's revenue for $\beta \leq 1$ $(1 \leq \beta \leq 3.75)$, and the neutral ranking maximizes it when the gains from advertisement slots are large enough (here $\beta \geq 3.75$).

  Therefore, if the potential revenues from its own content are large with respect to those from advertisement slots, the search engine is incentivized to favor a non-neutral stance in order to maximize revenue. In this example, this is at the expense of users (who are provided with less relevant results), and even of the aggregated revenues from organic results, illustrating that some regulatory intervention may be needed.

7.4.5 The case of a general set of keywords

In practice, users will not change search engine for every request they want to run. Rather, they will build a general opinion about the performance (relevance) of the search engines over time, depending on the results they obtain for a large set of requests. When they have a new search request to perform, they will then choose a search engine that they believe likely to provide relevant results; i.e., they will use that opinion.

We propose here a model for that case. Now $\lambda(r)$ represents the total rate of requests, not just for a single keyword, but for all keywords. Then, when the search engine receives a request, we assume it is capable of computing the relevance and gain values of the candidate results for that specific request. But, of course, those values vary from one request to another. A ranking policy then corresponds to a permutation of all links/ads associated with each potential keyword.

More specifically, we assume that, for a given request, the set of possible results is represented by two random vectors R and G, where R_i, the ith component of R (G_i, the ith component of G), is the relevance (expected revenue) of a displayable result labeled i. We do not focus on the computation of relevance or revenue scores here; instead their distributions can be considered as inputs of our model.

To each request there will correspond a realization of (R, G), for which a ranking decision must be taken to optimize the trade-off between attractiveness (through relevance values (R_i)) and revenue (through the values G_i). We suggest that it is valid to assume that the number of potential results is constant[12] and denoted by m. For example, when the components of R are independent and identically distributed with distribution function F and density f, then the average relevance r of the neutral ranking (i.e., that which is based solely on relevance) can be derived from order statistics [78]:

$$r = \sum_{j=1}^{m} \theta_j \int y \frac{m!}{(m-j-1)!\, j!} F(y)^{m-j-1} (1 - F(y))^j f(y) \mathrm{d}y.$$

Recall that a ranking policy is now a function that associates a permutation $\boldsymbol{\pi}(R, G)$ with each pair of vectors (R, G). Therefore, for a given policy, the average relevance r becomes

$$r = \mathbb{E}\left[\sum_{i=1}^{m} \theta_{\pi_i(R,G)} R_i\right],$$

with $\pi_i(R, G)$ the position of result i in the ranking.

In terms of the search engine's revenue, the counterpart of (7.7) when the search engine gets a proportion α of the gains of all results can thus be written as (removing the dependence on (R, G) to simplify the notation)

$$\begin{aligned} u_{\mathrm{SE}} &= \lambda(r) \cdot \left(\beta + \alpha \mathbb{E}\left[\sum_{i=1}^{m} \theta_{\pi_i} G_i\right]\right) \\ &= \lambda\left(\mathbb{E}\left[\sum_{i=1}^{m} \theta_{\pi_i} R_i\right]\right) \cdot \left(\beta + \alpha \mathbb{E}\left[\sum_{i=1}^{m} \theta_{\pi_i} G_i\right]\right). \end{aligned} \tag{7.8}$$

Determining a revenue-maximizing policy is a complex task because we have a product of two expected values, hence the classical dynamic programming tools cannot be used directly. One way to address it could be to consider a continuum of keywords, and derive the optimal policy thanks to first-order derivations with respect to variations in expected relevance r and revenue g, following the standard exchange arguments used

[12] This could be generalized to a random m, but we keep it like this for the sake of simplicity.

in dynamic programming. However, we did not intend to solve that problem here, but simply to present how the model can be built; in particular the product form in (7.8) shows the trade-off between relevance (attracting more searches) and revenue per search.

7.4.6 Personalization of search results: what I want to see versus what I need to see

Closely linked to the notion of search neutrality is that of personalization of search results, which has been evoked before. It is clear that, by using all the information available, a search engine can deliver more relevant results; however, if the algorithms used are not transparent there always remains some doubt as to whether those data are being used in the interest of the user or of the ad revenues that will accrue to the search engine.

In addition, asking whether search engines should be allowed to impose what they think a given individual should see raises legal as well as ethical questions. Even if it is not revenue-oriented, personalization can lead to users seeing always the same kinds of results: those that they like and tend to click on. But the risk is that each user may eventually acquire a very biased and limited version of the world, including only what he or she *wants* to see, and possibly missing very important information that he or she *needs* to see. Given that nowadays the influence of search engines is crucial (for example, they are extensively used when a society is faced with a cultural question), such a lack of objectivity and diversity in the results may be a danger, for example to good understanding among people and in terms of facilitating clannishness. This has to be kept in mind, especially since personalization is increasingly being implemented by default by search engines, and is becoming ever more complex to switch off.

References

[1] H. Z. Aashtiani and T. L. Magnanti. Equilibria on a congested transportation network. *SIAM Journal of Algebraic and Discrete Methods*, 2:213–226, 1981.

[2] V. Abhishek and B. Hajek. On the incentive to deviate in core selecting combinatorial auctions. In *Proceedings of Workshop on Telecom Economics, Engineering and Policy*, Kraków, 2012.

[3] D. Acemoglu, K. Bimpikis, and A. Ozdaglar. Price and capacity competition. *Games and Economic Behavior*, 66:1–26, 2009.

[4] E. Adar and B. Huberman. Free riding on Gnutella. *First Monday*, 5(10), October 2000.

[5] M. H. Afrasiabi and R. Guérin. Exploring user-provided connectivity: A simple model. In *Proceedings of 7th International Workshop on Advanced Internet Charging and QoS Technology (ICQT)*, Paris, 2011.

[6] G. Aggarwal, J. Feldman, S. Muthukrishnan, and M. Pál. Sponsored search auctions with Markovian users. In *Workshop on Internet and Network Economics (WINE)*, pages 621–628, 2008.

[7] A. Al Daoud, M. Alanyali, and D. Starobinski. Pricing strategies for spectrum lease in secondary markets. *IEEE/ACM Transactions on Networking*, 18(2):462–475, 2010.

[8] T. Alford and G. Morton. The Economics of Cloud Computing Addressing the Benefits of Infrastructure in the Cloud. Technical report, Booz Allen Hamilton, 2010.

[9] G. Allon and A. Federgruen. Service competition with general queueing facilities. *Operations Research*, 56(4):827–849, 2008.

[10] T. Alpcan and T. Başar. A game theoretic approach to decision and analysis in network intrusion detection. In *Proceedings of 42nd IEEE Conference on Decision and Control*, Maui, HI, 2003.

[11] T. Alpcan, T. Başar, R. Srikant, and E. Altman. CDMA uplink power control as a noncooperative game. *Wireless Networks*, 8(6):659–670, 2002.

[12] E. Altman, M. K. Hanawal, and R. Sundaresan. Nonneutral network and the role of bargaining power in side payments. Available at http://arxiv.org/abs/1204.4647, 2012.

[13] E. Altman, V. Kamble, and H. Kameda. A Braess type paradox in power control over interference channels. In *Proceedings of Physicomnet, WiOpt*, Berlin, 2008.

[14] E. Altman, A. Legout, and Y. Xu. Network non-neutrality debate: An economic analysis. In *Proceedings of Networking 2011*, pages 68–81, Barcelona, 2011.

[15] L. Anderegg and S. Eidenbenz. Ad hoc-VCG: A truthful and cost-efficient routing protocol for mobile ad hoc networks with selfish agents. In *Proceedings of 9th Annual International Conference on Mobile Computing and Networking (MobiCom 2003)*, pages 245–259, San Diego, CA, 2003.

[16] B. Andersen and M. Frenz. The impact of music downloads and P2P file-sharing on the purchase of music: A study for industry Canada. http://strategis.ic.gc.ca/epic/site/ippd-dppi.nsf/en/h_ip01456e.html, 2007.

[17] B. Andersen and M. Frenz. Don't blame the P2P file-sharers: The impact of free music downloads on the purchase of music CDs in Canada. *Journal of Evolutionary Economics*, 20:715–740, 2010.

[18] N. Andréasson, A. Evgrafov, and M. Patriksson. *An Introduction to Continuous Optimization*. Lund: Studentlitteratur, 2007.

[19] P. Antoniadis. *Economic Modelling and Incentive Mechanisms for Efficient Resource Provision in Peer-to-Peer Systems*. PhD thesis, Athens University of Economics and Business, 2006.

[20] ARCEP. Neutralité de l'internet et des réseaux: propositions et orientations. http://www.arcep.fr/uploads/tx_gspublication/net-neutralite-orientations-sept2010.pdf, 2010.

[21] ARCEP. ARCEP publishes the results of the awards procedure for 4G licences in the 2.6 GHz frequency band. http://www.arcep.fr/index.php?id=8571&L=1&tx_gsactualite_pi1\%5Buid\%5D=1431\&tx_gsactualite_pi1\%5BbackID\%5D=1\&cHash=136860fe4e, 2011.

[22] A. Arlandis and E. Baranes. Interactions between network operators, content producers and internet intermediaries: Empirical implications of network neutrality. *Intereconomics*, 46(2):98–105, 2011.

[23] M. Armbrust, A. Fox, R. Griffith *et al.* A view of cloud computing. *Communications of the ACM*, 53(4):50–58, 2010.

[24] I. Ashlagi, D. Monderer, and M. Tennenholtz. Competing ad auctions. In *Proceedings of the 4th Workshop on Ad Auctions*, Chicago, IL, 2008.

[25] R. Aumann. Subjectivity and correlation in randomized strategies. *Journal of Mathematical Economics*, 1(1):67–96, 1974.

[26] R. Aumann and L. Shapley. Long term competition: A game theoretic analysis. Mimeo, Hebrew University, 1976.

[27] R. J. Aumann. What is game theory trying to accomplish? In K. Arrow and S. Honkapohja, editors, *Frontiers in Economics*, pages 28–76. Oxford: Blackwell, 1985.

[28] R. J. Aumann. Rationality and bounded rationality. *Games and Economic Behavior*, 21:2–14, 1997.

[29] D. Austin. How Google finds your needle in the web's haystack. *American Mathematical Society Feature Column*, 10(12), 2006. http://www.ams.org/samplings/feature-column/fcarc-pagerank.

[30] K. Avrachenkov and N. Litvak. Decomposition of the Google PageRank and Optimal Linking Strategy. Rapport de recherche RR-5101, INRIA, 2004.

[31] Y. Bakos and E. Brynjolfsson. Bundling information goods: Pricing, profits, and efficiency. *Management Science*, 45(12):1613–1630, 1999.

[32] A. Banerjee and C. M. Dippon. Voluntary relationships among mobile network operators and mobile virtual network operators: An economic explanation. *Information Economics and Policy*, 21:72–84, 2009.

[33] M. Beckmann, C. B. McGuire, and C. B. Winsten. *Studies in the Economics of Transportation*. New Haven, CN: Yale University Press, 1956.

[34] M. Ben-Akiva and S. Lerman. *Discrete Choice Analysis*. Cambridge, MA: MIT Press, 1985.

[35] C. Berge. *Espaces topologiques. Fonctions multivoques*. Paris: Dunod, 1959.

[36] U. Berger. Brown's original fictitious play. *Journal of Economic Theory*, 135(1):572–578, 2007.

[37] D. P. Bertsekas. *Dynamic Programming and Optimal Control*. Belmont, MA: Athena Scientific, 1995.

[38] D. P. Bertsekas. *Nonlinear Programming*. Belmont, MA: Athena Scientific, 2nd edition, 1999.

[39] S. Bistarelli, M. Dall'Aglio, and P. Peretti. Strategic games on defense trees. In *Proceedings of 4th International Workshop on Formal Aspects in Security and Trust (FAST '06), LNCS 4691*, pages 1–15, Hamilton, Ontario, 2006.

[40] C. Blumrosen, J. D. Hartline, and S. Nong. Position auctions and non-uniform conversion rates. In *Proceedings of 4th Workshop on Ad Auctions*, Chicago, IL, 2008.

[41] N. Bohacek, J. P. Hespanha, J. Lee, C. Lim, and K. Obraczka. Game theoretic stochastic routing for fault tolerance and security in computer networks. *IEEE Transactions on Parallel and Distributed Systems*, 18(9):1227–1240, 2007.

[42] R. Bohn, H. W. Braun, K. C. Claffy, and S. Wolff. Mitigating the Coming Internet Crunch: Multiple Service Levels via Precedence. Technical report, University of California, San Diego, 1993.

[43] F. Boussion, P. Maillé, and B. Tuffin. Net neutrality debate: Impact of competition among ISPs. In *Proceedings of the Fourth International Conference on COMmunication Systems and NETworkS (COMSNETS)*, Bangalore, 2012.

[44] D. Braess. Über ein Paradoxon aus der Verkehrsplanung. *Unternehmensforschung*, 12:258–268, 1968.

[45] J. Bredin, R. T. Maheswaran, Ç. Imer *et al.* A game-theoretic formulation of multi-agent resource allocation. In *Proceedings of the 4th International Conference on Autonomous Agents*, 2000.

[46] S. Buehler, R. Dewenter, and J. Haucap. Mobile number portability in Europe. *Telecommunications Policy*, 30(7):385–399, 2006.

[47] C. Buragohain, D. Agrawal, and S. Suri. A game theoretic framework for incentives in P2P systems. In *Proceedings of the 3rd International Conference on Peer-to-Peer Computing, P2P '03*, pages 48–56, Washington, DC, 2003.

[48] L. Buttyán and J.-P. Hubaux. Enforcing service availability in mobile ad-hoc WANs. In *Proceedings of 1st IEEE/ACM Workshop on Mobile Ad Hoc Networking and Computing (MobiHOC)*, Boston, MA, 2000.

[49] L. Buttyán and J.-P. Hubaux. Stimulating cooperation in self-organizing mobile ad hoc networks. *Mobile Networks and Applications*, 8(5):579–592, 2003.

[50] S. Caron, G. Kesidis, and E. Altman. Application neutrality and a paradox of side payments. In *Proceedings of the Third International Workshop on Re-Architecting the Internet (ReArch 2010)*, Philadelphia, PA, 2010.

[51] D. Chakrabarty, Y. Zhou, and R. Lukose. Budget constrained bidding in keyword auctions and online knapsack problems. In *Workshop on Internet and Network Economics (WINE)*, pages 566–576, 2008.

[52] R. Chandramouli. Economics of security: Research challenges. In *Proceedings of the 16th International Conference on Computer Communications and Networks (ICCCN '2007)*, Hawaii, HI, 2007.

[53] C. K. Chau and K. M. Sim. The price of anarchy for non-atomic congestion games with symmetric cost maps and elastic demands. *Operations Research Letters*, 31:327–334, 2003.

[54] C.-K. Chau, Q. Wang, and D.-M. Chiu. On the viability of Paris metro pricing for communication and service networks. In *Proceedings of the 29th Conference on Information Communications*, INFOCOM '10, pages 929–937, Piscataway, NJ, 2010.

[55] H. Chen and Y. Li. Intelligent flow control under game theoretic framework. In D. W. Come, G. D. Smith, and M. J. Oats, editors, *Telecommunications Optimization: Heuristic and Adaptive Techniques*. New York: John Wiley & Sons, 2000.
[56] D. Clark, B. Lehr, S. Bauer *et al.* The growth of internet overlay networks: Implications for architecture, industry structure and policy. In *Proceedings of Telecommunications Policy Research Conference (TPRC-05)*, 2005.
[57] D. D. Clark. Internet cost allocation and pricing. In Lee W. McKnight and Joseph P. Bailey, editors, *Internet Economics*, pages 215–252. Cambridge, MA: MIT Press, 1997.
[58] E. H. Clarke. Multipart pricing of public goods. *Public Choice*, 11:17–33, 1971.
[59] R. Cocchi, D. Estrin, S. Shenker, and L. Zhang. A study of priority pricing in multiple service class networks. In *Proceedings of SIGCOMM '91*, pages 123–130, 1991.
[60] Federal Communications Commission. In the matter of expanding the economic and innovation opportunities of spectrum through incentive auctions. Notice of proposed rulemaking. Available at http://www.fcc.gov/document/broadcast-television-spectrum-incentive-auction-nprm, 2012.
[61] W. J. Cook, W. H. Cunningham, and W. R. Pulleyblank. *Combinatorial Optimization*. New York: John Wiley & Sons, 1998.
[62] J. R. Correa, A. S. Schulz, and N. E. Stier-Moses. Selfish routing in capacitated networks. *Mathematics of Operations Research*, 29(4):961–976, 2004.
[63] P. Coucheney, G. D'Acquisto, P. Maillé, M. Naldi, and B. Tuffin. Influence of Search Neutrality on the Economics of Advertisement-Financed Content. Working paper, 2013.
[64] P. Coucheney, P. Maillé, and B. Tuffin. Comparison of search engines non-neutral and neutral behaviors. *SIGMETRICS Performance Evaluation Review*, 40(2):14–17, 2012.
[65] P. Coucheney, P. Maillé, and B. Tuffin. Impact of Reputation-Sensitive Users and Competition between ISPs on the Net Neutrality Debate. Rapport de recherche, INRIA, 2012.
[66] P. Coucheney, P. Maillé, and B. Tuffin. Network Neutrality Debate and ISP Inter-relations: Traffic Exchange, Revenue Sharing, and Disconnection Threat. Working paper, available at http://hal.inria.fr/hal-00659278/fr/, 2012.
[67] C. Courcoubetis, M. P. Dramitinos, and G. D. Stamoulis. An auction mechanism for bandwidth allocation over paths. In *Proceedings of the 17th International Teletraffic Congress (ITC)*, 2001.
[68] P. Cramton. The FCC spectrum auctions: An early assessment. *Journal of Economics and Management Strategy*, 6(3):431–495, 1997.
[69] P. Cramton. Lessons from the United States spectrum auctions. Prepared testimony before the United States Senate Budget Committee, 2000. Available at http://www.cramton.umd.edu/papers2000-2004/00jre-collusive-bidding-lessons.pdf.
[70] P. Cramton. Simultaneous ascending auctions. In P. Cramton, Y. Shoham, and R. Steinberg, editors, *Combinatorial Auctions*, pages 99–114. Cambridge, MA: MIT Press, 2006.
[71] P. Cramton. Spectrum auction design. *Review of Industrial Organization*, 42(2):161–190, 2013.
[72] P. Cramton and J. A. Schwartz. Collusive bidding: Lessons from the FCC spectrum auctions. *Journal of Regulatory Economics*, 17:229–252, 2000.
[73] L. Cricelli, M. Grimaldi, and N. L. Ghiron. The competition among mobile network operators in the telecommunication supply chain. *International Journal on Production Economics*, 131(1):22–29, 2011.
[74] J. Crowcroft. Net neutrality: The technical side of the debate: A white paper. *Computer Communication Review*, 37(1):49–56, 2007.
[75] J. Crowcroft, R. Gibbens, F. Kelly, and S. Östring. Modelling incentives for collaboration in mobile ad hoc networks. In *Proceedings of WiOpt '03*, 2003.

[76] R. Dai and S.-L. Tang. Interconnection discrimination: A two-sided markets perspective. In *Proceedings of Fifth Hot Topics in Networks (HotNets-V '06)*, 2006.

[77] L. A. DaSilva. Pricing of QoS-enabled networks: A survey. *IEEE Communications Surveys & Tutorials*, 3(2):2–8, 2000.

[78] H. A. David. *Order Statistics*. New York: John Wiley, 2nd edition, 1981.

[79] R. W. Day and P. Cramton. The quadratic core-selecting payment rule for combinatorial auctions. *Operations Research*, 60(3):588–603, 2012.

[80] R. W. Day and P. Milgrom. Core-selecting package auctions. *International Journal of Game Theory*, 36(3–4):393–407, 2008.

[81] S. de Vries and R. Vohra. Combinatorial auctions: A survey. *INFORMS Journal on Computing*, 15(3):284–309, 2003.

[82] C. Dellarocas and S. Viswanathan. The holy grail of advertising? Allocative efficiency and revenue implications of "pay-per-action" advertising in environments with quality uncertainty. In *Proceedings of 4th Workshop on Ad Auctions*, Chicago, IL, 2008.

[83] X. Deng and J. Yu. A new ranking scheme of the GSP mechanism with Markovian users. In *Workshop on Internet and Network Economics (WINE)*, 2009.

[84] R. Dewenter and J. Haucap. Incentives to licence mobile virtual network operators (MVNOs). In R. Dewenter and J. Haucap, editors, *Access Pricing: Theory and Practice*, pages 305–325. Amsterdam: Elsevier Science, 2006.

[85] P. Dolan. Internet Pricing. Is the end of the World Wide Wait in view? *Communications & Strategies*, 37:15–46, 2000.

[86] L. Duan, J. Huang, and B. Shou. Competition with dynamic spectrum leasing. In *Proceedings of Dyspan 2010*, Singapore, 2010.

[87] P. K. Dutta. *Strategies and Games: Theory and Practice*. Cambridge, MA: MIT Press, 1999.

[88] D. Easley and J. Kleinberg. Evolutionary game theory. In *Networks, Crowds, and Markets: Reasoning about a Highly Connected World*. Cambridge: Cambridge University Press, 2010.

[89] R. Edell and P. Varaiya. Providing internet access: What we learn from INDEX. *IEEE Network*, 13(5):18–25, 1999.

[90] B. Edelman and B. Lockwood. Measuring bias in "organic" web search. http://www.benedelman.org/searchbias/, 2011.

[91] B. Edelman, M. Ostrovsky, and M. Schwarz. Internet advertising and the generalized second-price auction: Selling billions of dollars worth of keywords. *American Economic Review*, 97(1):242–259, 2006.

[92] S. Eidenbenz, G. Resta, and P. Santi. COMMIT: A sender-centric truthful and energy-efficient routing protocol for ad hoc networks. In *Proceedings of 5th IEEE International Workshop on Algorithms for Wireless, Mobile, Ad Hoc and Sensor Networks (WMAN '05)*, 2005.

[93] F. Etro. Advertising and search engines. A model of leadership in search advertising. *Research in Economics*, 67:25–38, 2013.

[94] M. Falkner, M. Devetsikiotis, and I. Lambadaris. An overview of pricing concepts for broadband IP networks. *IEEE Communications Surveys & Tutorials*, 3(2):2–13, 2000.

[95] A. Farahat and G. Perakis. Profit loss in differentiated oligopolies. *Operations Research Letters*, 37(1):43–46, 2009.

[96] P. Faratin. Economics of overlay networks: An industrial organization perspective on network economics. In *Proceedings of the ACM Conference on Electronic Commerce (EC '07)*, San Diego, CA, 2007.

[97] Federal Communications Commission. Appropriate Framework for Broadband Access to the Internet over Wireline Facilities. Technical Report FCC 05-151, CC Docket No. 02-33, 2005.

[98] M. Félegyházi, L. Buttyán, and J.-P. Hubaux. Equilibrium Analysis of Packet Forwarding Strategies in Wireless ad hoc Networks – the Dynamic Case. Technical Report IC/2003/68, École Polytechnique Fédérale de Lausanne (EPFL), 2003.

[99] M. Félegyházi, L. Buttyán, and J.-P. Hubaux. Equilibrium analysis of packet forwarding strategies in wireless ad hoc networks – the static case. In *Proceedings of the 8th International Conference on Personal Wireless Communications (PWC)*, 2003.

[100] H. Feng, S. Zhang, C. Liu, J. Yan, and M. Zhang. P2P incentive model on evolutionary game theory. In *Proceedings of the 4th International Conference on Wireless Communications, Networking and Mobile Computing (WiCOM)*, pages 1–4, 2008.

[101] K. Fjell. Online advertising: Pay-per-view or pay-per-click – a comment. *Journal of Revenue and Pricing Management*, 8(2/3):200–206, 2009.

[102] K. Fjell. Online advertising: Pay-per-view versus pay-per-click with market power. *Journal of Revenue and Pricing Management*, 9(3):198–203, 2010.

[103] R. Frieden. Internet Packet Sniffing and Its Impact on the Network Neutrality Debate and the Balance of Power between Intellectual Property Creators and Consumers. Technical report, SSRN, 2007. http://papers.ssrn.com/sol3/papers.cfm?abstract_id=995273.

[104] J. Friedman. A noncooperative equilibrium for supergames. *Review of Economic Studies*, 38:1–12, 1971.

[105] D. Fudenberg and E. Maskin. The folk theorem in repeated games with discounting or with incomplete information. *Econometrica*, 54(3):533–554, 1986.

[106] D. Fudenberg and J. Tirole. *Dynamic Models of Oligopoly*. Newark, NJ: Harwood Academic, 1986.

[107] D. Fudenberg and J. Tirole. *Game Theory*. Cambridge, MA: MIT Press, 1991.

[108] A. Ganesh, D. Gunawardena, P. Jey, L. Massoulié, and J. Scott. Efficient quarantining of scanning worms: Optimal detection and co-ordination. In *Proceedings of IEEE INFOCOM 2006*, Barcelona, 2006.

[109] J. Genachowski. Expanding the economic and innovation opportunities of spectrum through incentive auctions. Statement before the Federal Communications Commission, available at http://hraunfoss.fcc.gov/edocs_public/attachmatch/FCC-12-118A3.pdf, 2012.

[110] A. Ghosh and A. Sayedi. Expressive auctions for externalities in online advertising. In *Workshop on the World Wide Web (WWW)*, Raleigh, NC, 2010.

[111] A. Gibbard. Manipulation of voting schemes: A general result. *Econometrica*, 41(4):587–601, 1973.

[112] R. J. Gibbens and F. P. Kelly. Measurement-based connection admission control. In *Proceedings of the 15th International Teletraffic Congress*, 1997.

[113] R. J. Gibbens and F. P. Kelly. Distributed connection acceptance control for a connectionless network. In *Proceedings of the 16th International Teletraffic Congress*, 1999.

[114] R. J. Gibbens and F. P. Kelly. Resource pricing and the evolution of congestion control. *Automatica*, 35:1969–1985, 1999.

[115] R. J. Gibbens, R. Mason, and R. Steinberg. Internet service classes under competition. *IEEE Journal on Selected Areas in Communications*, 18(12):2490–2498, 2000.

[116] G. T. Gilbert. Positive definite matrices and Sylvester's criterion. *The American Mathematical Monthly*, 98(1):44–46, 1991.

[117] S. Goel, S. Lahaie, and S. Vassilvitskii. Contract auctions for sponsored search. In *Workshop on Internet and Network Economics (WINE)*, 2009.

[118] J. K. Goeree and Y. Lien. On the Impossibility of Core-Selecting Auctions. IEW Working Paper 452, Institut für Volkswirtschaftslehre, Wirtschaftswissenschaftliche Fakultät, Universität Zürich, 2009.

[119] R. Gonen. Characterizing optimal syndicated sponsored search market design. In *Proceedings of the 5th Workshop on Ad Auctions*, Stanford, CA, 2009.

[120] L. A. Gordon and M. P. Loeb. The economics of information security investment. *ACM Transactions on Information System Security*, 5(4):438–457, 2002.

[121] N. Gould. An introduction to algorithms for continuous optimization. Oxford University Computing Laboratory Notes, 2006.

[122] J. Green and J.-J. Laffont. Characterization of satisfactory mechanisms for the revelation of preferences for public goods. *Econometrica*, 45(2):427–438, 1977.

[123] T. Groves. Incentives in teams. *Econometrica*, 41(3):617–631, 1973.

[124] V. Guijarro, L. Pla, B. Tuffin, P. Maillé, and J. R. Vidal. Competition and bargaining in wireless networks with spectrum leasing. In *Proceedings of IEEE GLOBECOM*, Houston, TX, 2011.

[125] A. Gupta, D. O. Stahl, and A. B. Whinston. Priority pricing of integrated services networks. In L. W. McKnight and J. P. Bailey, editors, *Internet Economics*, pages 323–352. Cambridge, MA: MIT Press, 1997.

[126] I. Hafalir, R. Ravi, and A. Sayedi. Sort-cut: A pareto-optimal and semi-truthful mechanism for multi-unit auctions with budget-constrained bidders. In *Proceedings of the 5th Workshop on Ad Auctions*, Stanford, CA, 2009.

[127] N. Haile and J. Altmann. Value Creation in IT Service Platforms through Two-Sided Network Effects. TEMEP Discussion Papers 201297, Seoul National University; Technology Management, Economics, and Policy Program (TEMEP), 2012.

[128] P. Hand, M. Chiang, R. Calderbank, and S. Rangan. Network rate allocation with content provider participation. In *Proceedings of IEEE INFOCOM*, pages 990–998, Rio de Janeiro, 2009.

[129] G. Hardin. The tragedy of the commons. *Science*, 162(3859):1243–1248, 1968.

[130] R. Harms and M. Yamartino. The Economics of the Cloud. Technical report, Microsoft Research, 2010.

[131] J. C. Harsanyi. Games with incomplete information played by "Bayesian" players, parts I–III. *Management Science*, 8:159–182, 320–334, 486–502, 1967–1968.

[132] J. C. Harsanyi. Games with randomly disturbed payoffs: A new rationale for mixed-strategy equilibrium points. *International Journal of Game Theory*, 2:1–23, 1973.

[133] Y. Hayel, D. Ros, and B. Tuffin. Less-than-best-effort services: Pricing and scheduling. In *Proceedings of IEEE INFOCOM*, Hong Kong 2004.

[134] G. He. *A Game-Theoretical Approach to Resource Allocation in Wireless Networks*. PhD thesis, Eurecom, 2010.

[135] T. Henderson, J. Crowcroft, and S. Bhatti. Congestion pricing. Paying your way in communication networks. *IEEE Internet Computing*, 5(5):85–89, 2001.

[136] R. A. Horn and C. R. Johnson. *Matrix Analysis*. Cambridge: Cambridge University Press, 1985.

[137] K. Hosanagar, R. Krishnan, M. Smith, and J. Chuang. Optimal pricing of content delivery network (CDN) services. In *Proceedings of the 37th Annual Hawaii International Conference on System Sciences*, page 10, 2004.

[138] D. A. Hyman and D. J. Franklyn. Search Neutrality and Search Bias: An Empirical Perspective on the Impact of Architecture and Labeling. Illinois Program in Law, Behavior

and Social Science Paper No. LE13-24; University of San Francisco Law Research Paper No. 2013-15. Available at http://ssrn.com/abstract=2260942, 2013.

[139] INRIA. INRIA's response to ARCEP consultation about network neutrality, 2012.

[140] Interactive Advertising Bureau. IAB internet advertising revenue report, 2008. http://www.iab.net/insightsresearch/530422/1357.

[141] V. Jacobson. Congestion avoidance and control. *ACM Computer Communication Review; Proceedings of SIGCOMM '88*, 18(4):314–329, 1988.

[142] L. Jiang, V. Anantharam, and J. Walrand. Efficiency of selfish investments in network security. In *Proceedings of 3rd Workshop on the Economics of Networks, Systems, and Computation*, Seattle, WA, 2008.

[143] R. Johari and J. N. Tsitsiklis. Efficiency loss in a network resource allocation game. *Mathematics of Operations Research*, 29(3):407–435, 2004.

[144] R. Johari, G. Y. Weintraub, and B. Van Roy. Investment and market structure in industries with congestion. *Operations Research*, 58(5):1303–1317, 2010.

[145] J. Jormakka and J. Mölsä. Modelling information warfare as a game. *Journal of Information Warfare*, 4(2):12–25, 2005.

[146] S. Kakutani. A generalization of Brouwer's fixed point theorem. *Duke Mathematical Journal*, 8:457–459, 1941.

[147] P. Kalmus and L. Wiethaus. On the competitive effects of mobile virtual network operators. *Telecommunications Policy*, 34:262–269, 2010.

[148] F. P. Kelly, A. K. Maulloo, and D. K. H. Tan. Rate control in communication networks: Shadow prices, proportional fairness and stability. *Journal of the Operational Research Society*, 49:237–252, 1998.

[149] F. P. Kelly. Note on effective bandwidths. In F. P. Kelly, S. Zachary, and I. B. Ziedins, editors, *Stochastic Networks: Theory and Applications*, pages 141–168. Oxford: Oxford University Press, 1996.

[150] F. P. Kelly. Charging and rate control for elastic traffic. *European Transactions on Telecommunications*, 8:33–37, 1997.

[151] F. P. Kelly. Mathematical modelling of the Internet. In *Proceedings of the Fourth International Congress on Industrial and Applied Mathematics*, 2000.

[152] F. P. Kelly. Models for a self-managed Internet. *Philosophical Transactions of the Royal Society*, A358:2335–2348, 2000.

[153] D. Kempe and M. Mahdian. A cascade model for externalities in sponsored search. In *Workshop on Internet and Network Economics (WINE)*, pages 585–596, 2008.

[154] S. Keshav. *An Engineering Approach to Computer Networking: ATM Networks, the Internet and the Telephone Network*. New York: Addison-Wesley, 1999.

[155] H.-S. Kim and C.-H. Yoon. Determinants of subscriber churn and customer loyalty in the Korean mobile telephony market. *Telecommunications Policy*, 28(9–10):751–765, 2004.

[156] W. A. Kirk and M. A. Khamsi. *An Introduction to Metric Spaces and Fixed Point Theory*. New York: John Wiley, 2001.

[157] M. Kodialam and T. V. Lakshman. Detecting network intrusions via sampling: A game theoretic approach. In *Proceedings of IEEE INFOCOM*, 2003.

[158] B. Korte and J. Vygen. *Combinatorial Optimization: Theory and Algorithms*. Berlin: Springer, 4th edition, 2008.

[159] E. Koutsoupias and C. Papadimitriou. Worst-case equilibria. In *Proceedings of 16th Annual Symposium on Theoretical Aspects of Computer Science (STACS 1999)*, pages 404–413. Berlin: Springer, 1999.

[160] B. Krebs. I'll take 2 Mastercards and a Visa, please. http://krebsonsecurity.com/2010/09/ill-take-2-mastercards-and-a-visa-please/, 2010.

[161] C. C. Krueger and P. M. C. Swatman. Who are the Internet content providers? Identifying a realistic taxonomy of content providers. In *Proceedings of IFIP Conference on e-Business, e-Services, and e-Society (I3E)*, São Paolo, 2003.

[162] J.-J. Laffont, S. Marcus, P. Rey, and J. Tirole. Internet interconnection and the off-net-cost pricing principle. *The RAND Journal of Economics*, 34:370–390, 2003.

[163] S. Lahaie and D. M Pennock. Revenue analysis of a family of ranking rules for keyword auctions. In *Proceedings of the ACM Conference on Electronic Commerce (EC '07)*, San Diego, CA 2007.

[164] T. Lan, D. Kao, M. Chiang, and A. Sabharwal. An axiomatic theory of fairness in network resource allocation. In *Proceedings of IEEE INFOCOM*, 2010.

[165] A. A. Lazar and N. Semret. Design and analysis of the progressive second price auction for network bandwidth sharing. *Telecommunication Systems – Special Issue on Network Economics*, 20:255–263, 1999.

[166] H. Le Cadre, M. Bouthou, and B. Tuffin. Competition for subscribers between mobile operators sharing a limited resource. In *Proceedings of GameNets*, Istanbul, 2009.

[167] H. Le Cadre, M. Bouthou, and B. Tuffin. A pricing model for a mobile network operator sharing limited resource with a mobile virtual network operator. In P. Reichl, B. Stiller, and B. Tuffin, editors, *Network Economics for Next Generation Networks. Proceedings of the Sixth International Workshop on Advanced Internet Charging and QoS technologies (ICQT '09)*. Berlin: Springer, 2009.

[168] T. M. Lenard and R. J. May, editors. *Net Neutrality or Net Neutering: Should Broadband Internet Services be Regulated?* Berlin: Springer, 2006.

[169] X. Lin and N. B. Shroff. Pricing-based control of large networks. In S. Palazzo, editor, *IWDC 2001*, pages 212–231. Berlin: Springer, 2001.

[170] D. Liu, J. Chen, and A. B. Whinston. Competing keyword auctions. In *Proceedings of the 4th Workshop on Ad Auctions*, Chicago, IL, 2008.

[171] J. Liu and D. M. Chiu. Mathematical modeling of competition in sponsored search market. In *Proceedings of NetEcon*, 2010.

[172] P. Liu, W. Zang, and M. Yu. Incentive-based modeling and inference of attacker intent, objectives, and strategies. *ACM Transactions on Information and System Security*, 8(1):78–118, 2005.

[173] T. Locher, P. Moor, S. Schmid, and R. Wattenhofer. Free riding in BitTorrent is cheap. In *Proceedings of HotNets*, Irvine, CA, 2006.

[174] R. T. B. Ma, D.-M. Chiu, J. C. S. Lui, V. Misra, and D. Rubenstein. Interconnecting eyeballs to content: A Shapley value perspective on ISP peering and settlement. In *Proceedings of the International Workshop on Economics of Networked Systems (NetEcon)*, pages 61–66, 2008.

[175] R. T. B. Ma, D.-M. Chiu, J. C. S. Lui, V. Misra, and D. Rubenstein. On cooperative settlement between content, transit and eyeball internet service providers. In *Proceedings of the ACM International Conference on Emerging Networking EXperiments and Technologies (CoNEXT)*, 2008.

[176] R. T. B. Ma, S. C. M. Lee, J. C. S. Lui, and D. K. Y. Yau. An incentive mechanism for P2P networks. In *Proceedings of the 24th International Conference on Distributed Computing Systems*, pages 516–523, 2004.

[177] J. K. MacKie-Mason and H. R. Varian. Pricing the internet. In B. Kahin and J. Keller, editors, *Public Access to the Internet*, pages 269–314. Cambridge, MA: MIT Press, 1995.

[178] M. Mahdian, S. M. Li, and R. P. McAfee. Value of learning in sponsored search auctions. In *Workshop on Internet and Network Economics (WINE)*, Stanford, CA, 2010.

[179] P. Maillé. Auctioning for downlink transmission power in CDMA cellular systems. In *Proceedings of 7th ACM International Symposium on Modeling, Analysis and Simulation of Wireless and Mobile Systems (MSWiM)*, Venice, 2004.

[180] P. Maillé, E. Markakis, M. Naldi, G. Stamoulis, and B. Tuffin. Sponsored search auctions: An overview of research with emphasis on game theoretic aspects. *Electronic Commerce Research*, 12:265–300, 2012.

[181] P. Maillé, M. Naldi, and B. Tuffin. Price war with migrating customers. In *Proceedings of 17th IEEE International Symposium on Modelling, Analysis and Simulation of Computer and Telecommunication Systems (MASCOTS 2009)*, London, 2009.

[182] P. Maillé, P. Reichl, and B. Tuffin. Of threats and costs: A game-theoretic approach to security risk management. In N. Gülpınar, P. Harrison, and B. Rüstem, editors, *Performance Models and Risk Management in Communication Systems*. Berlin: Springer, 2010.

[183] P. Maillé, P. Reichl, and B. Tuffin. Interplay between security providers, consumers, and attackers: A weighted congestion game approach. In *Proceedings of the 2nd Conference on Decision and Game Theory for Security (GameSec 2011)*, College Park, MD, 2011.

[184] P. Maillé and B. Tuffin. Multi-bid auctions for bandwidth allocation in communication networks. In *Proceedings of IEEE INFOCOM*, 2004.

[185] P. Maillé and B. Tuffin. Multi-bid versus progressive second price auctions in a stochastic environment. In *Proceedings of ICQT '04, LNCS 3266*, pages 318–327, Barcelona, Berlin: Springer, 2004.

[186] P. Maillé and B. Tuffin. Pricing the internet with multibid auctions. *IEEE/ACM Transactions on Networking*, 14(5):992–1004, 2006.

[187] P. Maillé and B. Tuffin. Why VCG auctions can hardly be applied to the pricing of inter-domain and ad hoc networks. In *Proceedings of 3rd EuroNGI Conference on Next Generation Internet Networks*, pages 36–39, Trondheim, 2007.

[188] P. Maillé and B. Tuffin. Price war with partial spectrum sharing for competitive wireless service providers. In *Proceedings of IEEE GLOBECOM*, Honolulu, HI, 2009.

[189] P. Maillé and B. Tuffin. On the interest of introducing randomness in ad-word auctions. In *Proceedings of the 1st IFIP International Conference on the Network of the Future*, Brisbane, 2010.

[190] P. Maillé and B. Tuffin. Price war in heterogeneous wireless networks. *Computer Networks*, 54(13):2281–2292, 2010.

[191] P. Maillé and B. Tuffin. Adword auction bidding strategies of budget-limited advertisers on competing search engines. In *Proceedings of the 7th International Conference on Network and Service Management (CNSM 2011)*, Paris, 2011.

[192] P. Maillé and B. Tuffin. On the ranking strategy in adword auctions. In *Proceedings of the 8th International Conference on Information Technology: New Generations (ITNG)*, Las Vegas, CA, 2011.

[193] P. Maillé and B. Tuffin. Sponsored search engines in competition: Advertisers behavior and engines optimal ranking strategies. In *Proceedings of the 19th Annual Meeting of the IEEE/ACM International Symposium on Modelling, Analysis and Simulation of Computer and Telecommunication Systems (MASCOTS 2011)*, 2011.

[194] P. Maillé and B. Tuffin. Competition among providers in loss networks. *Annals of Operations Research, Special Volume on Operations Research and Systems (ALIO-INFORMS)*, 199:3–22, 2012.
[195] P. Maillé, B. Tuffin, and J.-M. Vigne. Technological investment games among wireless telecommunications service providers. *International Journal on Network Management – Special Issue: Economic Traffic Management*, 21(1):65–82, 2011.
[196] P. Maillé, M. Naldi, and B. Tuffin. Competition for migrating customers: A game-theoretic analysis in a regulated regime. In *Proceedings of IEEE GLOBECOM*, New Orleans, LA, 2008.
[197] M. Mandjes. Pricing strategies under heterogeneous service requirements. In *IEEE INFOCOM*, 2003.
[198] A. Mangani. Online advertising: Pay-per-view versus pay-per-click. *Journal of Revenue and Pricing Management*, 2:295–302, 2004.
[199] M. H. Manshaei, J. Freudiger, M. Félegyházi, P. Marbach, and J.-P. Hubaux. On wireless social community networks. In *Proceedings of IEEE INFOCOM*, Phoenix, AZ, 2008.
[200] M. H. Manshaei, P. Marbach, and J.-P. Hubaux. Evolution and market share of wireless community networks. In *Proceedings of IEEE International Conference on Game Theory for Networks (GameNets)*, Istanbul, 2009.
[201] P. Marbach. Pricing differentiated services networks: Bursty traffic. In *Proceedings of IEEE INFOCOM 2001*, 2001.
[202] P. Marbach. Priority service and max–min fairness. In *Proceedings of IEEE INFOCOM*, 2002.
[203] P. Marbach. Analysis of a static pricing scheme for priority services. *IEEE/ACM Transactions on Networking*, 12(2):312–325, 2004.
[204] P. Marbach and R. Berry. Downlink resource allocation and pricing for wireless networks. In *Proceedings of IEEE INFOCOM*, 2002.
[205] S. Marti, T. J. Giuli, K. Lai, and M. Baker. Mitigating routing misbehavior in mobile ad hoc networks. In *Proceedings of the Sixth ACM/IEEE International Conference on Mobile Computing and Networking*, pages 255–265, 2000.
[206] J. Maynard Smith and G. R. Price. The logic of animal conflict. *Nature*, 246(5427):15–18, 1973.
[207] D. McAdams. Multiproduct Monopoly Bundling. Technical report, Graduate School of Business, Stanford University, 1999.
[208] R. P. McAfee and J. McMillan. Auctions and bidding. *Journal of Economic Literature*, 25(2):699–738, 1987.
[209] R. P. McAfee, J. McMillan, and M. D. Whinston. Multiproduct monopoly, commodity bundling, and correlation of values. *The Quarterly Journal of Economics*, 114:371–383, 1989.
[210] J. K. MacKie-Mason and H. R. Varian. Some Economics of the Internet. Technical report, University of Michigan, November 1993. http://wueconb.wustl.edu:8089/eps/comp/papers:9401/9401001.pdf.
[211] J. K. MacKie-Mason and H. R. Varian. Pricing congestible network resources. *IEEE Journal on Selected Areas in Communications*, 13:1141–1149, 1995.
[212] L. McKnight, R. Solomon, J. Reagle *et al.* Information security for internet commerce. In L. W. McKnight and J. P. Bailey, editors, *Internet Economics*, pages 435–452. Cambridge, MA: MIT Press, 1997.
[213] I. Milchtaich. Congestion games with player-specific payoff functions. *Games and Economic Behavior*, 13(1):111–124, 1996.

[214] I. Milchtaich. Weighted congestion games with separable preferences. *Games and Economic Behavior*, 67(2):750–757, 2009.

[215] P. Milgrom. Putting auction theory to work: The simultaneous ascending auction. *Journal of Political Economy*, 108(2):245–272, 2000.

[216] P. Milgrom. *Putting Auction Theory to Work*. Cambridge: Cambridge University Press, 2004.

[217] J. Mo, W. Kim, and D. Lee. Impacts of universal service regulation for broadband internet services. In J. Cohen, P. Maillé, and B. Stiller, editors, *Economics of Converged, Internet-Based Networks*, pages 14–25. Berlin: Springer, 2011.

[218] J. Mo and J. Walrand. Fair end-to-end window-based congestion control. *IEEE/ACM Transactions on Networking*, 8(5):556–567, 2000.

[219] D. Monderer and L. S. Shapley. Fictitious play property for games with identical interests. *Journal of Economic Theory*, 68:258–265, 1996.

[220] D. Monderer and L. S. Shapley. Potential games. *Games and Economic Behaviour*, 14:124–143, 1996.

[221] I. Mundell. Piracy in Europe costs $13.7 billion. *Variety*, March 2010. http://www.variety.com/article/VR1118016618/?categoryId=1338&cs=1.

[222] J. Musacchio, J. Walrand, and G. Schwartz. Network neutrality and provider investment incentives. In *Conference Record of the Forty-First Asilomar Conference on Signals, Systems and Computers, 2007. ACSSC 2007*, pages 1437–1444, 2007.

[223] S. Muthukrishnan, M. Pal, and Z. Svitkina. Stochastic models for budget optimization in search-based advertising. In *Workshop on Internet and Network Economics (WINE)*, pages 131–142, 2007.

[224] H. Mutlu, M. Alanyali, and D. Starobinski. Spot pricing of secondary spectrum usage in wireless cellular networks. In *Proceedings of IEEE INFOCOM*, pages 682–690, Phoenix, AZ, 2008.

[225] R. B. Myerson. Incentive compatibility and the bargaining problem. *Econometrica*, 47(1):61–74, 1979.

[226] R. B. Myerson. Optimal auction design. *Mathematics of Operations Research*, 6(1):58–73, 1981.

[227] P. Nabipay, A. Odlyzko, and Z. L. Zhang. Flat versus metered rates, bundling, and bandwidth hogs. In *Proceedings of NetEcon 11: 6th Workshop on the Economics of Networks, Systems, and Computation*, 2011.

[228] A. Nagurney. *Supply Chain Network Economics: Dynamics of Prices, Flows and Profits*. Cheltenham: Edward Elgar Publishing, 2006.

[229] M. Naldi. A simple model for the effectiveness of delaying strategies for telecommunications churn reduction. In *Proceedings of 10th International Conference on Computer Modelling and Simulation (EUROSIM)*, Cambridge, 2008.

[230] J. Nash. Two-person cooperative games. *Econometrica: Journal of the Econometric Society*, 21(1):128–140, 1953.

[231] J. F. Nash. Equilibrium points in n-person games. In *Proceedings of the NAS*, 36:48–49, 1950.

[232] N. Nisan, T. Roughgarden, É. Tardos, and V. Vazirani, editors. *Algorithmic Game Theory*. Cambridge: Cambridge University Press, 2007.

[233] P. Njoroge, A. Ozdaglar, N. Stier-Moses, and G. Weintraub. Investment in Two Sided Markets and the Net Neutrality Debate. Technical Report DRO-2010-05, Columbia University, Decision, Risk and Operations Working Papers Series, 2010.

[234] J. R. Norris. *Markov Chains*. Cambridge: Cambridge University Press, 1998.

[235] F. Oberholzer and K. Strumpf. The effect of file sharing on record sales: An empirical analysis. *Journal of Political Economy*, 1(115):1–42, 2007.

[236] A. Odlyzko. Paris Metro pricing for the Internet. In *ACM Conference on Electronic Commerce (EC '99)*, pages 140–147, 1999.

[237] A. Odlyzko. Content Is Not King. Technical report, AT&T Labs, 2000.

[238] A. Odlyzko. The History of Communications and Its Implications for the Internet. Technical report, AT&T Labs, 2000.

[239] A. Odlyzko. The Internet and Past and Future Communications Revolutions. Technical report, School of Mathematics, University of Minnesota, 2009.

[240] A. Odlyzko. Network neutrality, search neutrality, and the never-ending conflict between efficiency and fairness in markets. *Review of Network Economics*, 8(1):40–60, 2009.

[241] A. Odlyzko, B. St. Arnaud, E. Stallman, and M. Weinberg. Know Your Limits: Considering the Role of Data Caps and Usage Based Billing in Internet Access Service. Technical report, Public Knowledge, May 2012.

[242] BEUC The European Consumer's Organization. Fair internet search – remedies in Google case. Position paper, available from www.beuc.org, 2013.

[243] M. J. Osborne and A. Rubinstein. *A Course in Game Theory*. Cambridge, MA: MIT Press, 1994.

[244] D. C. Parkes. *Iterative Combinatorial Auctions: Achieving Economic and Computational Efficiency*. PhD thesis, University of Pennsylvania, 2001.

[245] I. Ch. Paschalidis and Y. Liu. Pricing in multiservice loss networks: Static pricing, asymptotic optimality, and demand substitution effects. *IEEE/ACM Transactions on Networking*, 10(3):425–438, 2002.

[246] I. Ch. Paschalidis and J. N. Tsitsiklis. Congestion-dependent pricing of network services. *IEEE/ACM Transactions on Networking*, 8(2):171–184, 2000.

[247] A. Patcha and J.-M. Park. A game theoretic formulation for intrusion detection in mobile ad hoc networks. *International Journal of Network Security*, 2(2):131–137, 2006.

[248] G. Perakis. The "Price of Anarchy" under nonlinear and asymmetric costs. *Mathematics of Operations Research*, 32(3):614–628, 2007.

[249] A. C. Pigou. *The Economics of Welfare*. London: Macmillan, 1920.

[250] D. Porter, S. Rassenti, A. Roopnarine, and V. Smith. Combinatorial auction design. *Proceedings of the NAS*, 100(19):11153–11157, 2003.

[251] M. E. Porter. What is strategy? *Harvard Business Review*, November–December:61–78, 1996.

[252] J. Qi, Y. Zhang, Y. Zhang, and S. Shi. Treelogit model for customer churn prediction. In *Proceedings of IEEE Asia–Pacific Conference on Services Computing*, pages 70–75, Washington, DC, 2006.

[253] P. Reichl, R. Schatz, and B. Tuffin. Logarithmic laws in service quality perception: Where microeconomics meets psychophysics and quality of experience. *Telecommunication Systems*, 48(1–2):1–14, 2011.

[254] P. Reichl and B. Stiller. *Nil nove sub sole*: Why internet charging schemes look like as they do. In *Proceedings of the 4th Berlin Internet Economic Workshop*, 2001.

[255] J. Riley and W. Samuelson. Optimal auctions. *The American Economic Review*, 71(3):381–392, 1981.

[256] J. W. Roberts. Quality of service guarantees and charging in multiservice networks. *IEICE Transactions on Communication*, E81(5):824–831, 1998.

[257] D. Ros and B. Tuffin. A mathematical model of the Paris Metro pricing scheme for charging packet networks. *Computer Networks*, 46(1):73–85, 2004.

[258] E. Rosen, A. Viswanathan, and R. Callon. Multiprotocol Label Switching Architecture. IETF report RFC 3031, 2001.

[259] R. W. Rosenthal. A class of games possessing pure-strategy Nash equilibria. *International Journal of Game Theory*, 2(1):65–67, 1973.

[260] T. Roughgarden. The price of anarchy is independent of the network topology. *Journal of Computer and System Sciences*, 67:341–364, 2003.

[261] A. Rubinstein. Equilibrium in supergames with the overtaking criterion. *Journal of Economic Theory*, 21:1–9, 1979.

[262] W. Rudin. *Functional Analysis*. New York: McGraw-Hill, 2nd edition, 1991.

[263] W. H. Sandholm. Potential games with continuous player sets. *Journal of Economic Theory*, 97(1):81–108, 2001.

[264] W. H. Sandholm. Evolutionary game theory. In R. A. Meyers, editor, *Encyclopedia of Complexity and Systems Science*, pages 3176–3205. Berlin: Springer, 2009.

[265] W. H. Sandholm. Large population potential games. *Journal of Economic Theory*, 144(4):1710–1725, 2009.

[266] W. H. Sandholm. *Population Games and Evolutionary Dynamics*. Cambridge, MA: MIT Press, 2010.

[267] C. U. Saraydar, N. B. Mandayam, and D. J. Goodman. Pricing and power control in a multicell wireless data network. *IEEE JSAC Wireless Series*, 19(2):277–286, 2001.

[268] C. U. Saraydar, N. B. Mandayam, and D. J. Goodman. Efficient power control via pricing in wireless data networks. *IEEE Transactions on Communications*, 50(2):291–303, 2002.

[269] P. S. Sastry, V. V. Phansalkar, and M. A. L. Thathachar. Decentralized learning of Nash equilibria in multi-person stochastic games with incomplete information. *IEEE Transactions on Systems, Man, and Cybernetics*, 24:769–777, 1994.

[270] M. J. Scott. Network Operators and Content Providers: Who Bears the Cost? Technical report 201297, SSRN, 2011.

[271] N. Semret. *Market Mechanisms for Network Resource Sharing*. PhD thesis, Columbia University, 1999.

[272] N. Semret, R. Liao, A. Campbell, and A. Lazar. Market Pricing of Differentiated Internet Services. Technical report CU/CTR/TR 503-98-37, Columbia University, 1998.

[273] N. Semret, R. R.-F. Liao, A. T. Campbell, and A. A. Lazar. Pricing, provisioning and peering: Dynamic markets for differentiated internet services and implications for network interconnections. *IEEE Journal on Selected Areas in Communications*, 18(12):2499–2513, 2000.

[274] S. Sengupta and M. Chatterjee. An economic framework for dynamic spectrum access and service pricing. *IEEE/ACM Transactions on Networking*, 17(4):1200–1213, 2009.

[275] D. H. Shin and M. Bartolacci. A study of MVNO diffusion and market structure in the EU, US, Hong Kong, and Singapore. *Telematics and Informatics*, 24:86–100, 2007.

[276] G. Singh, C. Kesselman, and E. Deelman. Adaptive pricing for resource reservations in shared environments. In *Proceedings of the 8th IEEE/ACM International Conference on Grid Computing, GRID '07*, pages 74–80, Washington, DC, 2007.

[277] V. A. Siris. Resource control for elastic traffic in CDMA networks. In *Proceedings of the 8th International Conference on Mobile Computing and Networking*, pages 193–204, Atlanta, GA, 2002.

[278] V. A. Siris and C. Courcoubetis. Resource control for loss-sensitive traffic in CDMA networks. In *Proceedings of IEEE INFOCOM 2004*, Hong Kong, 2004.
[279] D. J. Songhurst, editor. *Charging Communication Networks*. Amsterdam: Elsevier, 1999.
[280] B. Stiller, P. Reichl, and S. Leinen. Pricing and cost recovery for Internet services: Practical review, classification, and application of relevant models. *Netnomics*, 3(2):149–171, 2001.
[281] J. Sun, E. Modiano, and L. Zheng. A novel auction algorithm for fair allocation of a wireless fading channel. In *Proceedings of 38th Annual Conference on Information Science and Systems*, Princeton, NJ, 2004.
[282] J. Sun, L. Zheng, and E. Modiano. Wireless channel allocation using an auction algorithm. In *Proceedings of Allerton Conference on Communications, Control and Computing*, 2003.
[283] K. Sydsaeter and P. Hammond. *Essential Mathematics for Economic Analysis*. Harlow: Prentice Hall, 3rd edition, 2008.
[284] Symantec. State of Enterprise Security. http://www.symantec.com/content/en/us/about/presskits/SES_report_Feb2010.pdf, 2010.
[285] K. T. Talluri and G. J. van Ryzin. *The Theory and Practice of Revenue Management*. Boston, MA: Kluwer, 2004.
[286] H. Tran Hoang and B. Tuffin. Inter-domain pricing: Challenges and possible approaches. *International Journal of Network Management*, 21(3):223–246, 2011.
[287] K. S. Trivedi. *Probability and Statistics with Reliability, Queuing, and Computer Science Applications*. New York: John Wiley & Sons, 2nd edition, 2002.
[288] B. Tuffin. Charging the Internet without bandwidth reservation: An overview and bibliography of mathematical approaches. *Journal of Information Science and Engineering*, 19(5):765–786, 2003.
[289] H. R. Varian. Position auctions. *International Journal of Industrial Organization*, 25:1163–1178, 2005.
[290] W. Vickrey. Counterspeculation, auctions, and competitive sealed tenders. *Journal of Finance*, 16(1):8–37, 1961.
[291] A. J. Viterbi. *CDMA. Principles of Spread Spectrum Communication*. New York: Addison-Wesley, 1995.
[292] X. Vives. *Oligopoly Pricing*. Cambridge, MA: MIT Press, 1990.
[293] H. von Stackelberg. *Market Structure and Equilibrium*. Berlin: Springer, 2010. Translation by D. Bazin, R. Hill, and L. Urch.
[294] E. Walker, W. Brisken, and J. Romney. To lease or not to lease from storage clouds. *Computer*, 43(4):44–50, 2010.
[295] J. Walrand and P. Varaiya. *High-Performance Communication Networks*. New York: Morgan Kaufmann Publishers, 2nd edition, 2000.
[296] G. R. Walsh. *Methods of Optimization*. New York: John Wiley & Sons, 1975.
[297] J. G. Wardrop. Some theoretical aspects of road traffic research. *Proceedings of the Institute of Civil Engineers*, 1:325–378, 1952.
[298] D. Waterman and S. Choi. Non-discrimination rules for ISPs and vertical integration: Lessons from cable television. *Telecommunications Policy*, 35(11):970–983, 2011.
[299] D. Waterman, R. Sherman, and S. W. Ji. Television: Revenue models, aggregation, and "TV everywhere." In *Research Conference on Communication, Information and Internet Policy (TPRC)*, 2012.
[300] C. Watkins and P. Dayan. *Q*-learning. *Machine Learning*, 8:279–292, 1992.
[301] K. Wieland. The customer retention challenge. *Telecommunications*, 40(10):14–17, 2006.
[302] L. A. Wolsey. *Integer Programming*. New York: John Wiley & Sons, 1998.

[303] S. Wong, J. Rojas-Mora, and E. Altman. Public Consultations on Net Neutrality 2010: USA, EU and France. Technical report, SSRN, 2010. http://papers.ssrn.com/sol3/papers.cfm?abstract_id=1684086.

[304] J. D. Wright. *Defining and Measuring Search Bias: Some Preliminary Evidence*. Fairfax, VA: George Mason University School of Law, 2012.

[305] S. Yang and B. Hajek. VCG–Kelly mechanisms for allocation of divisible goods: Adapting VCG mechanisms to one-dimensional signals. *IEEE Journal on Selected Areas in Communications,* 25(6):1237–1243, 2007.

[306] C. S. Yoo. Vertical integration and media regulation in the new economy. *Yale Journal on Regulation*, 19(1):171–300, 2002.

[307] S. Zamir. Bayesian games: Games with incomplete information. In R. A. Meyers, editor, *Encyclopedia of Complexity and Systems Science*, pages 426–441. Berlin: Springer, 2009.

[308] T. Zeller. Black market in stolen credit card data thrives on Internet. *The New York Times*, June 21, 2005.

[309] L. Zhang, S. Deering, D. Estrin, S. Shenker, and D. Zappala. RSVP: A new resource reservation protocol. *IEEE Network*, 7(9):8–18, 1993.

[310] B. Q. Zhao, J. C. S. Lui, and D.-M. Chiu. Mathematical modeling of incentive policies in P2P systems. In *Proceedings of the 3rd International Workshop on Economics of Networked Systems, NetEcon '08*, pages 97–102, New York, 2008.

[311] S. Zhong, J. Chen, and Y. R. Yang. Sprite: A simple, cheat-proof, credit-based system for mobile ad-hoc networks. In *Proceedings of IEEE INFOCOM*, 2003.

Index